PACIFIC NEOGENE

PACIFIC NEOGENE

Environment, Evolution, and Events

Edited by Ryuichi Tsuchi and James C. Ingle Jr.

UNIVERSITY OF TOKYO PRESS

CONTENTS

Preface

List of Contributors

I. PALEOENVIRONMENT OF THE PACIFIC

II. TECTONIC EVOLUTION OF THE PACIFIC

III. PACIFIC NEOGENE EVENTS IN TIME AND SPACE

APPENDICES

PREFACE

This volume presents contributions stemming from the Fifth International Congress on Pacific Neogene Stratigraphy and associated meetings of IGCP Project 246 (Pacific Neogene Events in Time and Space). The congress was held October 6–10, 1991, in Shizuoka, Japan, and was attended by 140 earth scientists from every quarter of the Pacific and elsewhere. The theme of this very successful gathering was "Pacific Neogene: Environment, evolution, and events". Some 82 papers and posters were presented, providing new evidence and syntheses of events in the Pacific region as recorded in Neogene deposits from high to low latitudes and in continental to deep-sea settings. Individual sessions focused on a wide spectrum of topics including the impact of Neogene global climate on the Pacific Ocean and biota, Pacific-wide effects associated with the tectonic closing of ocean gateways and mountain building, paleoceanography, isotopic events and Neogene chronology, biostratigraphy, paleobiogeography, and paleoenvironments and event stratigraphy. Lesser known areas of the Pacific rim were also highlighted, with a special session devoted to the Neogene of Central and South America. Similarly, a number of papers reviewed studies of Neogene marine and continental deposits in the northwestern Pacific, including Kamchatka and Sakhalin, with an equal number of reports detailing new insights into the Neogene of mainland Asia and the peninsular and insular areas of Southeast Asia. The organization and logistics of the meeting and field trips were superbly handled by a very effective committee of Japanese colleagues.

Although space limitations prevented all of the papers presented at the Congress from being included in this volume, the Editorial Board attempted to select contributions which highlight the Congress themes from global, regional, and local perspectives. The papers are organized in a manner reflecting three key topics of Neogene research in the Pacific region—(a) paleoenvironments, Neogene global climate, and the paleoceanography of the Pacific, (b) Neogene tectonic evolution of the Pacific, and (c) Pacific Neogene biotic and physical events in time and space. Although the Pacific region encompasses half of the world, the results of many of the studies presented herein illustrate that the major tectonic, depositional, eustatic, and oceanographic events marking the Neogene are clearly expressed in stratigraphic sequences around the entire Pacific rim as well as in the deep Pacific Ocean. The apparent synchroneity of these events enhances Pacific Neogene correlations but at the same time poses fundamental questions about the nature of local versus regional and ocean-wide controls on provincial stratigraphies. Indeed, studies of local stratigraphic sequences form the foundation for deciphering the Neogene history of the Pacific region as a whole. Data gathered from a small exposure high in the Himalayas or from an uplifted neritic sequence in Peru may in fact aid interpretation of isotopic evidence of a major Neogene paleoclimatic event garnered from a deep sea core in the central Pacific Ocean. Hence, it is critically important for individual workers to gain a larger view through timely forums such as that presented by the Fifth I.C.P.N.S. in Shizuoka. The papers in this volume offer solid evidence of recent progress in understanding the Neogene evolution of the Pacific region and the power of both local and regional approaches to reading the stratigraphic record.

Finally, and on behalf of all who attended the Fifth I.C.P.N.S., we would like to dedicate this volume to two pioneers of Pacific Neogene study—Professor Nobuo Ikebe of Osaka, Japan, and Dr. Norcott de B. Hornibrook of Lower Hutt, New Zealand. The I.C.S. Regional Committee on Pacific Neogene Stratigraphy and IGCP Project 114 (Biostratigraphic Datum Planes of the Pacific Neogene) and its successor, IGCP Project 246 (Pacific Neogene Events in Time and Space), have formed a remarkably vigorous scientific alliance over the past 14 years to investigate the Neogene history of the Pacific region. The success of the various endeavors undertaken by these organizations can be traced directly to the early efforts of Professor Ikebe and Dr. Hornibrook, who continue to be enthusiastic supporters of Pacific Neogene research.

September 1992

Ryuichi Tsuchi
Leader of IGCP Project 246
Chairman, RCPNS (1992-   )

James C. Ingle Jr.
Chairman, Regional Committee on
Pacific Neogene Stratigraphy (1987-91)

List of Contributors

C. Geoffrey ADAMS, Department of Palaeontology, British Museum (Natural History), London SW7 5BD, U.K.

John A. BARRON, Branch of Paleontology & Stratigraphy, U.S. Geological Survey, MS 915, 345 Middlefield Rd., Menlo Park, CA 94025, U.S.A.

Robert E. GARRISON, Earth Sciences Department, University of California, Santa Cruz, California 95064, U.S.A.

Joseph T. GASPERI, Department of Geological Sciences and Marine Science Institute, University of California, Santa Barbara, CA 93106, U.S.A.

Yuri B. GLADENKOV, Geological Institute of the Russian Academy of Sciences, Moscow, Russia

Linda E. HEUSSER, Lamont-Deherty Geological Observatory of Columbia University, Palisades, New York 10964, U.S.A.

Kimio HIROOKA, Department of Earth Sciences, Faculty of Science, Toyama University, Toyama 930, Japan

Norcott de B. HORNIBROOK, Research Associate, DSIR Geology and Geophysics, Lower Hutt, New Zealand

Masako IBARAKI, Geoscience Institute, Faculty of Science, Shizuoka University, Shizuoka 422, Japan

James P. KENNETT, Department of Geological Sciences and Marine Science Institute, University of California, Santa Barbara, CA 93106, U.S.A.

Itaru KOIZUMI, Department of Geology and Mineralogy, Faculty of Science, Hokkaido University, Sapporo 060, Japan

Susumu NISHIMURA, Department of Geology and Mineralogy, Faculty of Science, Kyoto University, Kyoto 606, Japan

Maureen E. RAYMO, Department of Earth, Atmospheric and Planetary Sciences, Massachusetts Institute of Technology, Cambridge, MA 02139, U.S.A.

D.K. SINHA, Department of Geology, Banaras Hindu University Varanasi-221 005, India

M.S. SRINIVASAN, Department of Geology, Banaras Hindu University Varanasi-221 005, India

Ryuichi TSUCHI, Shizuoka University, Shizuoka 422, Japan

Kuo-Yen WEI, Department of Geology and Geophysics, Yale University, New Haven, CT 06511-8130, U.S.A.

I
PALEOENVIRONMENT OF THE PACIFIC

Stratigraphic and Paleoenvironmental Implications of Neogene Palynology of ODP Sites 794, and 797 in the Sea of Japan

L. E. HEUSSER

Lamont-Doherty Geological Observatory of Columbia University, Palisades, New York 10964, U.S.A.

Abstract

Preliminary pollen analyses of cores taken on ODP Legs 127 and 128 provide a consistent Neogene pollen stratigraphy characterized by a late Miocene decline in deciduous and Tertiary types, a subsequent rise in boreal components, a mid-Pliocene reversal of these trends centered at 4 Ma, followed by late Pliocene/Pleistocene high-frequency, high-amplitude fluctuations of temperate and boreal pollen types. Late Miocene vegetation inferred from these pollen data, a mix of conifer and broad-leaf elements with now-extinct Tertiary elements well-represented, appears similar to Aniai-type floras of Japan. During the Late Miocene through Early Pliocene, as Tertiary types declined, conifers became more prominent than broadleaf elements, and herbs increased. Mid-Pliocene pollen assemblages imply significant changes in forest composition. In the 500 k.y. interval centered at ~4 Ma, Tertiary and warm temperate deciduous types were comparable to or greater than Late Miocene levels, while temperate and cold temperate conifers were minimal. In Late Pliocene forests, Tertiary and deciduous components decreased, conifers were again prominent, and herbs increased. Over the last ~2 m.y., the very large and frequent changes in forest composition inferred from pollen in the Sea of Japan correspond with forest dynamics inferred from changes in pollen and floral assemblages throughout Japan.

Broad trends in Neogene climate inferred from these preliminary pollen data are decreasing temperature and increasing amplitude and frequency of climatic change. Two warm events, centered at ~9 Ma and ~ 4 Ma, punctuate the gradual deterioration of the equable warm humid subtropical/warm temperate Late Miocene and Early Pliocene climates. The first indication of conditions comparable to those of Pleistocene glacial intervals occurs ~ 3 Ma. Subsequently, regional climates oscillated rapidly between climatic extremes which were never great enough to displace warm-temperate and temperate forests from Honshu nor to produce arctic climates on the west coast of Japan.

Introduction

Japan is a dynamic country with an exciting Neogene tectonic history (Tamaki, 1988). Equally dramatic Neogene paleoenvironmental changes range from the Miocene climatic optimum to climatic mimina of the "ice ages". Investigations of pollen and plant macrofossils in Neogene sediments in Japan suggest that the composition of vegetation in the Japanese island arc has changed repeatedly in response to these climatic variations (Fuji, 1988; Hase and Hatanaka, 1984; Igarashi et al., 1988; Numata, 1974; Tai, 1973; Yamanoi, 1989). Unfortunately, prior to ~30,000 yr, lengthy, continuous terrestrial records are rare, and the precise ages of many records are not firmly established. Pollen in marine sediments offshore provides a means of directly correlating long, continuous records of Japanese vegetation with high-resolution regional and global marine chronologies (Heusser, 1989; Ishii et al., 1981; Koroneva, 1961). This paper presents the results of the initial pollen analyses of Neogene sediments from ODP Sites 794 and 797 in the Sea of Japan (Fig. 1).

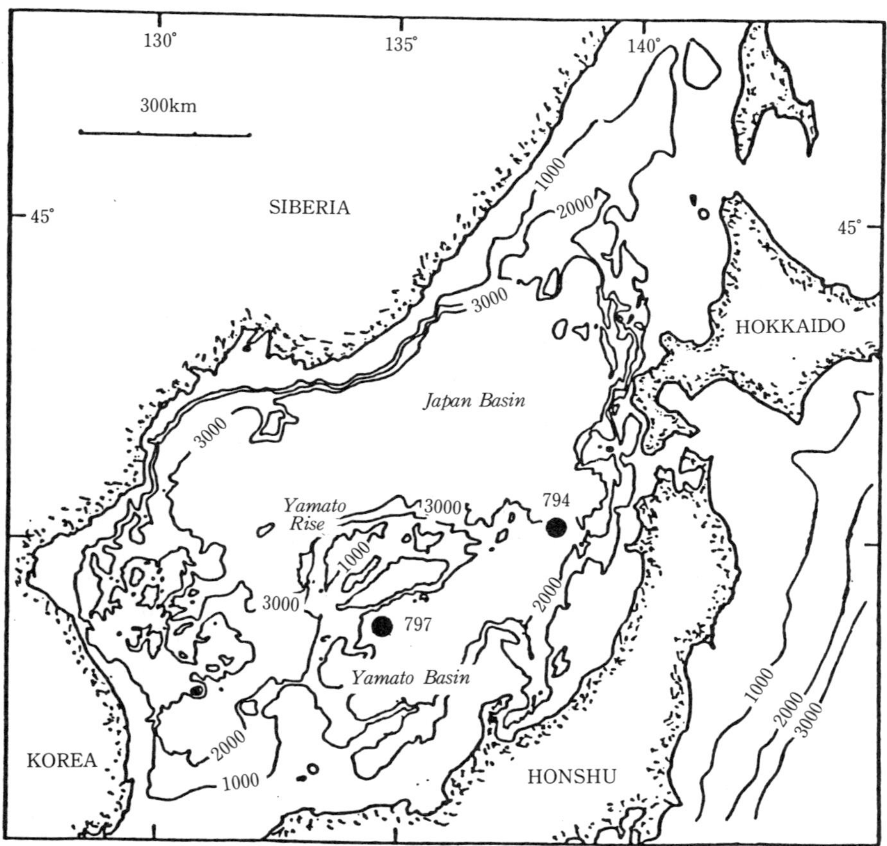

Fig. 1. Map of the Japan Sea and northwest Pacific Ocean showing the locations of ODP Sites 794 and 797 (from Tamaki, Pisciotto, Allan, et al., 1990).

Methods

Samples for pollen analysis were taken from the upper 345 m from Hole 794A (40°11.41′N, 138°13.86′E, 2821 mbsf) and the upper 379 m from Hole 797B (40°11.41′N, 138°13.86′E, 2821 mbsf). Preparation of these fine-grained hemipelagic, diatomaceous, and terrigenous sediment samples followed standard techniques (Heusser, in press). Reference collections of modern Japanese pollen and spore taxa along with published morphological descriptions and keys were used in pollen and spore identification. Taxonomic classification of pollen was usually conservative because criteria sometimes used to subdivide *Quercus* and *Pinus*, for example, were not consistently observed (Heusser, in press). Botanical nomenclature follows Ohwi (1984).

Pollen and spores were analyzed from 192 samples from Site 794 and from 122 samples from Site 797. Pollen percentages were calculated from the total number of arboreal and nonarboreal pollen identified. In the limited time available for the initial survey of pollen in cores taken on ODP Legs 127 and 128, a minimum of 100 pollen grains per sample were counted (Heusser, in press).

Age control for the intervals analyzed for pollen from Holes 794A and 797B derives from paleomagnetic measurements and siliceous and calcareous biostratigraphic datums (Heusser, in press; Tamaki, Pisciotto, Allan et al., 1990). Identification of the Aso 4 ash (with an estimated age of ~0.7-0.8 Ma; R. Tada, pers. comm., 1991) and correlation between late Quaternary terrestrial and marine pollen data provides additional chronostratigraphic control (Heusser, 1990). Using the sedimentation rates determined from these age models, intervals between productive pollen samples range from 0.04 m.y. in Hole 794A to ~2 m.y. in Hole 797B.

Results and Discussion

In most samples, pollen from the diagnostic taxa of vegetation/climate zones of Japan was relatively abundant and well preserved. Over the 8 m.y. of record from Site 794, pollen concentration ranged from < 100 to > 100,000 pollen grains/gmdws (Fig. 2). Highest amounts of pollen, which occur in late Quaternary sediments, are comparable with pollen concentration in Japanese bogs, and exceed that reported in marine cores elsewhere (Heusser, in press).

For ease of presentation, pollen taxa are combined into groups. The Taxodiaceae group includes *Cryptomeria,* other Taxodiaceae, and representatives of the Cupressaceae. Herbs consist of representatives of the Compositae; Chenopodiaceae, Gramineae, and Cyperaceae; and the Tertiary pollen group includes plants now extinct in Japan, such as *Carya, Liquidambar, Nyssa,* and *Sequoia/Metasequoia.*

The *Picea + Abies + Tsuga* group gradually decreases from a basal peak at ~275 mbsf to minimal values at ~125 mbsf, increases to a peak ~95mbsf, and then shifts rapidly back and forth in the upper ~50 m (Fig. 3a). Percentages of Taxodiaceae and herbs also fluctuate rapidly between extreme values in the upper part of the record.

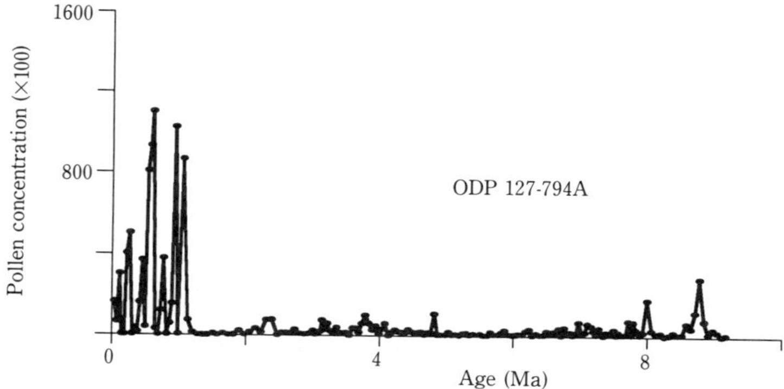

Fig. 2. Concentration of pollen (pollen/gwds) from Hole 794A plotted against age.

In the basal 300 mbsf, Taxodiaceae percentages are comparatively low. Overall, the relative contribution of herbs increases upwards, and mean values of *Quercus*, Tertiary, and warm temperate types decrease upwards in younger sediments (Fig. 3). Patterns of Tertiary types and of warm temperate types are marked by two robust peaks centered at ~325 mbsf and at ~125 mbsf.

Generally, pollen records from Hole 797A resemble those of 794. The relative abundance of Taxodiaceae remains low until the upper ~75 mbsf, where percentages fluctuate rapidly back and forth from a high of 60% to a low of <5% (Fig. 4a). Herbs and the *Picea + Tsuga + Abies* group show similar oscillations in the uppermost samples. In contrast, percentages of *Quercus*, Tertiary, and warm temperate types (Fig. 4b) display an overall upward decrease from basal maxima.

When plotted with age on the Y-axis, the pollen record from Hole 794 shows that the late Miocene peak in Tertiary and warm temperate types is succeeded by an extended interval characterized by *Picea + Abies + Tsuga* (Fig. 5). From ~6 Ma to shortly before ~4 Ma, conifer pollen dominates, with the Taxodiaceae becoming more important in the early Pliocene. A significant exception to the overall increase in conifer and herb pollen and decline in pollen from broadleafed trees — a reversal in these trends — occurs in the middle Pliocene. During this interval, centered at ~ 4 Ma, a lengthy resurgence of Tertiary and warm temperate pollen occurs. This is succeeded by a period of conifer dominance between ~3 and 2 Ma, which is followed by the distinctive high-frequency, high-amplitude fluctuations of the last ~million years.

These pollen data suggest that during the late Miocene, mesic, warm-temperate mixed conifer and broadleaf forests, with Tertiary types well represented, were replaced by forests in which Taxodiaceae and cold temperate/boreal types were relatively more important. Mid-Pliocene pollen assemblages imply significant changes in forest development in western Japan. In the half-million-year interval centered at ~4 Ma, Tertiary and warm temperate deciduous types re-expanded and were comparable to or greater than later Miocene levels of these same taxa. In middle Pliocene forests, representation of cold temperate and boreal conifers (*Picea + Abies + Tsuga*) was apparently minimal. Similar change at ~4 Ma, a re-expansion of mixed forest with Tertiary type conifers and broad-leafed trees, with appreciable *Fagus* have been described in northeastern Japan (Yamanoi, 1989).

After ~4 Ma, basic themes which began in the Late Miocene, overlain by high frequency variations, continue. In late Pliocene Japanese forests, Tertiary and warm deciduous types, *Quercus* and Taxodiaceae, were relatively unimportant. *Picea* + *Abies* + *Tsuga* were prominent, and herbs apparently played a larger role in the vegetation than in the early Pliocene. Similar trends in the vegetation of Hokkaido have been observed by Igarashi et al. (1988) and Yahata et al. (1989). In the last ~2 m.y., the very large and frequent changes in forest composition reconstructed from pollen assemblages from the Sea of Japan correspond with forest dynamics inferred from changes in pollen and floral assemblages throughout Japan (Heusser, in press).

Distribution of vegetation in Japan is intimately related to North Pacific and northeast Asian climatic processes (Fukui, 1977). For example, the boundary between warm temperate and temperate forests on Japan coincides with the general position of the atmospheric and marine fronts, and the abundance of Taxodiaceae (e.g., *Cryptomeria*) is strongly correlated with the high precipitation levels associated with the summer monsoon (Heusser, 1989; Tsukada, 1988). *Cryptomeria* and *Picea* + *Tsuga* + *Abies* are frequently used as temperature indicators. *Cryptomeria* is usually interpreted as indicative of moist temperate and warm-temperate conditions, and *Picea* and *Tsuga* are regarded as good indicators of subalpine/temperate environments (Tsukada, 1988).

Using these vegetation/climate relationships as guides, the major features of Neogene climate inferred from the preliminary pollen data from Site 794A are decreasing temperatures, increasing seasonality in temperatures and precipitation, and increasing amplitude and frequency of climatic change. A warming peak, centered at ~4 Ma, punctuated the gradual deterioration of the equable warm humid subtropical/ warm temperate Late Miocene and Early Pliocene climates. Following the mid-Pliocene warm humid interval, a rapid change to cooler and drier Late Pliocene environments began. In the pollen record from Hole 794A, the first indication of cold temperate conditions comparable to Pleistocene glacial intervals occurs ~3 Ma. Subsequently, regional Pleistocene temperate and cold temperate climates oscillated rapidly between extremes of temperature and precipitation which appear to exceed those of the previous part of the record.

These climatic extremes, however, were never great enough to totally displace subtropical components of Japanese vegetation nor to displace warm-temperate and temperate forests from Honshu. On the west coast of Japan, temperate and cold temperate climates, which supported conifer and mixed broadleaf forests, apparently prevailed through the late Pliocene and Quaternary, and there is no evidence of arctic climates on the west coast of Japan, i.e., of climates similar to those which now prevail around the Sea of Okhotsk (Heusser, in press).

The pollen data and inferred climatic changes imply that, over the time of record and within the time-span of each sample, winter sea surface temperatures off the west

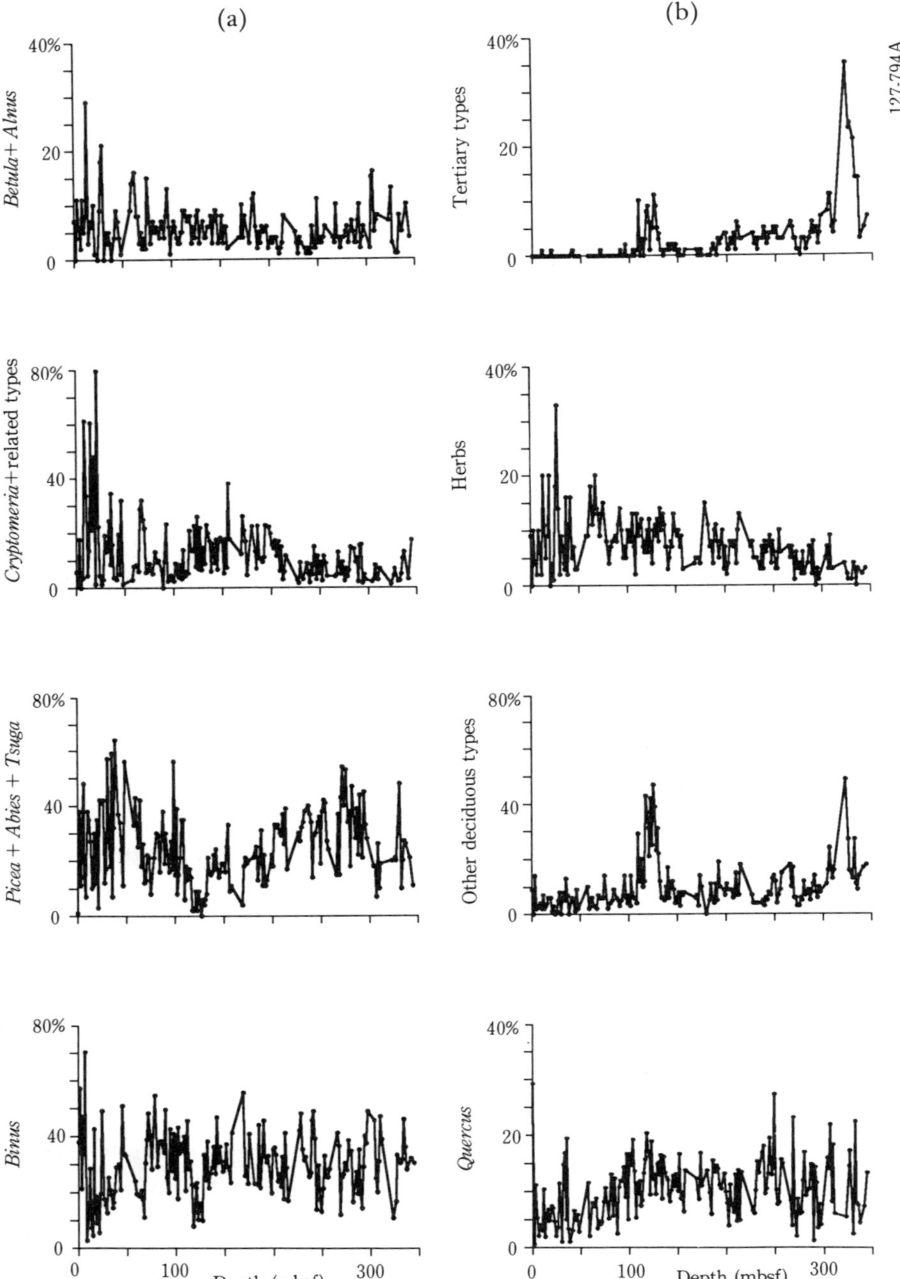

Fig. 3a, b. Percentages of selected pollen taxa from Hole 794A plotted against depth.

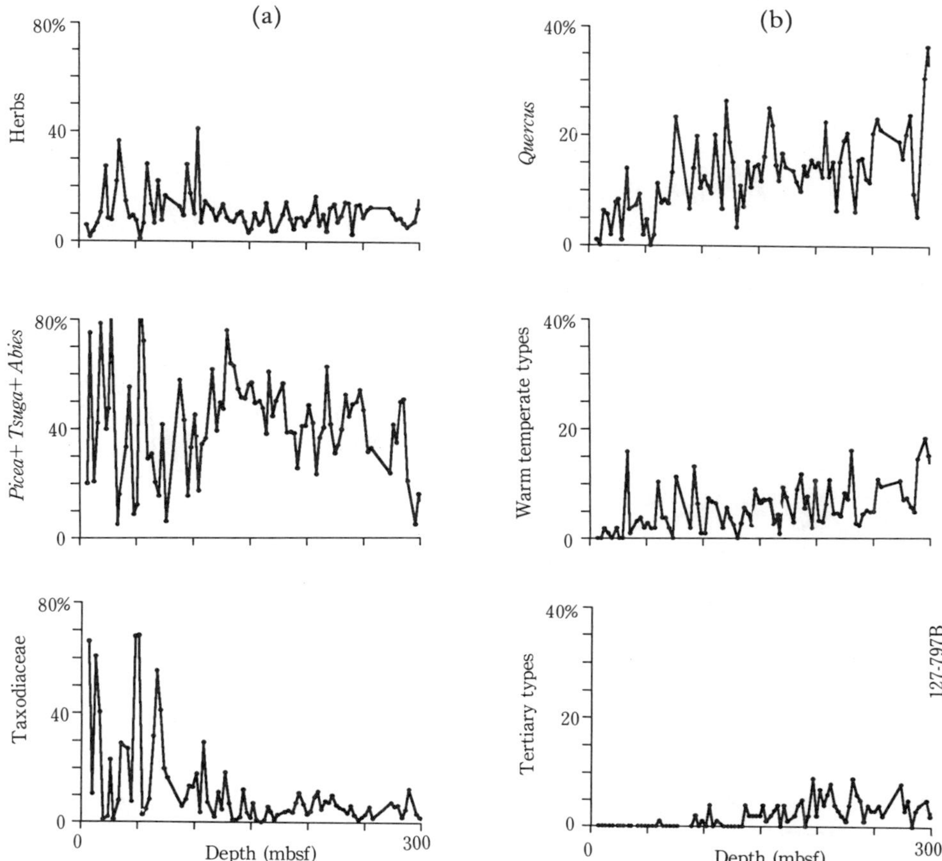

Fig. 4a, b. Percentages of selected pollen taxa from 797B plotted against depth.

coast of Japan did not vary beyond present extremes in the Sea of Japan for any extended interval. The pollen data from ODP cores and from piston cores also imply that, even during Quaternary eustatic sea-level depressions, warm North Pacific waters must have modified the effects of the enhanced Siberian lows during winter (Morley et al., 1986; Heusser, in press).

Paleoenvironmental reconstructions from these preliminary pollen analyses from ODP Holes 794 and 797 correspond with similar changes in Neogene environments in the Sea of Japan and elsewhere in the world. The overall trend of aridification and cooling in Japan apparently reflects uplift of the Tibetan Plateau, the major climate determinant in northeast Asia, and the ~6 Ma cooling may correspond with the expansion of Antarctic ice and increased aridity in Africa and Australia. An apparent exception to the overall trend of cooling and aridification in Japan is the warming at ~4 Ma.

The middle Pliocene warming implied by pollen data from Site 794 agrees with

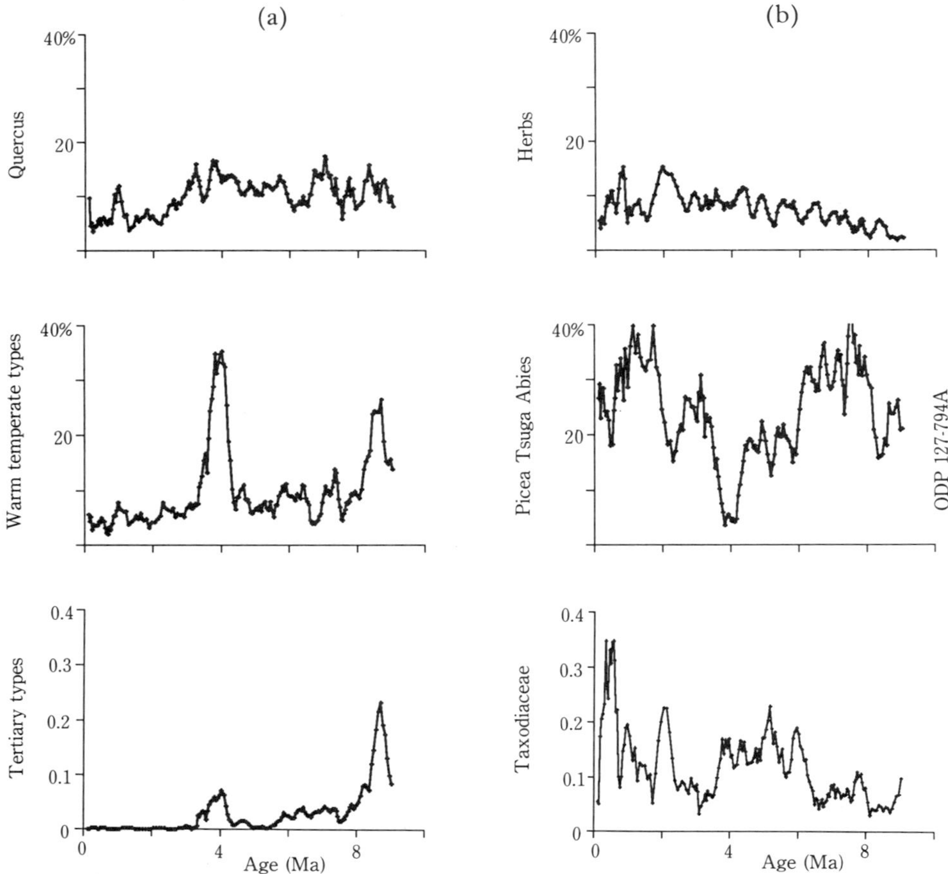

Fig. 5a, b. Percentages of selected pollen taxa from Hole 794A plotted against age. (For ease of presentation, data have been smoothed with a 5-point running mean).

temperature reconstructions from detailed analyses of onshore Neogene sequences in Japan (e.g., Asano et al., 1969) from the northeast Pacific (Barron, 1973; Ingle, 1973) and from the southern hemisphere (Elmstrom and Kennett, 1986). The ~4 Ma warming in Japan also conforms with global temperatures, as reflected in the composite oxygen isotope curve shown in Figure 6. The protracted warming implied by the increase in warm temperate pollen types and now-extinct Tertiary types corresponds with an extended interval when global temperatures, inferred from oxygen isotope variations in the Atlantic and Pacific Oceans, were higher and more stable than in the late Pleistocene (Shackleton and Cita, 1979; Shackleton and Hall, 1985; Keigwin, 1987; Keigwin et al., 1987).

Terrestrial data – pollen data from Japan – and marine data from the north and south Pacific and the south Atlantic (Venz and Hodell) clearly indicate a protracted warming event centered at 4 Ma. In contrast with paleoclimatic data from the North Atlantic, pollen data from Japan do not reflect a 3 or 3.5 Ma. warming event.

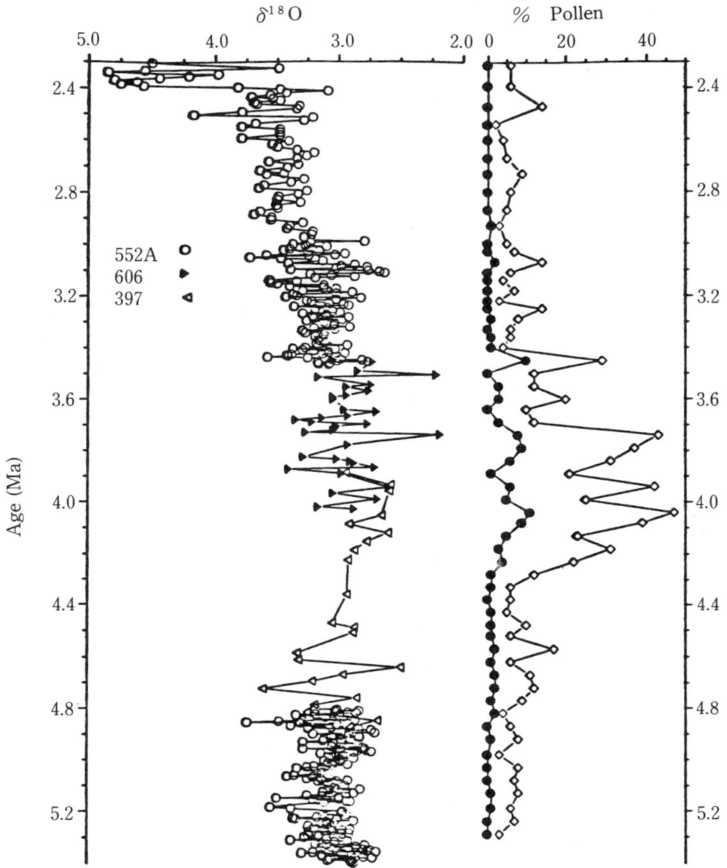

Fig. 6. Early and middle Pliocene paleoclimatic variations as expressed by benthic foraminifera δ^{18}O from ODP Sites 552A, 606, and 397, and by pollen indicator taxa (percentages of now-extinct Tertiary types) from Japan Sea ODP Site 794A. The composite oxygen isotope curve derives from the work of Shackleton and Cita (1979) (ODP Site 397), Shackleton and Hall (1985) (ODP Site 552A), Keigwin (1987) (ODP Site 606), and Keigwin et al. (1987) (ODP Site 552A).

Interpreting and modeling the causal mechanisms of these Pliocene climatic variations presents an exciting challenge.

Acknowledgements

I would like to thank Professor R. Tsuchi and the organizers of the Neogene conference and those who participated in ODP Legs 127/128, particularly Professor James C. Ingle.

References

Asamo, K., Ingle, J.C., and Takayamaga, Y., 1969. Neogene planktonic foraminiferal sequence ion northeastern Japan. *In:* P. Bronniman and H. Rehz (Eds.), *Proc. First Internat. Conf. Planktonic Microfossils*, 5, Leiden (J. Brill), 12–25.

Barron, J., 1973. Late Miocene-Early Pliocene paleotemperatures for California from marine diatom evidence. *Palaeogeogr., Palaeoclimatol., Palaeoecol.*, 14: 271–291.

Elstrom, K.M. and Kennett, J.P., 1986. Late Neogene paleoiceanographic evolution of Site 590: southwest Pacific. *In:* J.P. Kennett, C.C. von der Borch, et al., *Initial Reports of the Deep Sea Drilling Project, 90*, U.S. Government Printing Office, Washington, pp. 1361–1381.

Fuji, N., 1988. Palaeovegetation and palaeoclimate changes around Lake Biwa, Japan during the last ca. 3 million years. *Quat. Sci. Rev., 7*:21–28.

Fukui, E., 1977. *The Climate of Japan*: New York (Elsevier).

Hase, Y. and Hatanaka, K., 1984. Pollen stratigraphical study of the Late Cenozoic sediments in southern Kyushu, Japan. *Quat. Res. Japan, 22*:1–20.

Heusser, Y., 1989. Northeast Asian Climatic Change over the last 140,000 years inferred from pollen in marine cores taken off the Pacific Coast of Japan. *In:* M. Leinen and M. Sarthein (Eds.), *Paleoclimatology and Paleometerology: Modern and Past Patterns of Global Atmospheric Transport*: Dordrecht (Kluwer), pp. 665–92.

Heusser, Y., 1990. Northeast Asian pollen records for the last 150,000 years from deep-sea cores V28-304 and RC14-99 taken off the Pacific coast of Japan. *Rev. Palaeobot. Palynol., 65*:1–8.

Heusser, Y., in press. Neogene palynology of Holes 794A, 795A, and 797B in the Sea of Japan: stratigraphic and paleoenvironmental implications of the preliminary results. *In:* K. Tamaki, K. Suyehiro, J. Allan, M. McWiliams, et al. (Eds.), *Proc. ODP, Scientific Results*, Vol. 127/128, Pt. 2.

Igarashi, Y., Tonosaki, T., and Yoshida, M., 1988. Pollen stratigraphy of the Honbetsu and Ashoro formation, Tokachi Group. *Earth Sci. (Japan)*, 42(5):277–289.

Ingle, J.C., Jr., 1973. Neogene foraminifera from the northeastern Pacific Ocean, Leg 18, Deep Sea Drilling Project. *In:* L.D. Kulm, R. von Huene, et al. (Eds.), *Initial Reports of the Deep Sea Drilling Proejct, XVIII*, U.S. Gov. Printing Office, Washington, D.C., pp. 517–567.

Ishii, J., Igarashi, Y., Sasaki, S., Mino, N., and Matsumoto, K., 1981. On the peats collected from the continental shelf of the Isikari Bay, Hokkaido, Japan. *Earth Science*, 35(5):231–239.

Keigwin, L.D., 1987a. Pliocene stable-isotope record of Deep Sea Drilling Project Site 606: sequential events of ^{18}O enrichment beginning at 3.1Ma. *In:* W.F. Ruddiman, R.B. Kidd, E. Rhomas, et. al., *Initial Reports of the Deep Sea Drilling Project, 94*, U.S. Government Printing Office, Washington, pp. 911–920.

Keigwin, L.D., Aubry, M.-P., and Kent, D.V., 1987b. North Atlantic Late Miocene stable-isotope stratigraphy, biostratigraphy, and magnetostratigraphy. *In:* W.F. Ruddiman, R.B. Kidd, E. Thomas, et al., *Initial Reports of the Deep Sea Drilling Project, 94*, U.S. Government Printing Office, Washington, pp. 936-963.

Koroneva, E.V., 1961. Investigation by method of pollen and spore analysis of two marine cores from the Sea of Japan. *Oceanog. T. Geol. Inst. Academy Sci. USSR*, 126(1):1–16.

Morley, J., Heusser, L., and Sarro, T., 1986. Latest Pleistocene and Holocene palaeoenvironment of Japan and its marginal Sea. *Palaeogeog., Palaeoclimatol., Palaeoecol.,* 53: 349–358.

Numata, M., 1974. *The Flora and Vegetation of Japan:* New York (Elsevier).

Ohwi, J., 1984. *Flora of Japan:* Washington (Smithsonian Inst.).

Shackleton, N.J. and Cita, M.B., 1979. Oxygen and carbon isotope stratigraphy of benthic foraminifers at Site 397: detailed history of climatic change during the Neogene. *In:* U. von Rad, W.B.F. Ryan, et al., *Initial Reports of the Deep Sea Drilling Project, 47, Part I,* U.S. Government Printing Office, Washington, pp. 433– 445.

Shackleton, N.J. and Hall, M.A., 1985. Oxygen and carbon isotope stratigraphy of Deep Sea Drilling Project hole 552A: Plio-Pleistocene glacial history. *In:* D.G. Roberts, D. Schnitker, et. al., *Initial Reports of the Deep Sea Drilling Project, 81,* U.S. Government Printing Office, Washington, pp. 599–609.

Tai, A., 1973. A study of the pollen stratigraphy of the Osaka Group, Pliocene-Pleistocene deposits in the Osaka Basin. *Mem. Faculty of Sci., Kyoto Univ., Ser. Geol. & Mineral,* 39(2):123–165.

Tamaki, K. 1988. Geological structure of the Japan Sea and its tectonic implications. *Bull. Geol. Survey Japan,* 39: 269–365.

Tamaki, K., Pisciotto, K., Allan, J., et al., 1990. *Proceedings of the Ocean Drilling Program, Initial Reports:* College Station, Texas (Ocean Drilling Program). College Station, Texas.

Tsukada, M., 1988. Japan. *In:* B. Huntley and T.III Webb (Eds.), *Vegetation History:* Dordrecht (Kluwer), 459–518.

Yahata, M., Igarashi, Y., Gautam, P., and Wada, N., 1989. Plio-Pleistocene in the eastern part of Toya Lake, southwest Hokkaido: sedimentary facies, pollen stratigraphy and magnetostratigraphy. *Earth Sci. (Japan),* 43(5):261–276.

Yamanoi, T., 1989. Neogene palynological zones and events in Japan. *Proc. Int. Symp. Pac. Neogene Continental and Marine Events*, pp. 83–90.

Biostratigraphy and Paleoceanography
of the Japan Sea Based on Diatoms: ODP Leg 127

Itaru KOIZUMI

Department of Geology and Mineralogy, Faculty of Science,
Hokkaido University, Sapporo, Japan

Abstract

A series of upper Miocene through Quaternary diatomaceous sequences was recovered at six sites in the Japan Sea during ODP Legs 127 and 128. Diatoms are mostly abundant and preserved above the diagenetic boundary opal-A/opal-CT, and are useful in dating carbonate concretions below the boundary. Forty diatom datum levels were evaluated biostratigraphically on the basis of the sediment accumulation rate curve, and several isochronous datum levels were newly proposed for the Japan Sea area. The diagenetic boundary between opal-A and opal-CT forms a distinctively diachronic plane from 8.50 Ma at Hole 797B of the Yamato Basin to 5.73 Ma at Hole 795A of the Japan Basin. A warm-water (paleo-Tsushima) current did not penetrate into the Japan Sea through the (paleo-) Tsushima strait during the late Miocene and Pliocene. Diatoms occur cyclically in variable abundance and states of preservation which are affected by eustatic sea level changes, and ultimately controlled by Milankovich cycles, during the late Quaternary sequences at all four sites.

Introduction

ODP Legs 127 and 128 were expected to provide complete sedimentary and volcanic sequences from the Japan Sea as reference sections for resolving geologic uncertainties in on-land sections (Fig. 1).

In order to understand how the Japan Sea was formed and how it changed, it is necessary to know the geo-history of the Japanese islands, including paleo-geography, paleoclimate, sedimentation, and volcanic and tectonic movement. The history of the Japan Sea has been clarified in detail over the last 20 years, and the results accumulated so far will be summarized, with more findings added by drilling in the Japan Sea which was carried out on ODP Leg 127 in the summer of 1989 (Tamaki, K., Pisciotto, K., Allan, J. et al., 1990).

On Leg 127, it was expected that fresh-water diatom fossils associated with the older green tuff which is older than 20 Ma and brackish ones associated with the Daijima-type floras would be recognized. This expectation was not fulfilled, but the existence of sediments of the Daijima stage in the Yamato basin was ascertained because the alternation of sand and mud facies with a basalt dated at 19 Ma by ^{40}Ar—

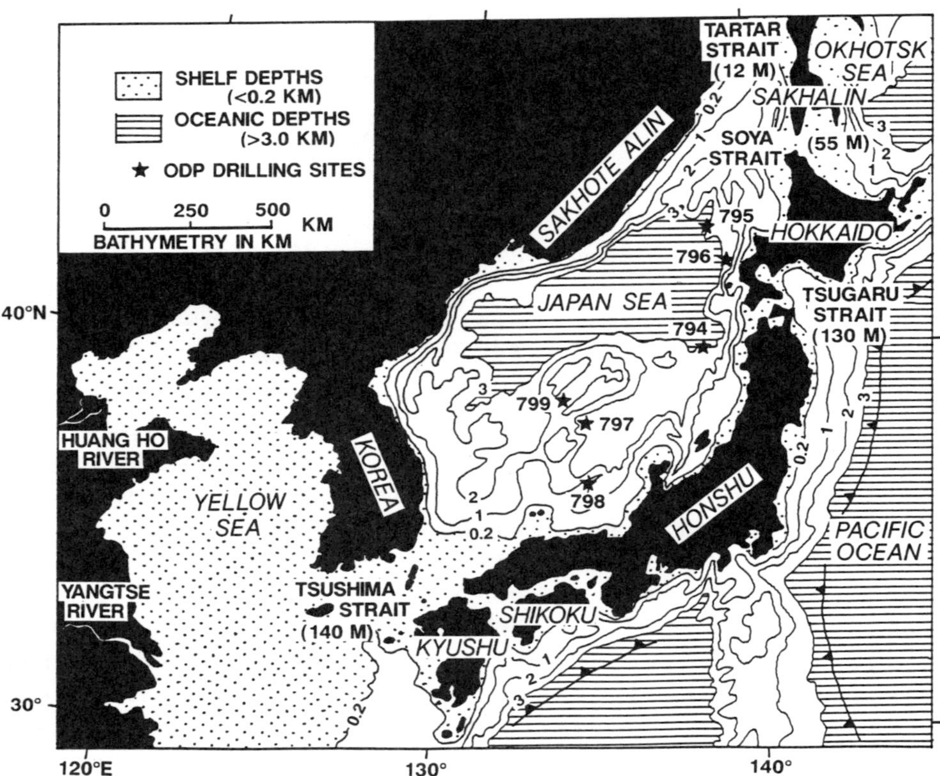

Fig. 1. Map of the Japan Sea showing the location of Leg 127 Sites 794 through 797 and Leg 128 Sites 794, 798 and 799. Bathymetry in kilometers. Isobaths representing water depths in excess of 3 km in the Pacific Ocean are not shown. Maximum water depths in the Japan Sea straits are shown in parentheses.

^{39}Ar method shows a deltaic stage of sedimentation at the base of the Site 797 located on the foot of Yamato Rise (Kaneoka et al., in press).

Methods of Study

Samples material was placed in an oven at 60°C for 24 hr, and 0.1g of dried-up materials was boiled in a 100 ml beaker with about 10 ml of hydrogen peroxide solution (15%) for several seconds and then left to stand for 24 hr after diluting with distilled water. After the suspension was poured off, the residue was diluted with 50 ml of distilled water and homogenized for about 3s in an ultrasonic washer (Clean Matic; 20W, 40kHz). Using a micropipette (Justor-Jv 500μl), 0.25 ml of this solution was placed on a cover glass (18×18 mm in size), dried on a hot plate at 50°C, and then mounted on a glass slide using Pleurax.

All diatoms were identified and counted until the number of individual specimens totaled 200, excluding *Bacteriastrum* spp. and *Chaetoceros* spp. Several samples did not contain enough diatoms to total 200 specimens.

The diatom zonation of Koizumi (1985) was adapted with minor changes to the samples of Leg 127 in the Japan Sea (Koizumi, in press).

Diatom Biostratigraphy

Nine diatom zones were recognized in Hole 794A, six zones in Hole 795A, seven zones in Hole 796A, and nine zones in Hole 797B (Table 1, Fig. 2). The lowest diatom zone is the middle Miocene *Denticulopsis praedimorpha* Zone, which contains *D. praedimorpha*, in the carbonate samples at 630.65 m, 608.18 m, 572.8 m, and 560.5 m below the sea-floor in Hole 795B and 350.22 m and 342.64 m below the sea-floor in Hole 797B. A carbonate layer 481.81 m below the sea-floor belongs to the upper Miocene *Denticulopsia dimorpha* Zone, because it contains valves of *D. dimorpha* and *Denticulopsis hustedtii*.

Diatoms decline rapidly in abundance below the diagenetic transition from opal-A to opal-CT. The age of the opal-A/opal-CT boundary at each site was estimated by extrapolating the sedimentation rates based on diatom datum levels above the boundary. The age is 8.3 Ma at 293 m below the sea-floor in Hole 794A and 8.5 Ma at 295 m below the sea floor, which belong to the *Denticulopsis katayamae* Zone, and 5.7 Ma at 325 m below the sea floor in Hole 795A and 6.07 Ma at 224 m below the sea floor in Hole 796A belonging to the *Neodenticula kamtschatica* Zone. The opal-A/opal-CT boundary is a distinctively diachronous plane (Table 1).

Diatoms occur cyclically with variable abundance and state of preservation through the Quaternary sequences at all four sites in response to glacial cycles in productivity and water mass conditions that are affected by eustatic sea level changes during the time. Therefore, the *Actinocyclus oculatus* Zone was not identified at either Site 797 or Site 796. Similarly, the *Pseudoeunotia doliolus* Zone could not be applied to the sediments of Leg 127, because such subtropical warm-water age-diagnostic species as *Pseudoeunotia doliolus*, *Nitzschia reinholdii*, *Rhizosolenia praebergonii*, and *Nitzschia jouseae* are scarce in the samples from Sites 794 to 797 of the Japan Sea. In place of *P. doliolus* in the Japan Sea area, the last occurrence of *Neodenticula koizumii* is very useful as a marker datum for the *A. oculatus/Neo. koizumii* zonal boundary. It falls just below the Pliocene/Pleistocene 2.0 Ma.

Forty stratigraphically useful upper Miocene to Quaternary diatom datum levels were discussed in Koizumi (in press) with ages based on (1) the paleomagnetic reference time scale and (2) sediment accumulation rate diagrams in the Japan Sea area.

Table 1. Diatom zonation of samples from ODP Leg 127. Numbers refer to core, section, and interval in cm.

Age	Diatom Zone	Core, Section, Interval (cm)			
		Hole 797B	Hole 794A	Hole 796A	Hole 795A
Quaternary	*Neodenticula seminae*	1-1, 13-15 2-5, 44-15	1-1, 43-44 2-3, 44-45	1-1, 13-14 2-6,110-111	1-1, 30-31 2-3,29-30
	Rhizosolenia curvirostris	3-3, 14-16	2-5,44-45 5-3,48-49	3-3, 74-75 5-3,84-85	3-1,29-30 5-5,30-31
	Actinocyclus oculatus	 9-5, 14-16	5-5,45-46 7-5,44-46	6-1, 26-27	6-2,30-31 11-5,30-31
Pliocene	*Neodenticula koizumii*	10-2,109-111 11-4, 43-45	7-7, 44-45 9-1, 43-44	10-1, 40-41	12-2, 29-30 14-6, 29-30
	Neodenticula koizumii *Neodenticula kamtschatica*	12-1,109-111 20-3, 38-39	9-3, 43-44 12-5, 43-44	11-1, 40-41 14-5, 40-41	15-4, 30-31 22-3, 29-30
	Thalassiosira oestrupii	21-1, 39-40 23-2, 40-41	13-1, 43-44 20-5, 43-44	15-1, 40-41 21-5, 24-25	22-7, 29-30
upper Miocene	*Neodenticula kamtschatica*	23-6, 40-41 26-5, 38-39	21-3, 44-45 26-3, 44-45	22-3, 23-24	
	Rouxia californica	27-1, 37-38 27-5, 38-39	27-1, 43-44 27-5, 43-44		
	Thalassionema schraderi	28-4, 38-39 31-4, 38-39	28-2, 44-45		
	Denticulopsis katayamae	32-2, 36-37			
	Denticulopsis dimorpha				
	Thalassiosira yabei				
middle Miocene	*Denticulopsis praedimorpha*	37-2, 87 38-1, 22			

Paleoenvironment–Paleogeography

Late Miocene to Pliocene diatomaceous sediments are distributed widely on the floor of the Yamato and Japan basins and form the topographical mounds named the Oki and Yamato Ridges. These sediments have the same ages as those distributed along the coastal area of the Japan Sea.

Among the diatom assemblages contained in the diatomaceous sediments sampled at Sites 794–797, cold-water species such as *Coscinodiscus marginatus* and *Neodenticula kamtschatica* dominate. On the other hand, the warm-water species *Hemidiscus cuneiformis* appears first at 3.5 Ma at Site 794 and 3.0 Ma at Site 797.

A paleo-Tsushima warm water current is supposed to have flown into the Japan Sea, passing Tsushima Island as does the modern Tsushima current. This supposition can be proved partly when the brackish-water species *Paralia sulcata*, which originates in the Yellow Sea, shows high frequency. At southern Sites 797 and 794, this species increases its frequency in sequences dated at 3.5 Ma and younger, while at northern Sites 795 and 796, *P. sulcata* shows increased abundances after 2.0 Ma.

Actinoptychus senarius, a coastal diatom which attaches to the surface of sea-floor sediments and decayed plants, occurs in abundance throughout the upper Miocene at the two southern sites, with cyclical decreases in abundance. But at the northern sites this species occurs less frequently.

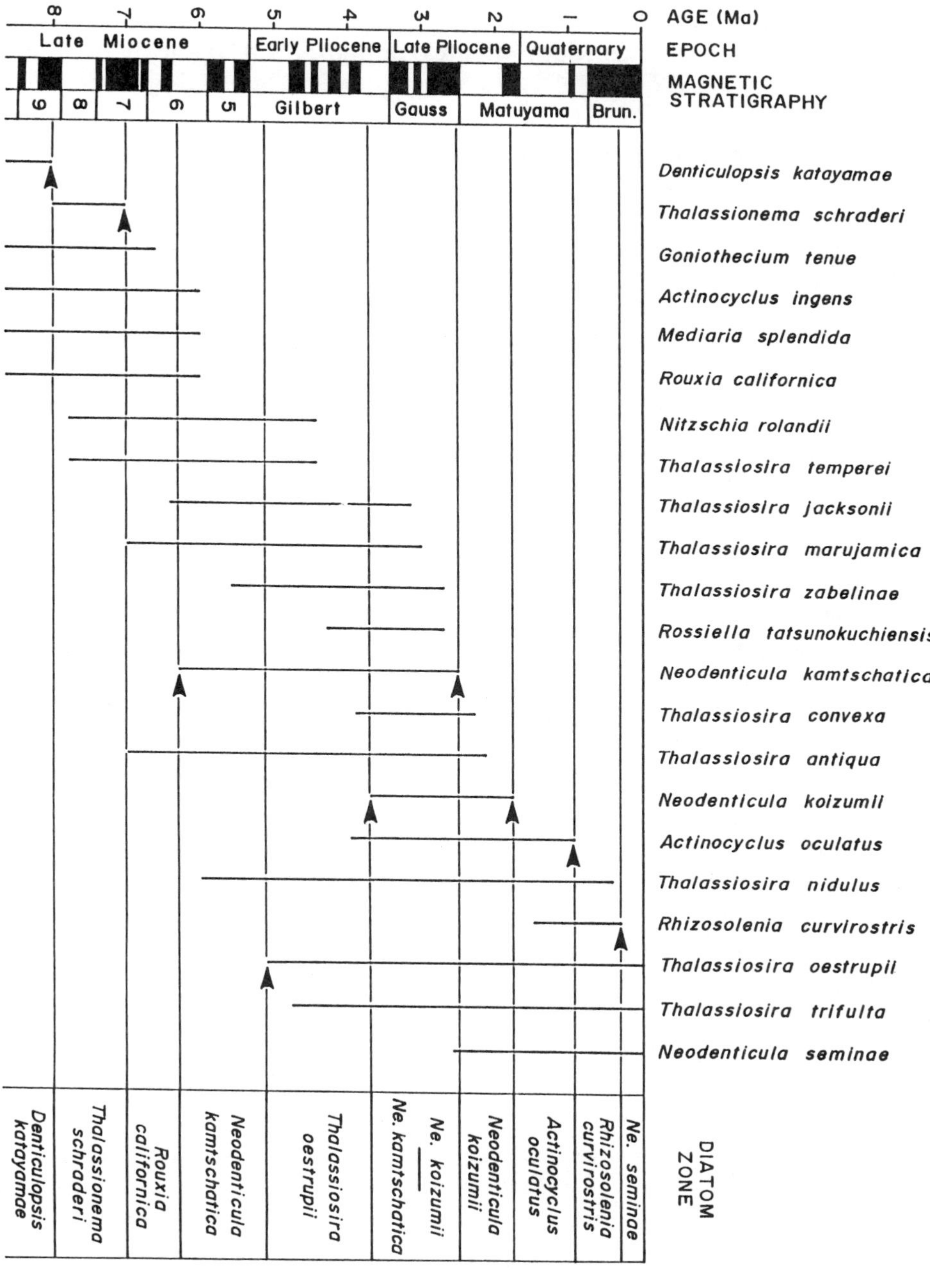

Fig. 2. Summary of stratigraphic ranges of selected diatom species and zonal subdivision encountered at ODP Leg 127 sites. Arrows indicate zonal markers.

Stephanopyxis turris, planktonic in shallow water for the specified period of lifetime and benthic as a spore for the rest of its lifetime, displays extremely high frequency after 3.5 Ma at the two southern sites, and after 4.0 Ma at the two northern sites (Figs. 3,4).

Presumably, these occurrences support the theory that from the late Miocene to the early Pliocene (8–3.5 Ma) the Japan Sea opened to the north and was almost closed to the south near the Tsushima strait. The floral and faunal fossil assemblages contained in the upper Miocene to early Pliocene in the Japan Sea coastal land-area do not suggest the flowing of a (paleo-) Tsushima warm-water current into the Japan Sea during this period (Iijima and Tada, 1990; Kano et al., 1991).

High Resolution Analyses of the Late Quaternary Diatom Assemblages

The late Quaternary diatom assemblages in 174 samples taken every 15 cm (representing 5 ky) between the sea-floor and the Brunhes/Matuyama paleomagnetic boundary (marking 730 ka) at Site 794 and in 148 samples taken every 30 cm (representing 6 ky) at Site 797 were analyzed.

The numbers of diatom valves per 1 g of the sample show changes in 100 ky cycles (Fig. 5). The changes in the order of 10 ky make a jagged pattern, which resemble closely the jagged line of late Quaternary oxygen isotope curves. Correlation between the peaks in both lines shows that diatom valves were less abundant during glacial periods and abundant during interglacial periods.

Major organic productivity, including diatoms, increases when nutrient-rich middle-deep sea water is provided to the surface as upwelling. During transgressive and interglacial periods, glaciers in the polar regions melt and cause a rise of sea level and active upwelling in the Japan Sea. Consequently productivity increases. During regressive and glacial periods productivity decreases when glaciers enlarge and cause lowering of sea level, causing the strait between the Japan Sea and outer oceans to become shallow, and retarding the flow of middle-deep sea water into the Japan Sea (Tada et al., in press).

The frequency of *Paralia sulcata* also shows a jagged pattern of stratigraphic changes more markedly than frequency of diatom valves. This is presumably because that branch of the Kuroshio current (Tsushima warm-water current) flowed northward, when sea level rose, into the Japan Sea, and carried with it many species living in the East China Sea and Yellow Sea when it passed through these areas.

Three cold-water diatoms, *Neodenticula seminae, Thalassiosira trifulta,* and *Rhizosolenia curvirostris,* also occur frequently through the periods of high sea level. The abundance pattern of these cold-water species downcore is a jagged and gradually decreasing one, unlike that of *P. sulcata.* Presumably, cold-water species increased in abundance as the Oyashio cold-water current flowed into the Japan Sea via the Tsugaru strait due to a rise of sea level during the shift from a glacial to an interglacial period. However, as the Tsushima warm-water current gradually became dominant, they began to decrease.

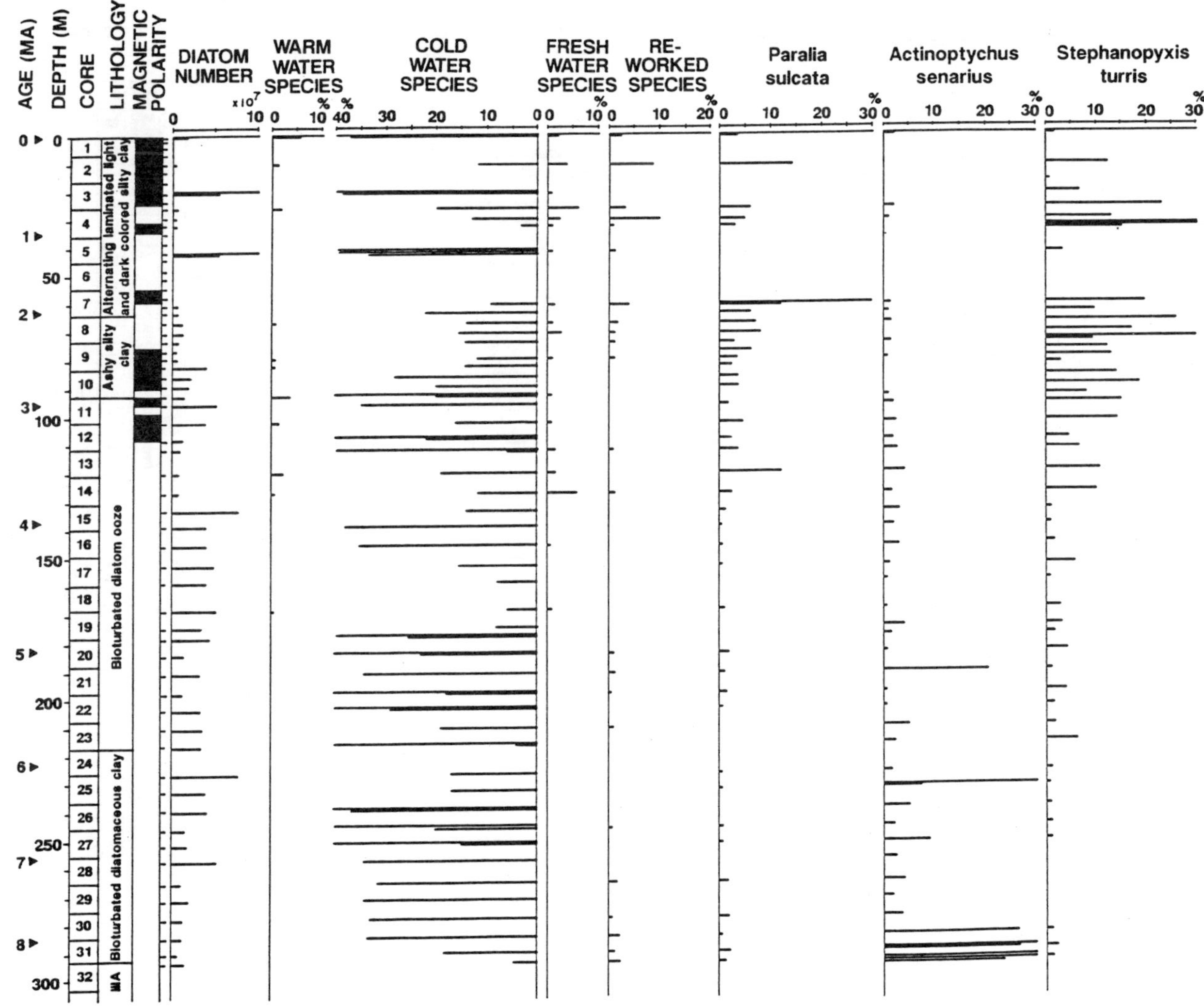

Fig. 3. Variation of diatom numbers (per g) and quantitative distribution of selected diatom species at ODP Leg 127 Site 794.

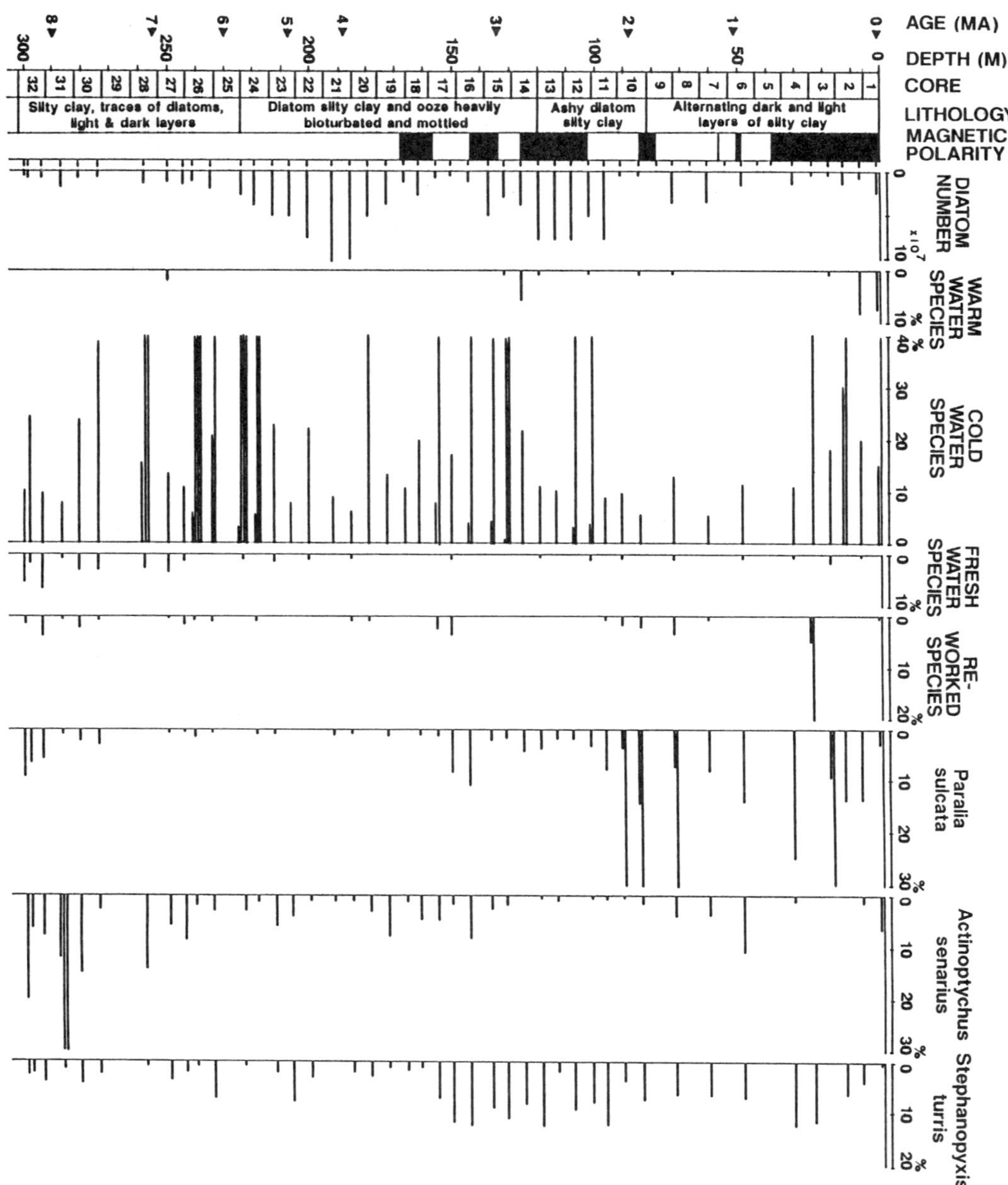

Fig. 4. Variation of diatom numbers (per g) and quantitative distribution of selected diatom species at ODP Leg 127 Site 797.

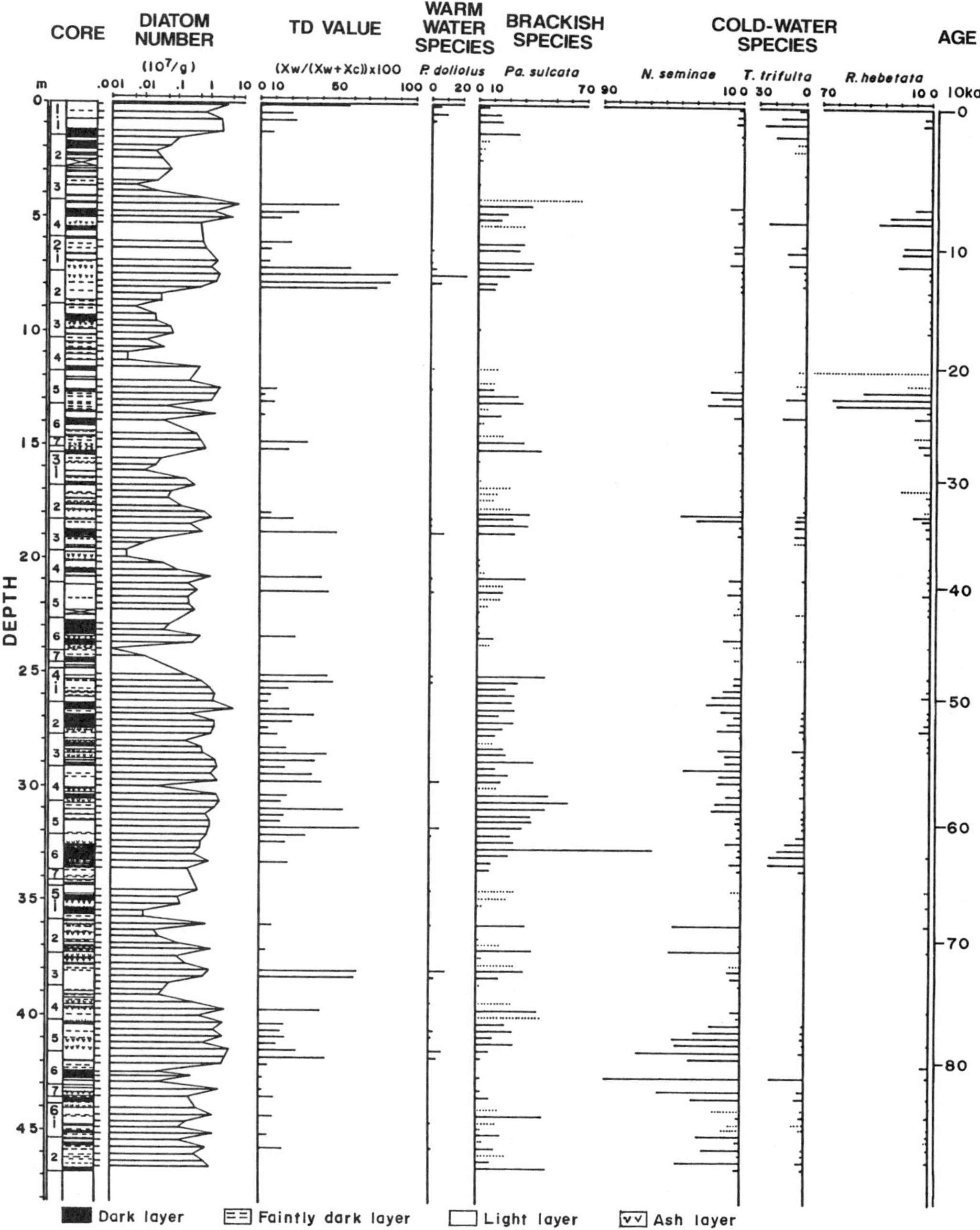

Fig. 5. Variation of diatom numbers (per g) and selected diatom species in the upper Quaternary sediments at ODP Leg 127 Site 797.

Future Research

ODP Legs 127 and 128 into the Japan Sea brought a great deal of material to be analyzed in order to learn more about the Japan Sea. Additional studies are needed now: (1) settling of sedimentary trap and piston corings at the drilling sites in order to understand better the circulation in the atmospher-ocean-sediment system and its mechanism, and (2) deeper drilling at the central area in the Japan Basin in order to know more about the establishment of the Japan Sea.

Acknowledgements

I am grateful to Dr. John A. Barron for a critical review of this manuscript. This study was partly supported by the Grant-in Aid for Scientific Research of the Ministry of Education, Science and Culture of Japan No. 04402014 provided to I. Koizumi.

References

Iijima, A. and Tada, R., 1990. Evolution of Tertiary sedimentary basins of Japan in reference to opening of the Japan Sea. *J. Fac. Sci. Univ. Tokyo, Sec. II,* 22: 121–171.

Kano, K., Kato, H., Yanagisawa, Y., and Yoshida, F. (Eds.), 1991. *Stratigraphy and geologic history of the Cenozoic of Japan.* Geol. Surv. Japan (Tsukuba), Rep. No. 274: 1–114.

Koizumi, I., in press. Diatom biostratigraphy of the Japan Sea: Leg 127. *In*: K. Tamaki, K. Suyehiro, J. Allan, M. McWilliams, et al., *Proc. ODP, Sci. Results,* College Station TX (Ocean Drilling program), 127–128.

Kaneoka, I., Takagi, Y., Takaoka, N., Yamashita, S., and Tamaki, K., in press. ^{40}Ar–^{39}Ar analyses of volcanic rocks recovered from the Japan Sea floor by Leg 127/128: constraint on the formation age of the Japan Sea. *In*: K. Tamaki, K. Suyehiro, J. Allan, M. McWilliams, et al., *Proc. ODP, Sci. Results,* College Station TX (Ocean Drilling Program), 127–128.

Tada, R., Koizumi, I., Cramp, A., and Rahman, A., in press. Correlation of dark and light layers, and the origin of their cyclicity in the Quaternary sediments from the Japan Sea. *In*: K. Tamaki, K. Suyehiro, J. Allan, M. McWilliams, et al., *Proc. ODP, Sci. Results,* College Station TX (Ocean Drilling Program), 127–128.

Tamaki, K., Pisciotto, K., Allan, J., et al., 1990. *Proc. ODP, Init. Repts.,* College Station TX (Ocean Drilling Parogram), 127: 1–844.

Paleoceanographic and Tectonic Controls on the Pliocene Diatom Record of California

J. A. BARRON

Branch of Paleontology & Stratigraphy, U.S. Geological Survey, MS 915, 345 Middlefield Rd., Menlo Park, CA 94025, U.S.A.

Abstract

The Pliocene diatom stratigraphy for California is refined for the interval between 5.0 and 2.8 Ma through study of diatoms from the Centerville Beach and Harris Grade sections and cores from Deep Sea Drilling Project Site 32. This refined diatom stratigraphy is applied along with tephrostratigraphy and magnetostratigraphy to correlate onshore and offshore Pliocene sections of California. These correlations suggest that tectonic and paleoceanographic events have exerted a major influence on deposition of sediments (and diatoms) throughout the California coastal region during the Pliocene.

The lithologic character and diatom content of the onshore Pliocene sections in California were strongly influenced by tectonic events. Tectonic reorganization of the California continental borderland began in the latest Miocene and was intensified during the Pliocene as the Coast Ranges of California were uplifted and adjoining basins underwent subsidence. An eastward jump in the San Andreas fault system at 5.5 Ma and a major change in the motion of the Pacific plate between 3.9 and 3.5 Ma appear to have been responsible for this tectonic reorganization. Increased deposition of clastic-rich sediments followed these tectonic events and also coincided with global falls in sea level at 5.5, 3.8, and 3.0 Ma. Basins formed during the Miocene were rapidly filled with Pliocene clastic sediments, and diatoms persisted as a major component only in sediments deposited in the center of basins and only until about 4 Ma.

In offshore DSDP sections, where terrigenous dilution is reduced, an interval of poor diatom preservation is present throughout the middle part of the Pliocene. At DSDP Sites 467 and 469 off southern California this dissolution interval begins about 4.8 Ma and persists until about 2.6 Ma. At DSDP Site 32 off central California, this interval of poor diatom preservation occurs between 3.7 and 3.0 Ma. This dissolution interval coincides with a period of high-latitude warming or deglaciation according to oxygen isotope studies, and it is taken as evidence of decreased southerly flow of the California Current and a reduction in upwelling off California during the middle part of the Pliocene. Intervals of diatom dissolution in these sections are inferred to correspond to August sea surface temperatures in excess of about 17°C.

Introduction–The Pliocene of California

Beginning in the latest Miocene (ca. 6.0 Ma), the detritus-poor, biogenic-rich sediments of the lower to upper Miocene Monterey formation of central and southern California were abruptly replaced by hemipelagic sediments containing an increased flux of terrigenous material in the form of fan turbidites and fine-grained detritus (Ingle, 1980; Barron, 1986a). This increasing dilution of the pelagic record effectively masked diatom sedimentation in onshore California basins (Barron, 1986a) obscuring the Pliocene and Quaternary diatom record.

Where present, early Pliocene diatom assemblages of onshore California basins are strongly provincial in character, making correlation with the geologic time scale very difficult (Barron, 1981, 1989). Neither standard North Pacific nor equatorial Pacific diatom biostratigraphies can be applied in dating lower Pliocene diatomaceous sediments in this region (Barron, 1981; Barron and Baldauf, 1986).

Offshore sites cored by DSDP Legs 18 and 63 provide Pliocene sections more distally removed from clastic sediment sources, but the Pliocene diatom record in these sites is also fragmented (Schrader, 1973; Barron, 1981, 1989). A hiatus between 4.8 and 2.8 Ma at DSDP Site 173 (Sarna-Wojcicki et al., 1987; Barron, 1989) and the poor preservational state of diatom assemblages between about 4.8 and 2.6 Ma at DSDP Sites 467 and 469 (Barron, 1981) (Fig. 1) severely limit our knowledge of the diatom record of the middle part of the Pliocene. About this poorly known interval, the well-established diatom stratigraphy of the northwest Pacific (see Koizumi and Tanimura, 1985, and references therein) can be readily applied to uppermost Pliocene (younger than 2.8 Ma) sediments of California (Schrader, 1973; Barron, 1981, 1989).

The purpose of this paper is to develop a workable diatom stratigraphy for the Pliocene of California and to apply that stratigraphy in correlating key onshore and offshore Pliocene sections of California. The resulting Pliocene sediment record will then be compared with known eustatic, tectonic, and paleoceanographic events. The geologic time scale of Berggren et al. (1985) will be used throughout this paper.

Pliocene Diatom Stratigraphy

Throughout coastal California, the diatom-rich Monterey Formation of the lower to upper Miocene is typically overlain by more terrigenous diatom-bearing rocks such as the Purisima, Sisquoc, and Capistrano Formations (Barron, 1986a). Diatom stratigraphy (Barron, 1986a, b) reveals that the lower parts of these overlying units contain latest Miocene to earliest Pliocene diatoms (typically 6.0 to 5.0 Ma), but that diatoms quickly disappear upsection as the terrigenous component increases. Diatoms persist only in sections deposited in the middle of basins such as the section of the Sisquoc Formation which is exposed along Harris Grade (old California Highway 1) north of Lompoc, California (Fig. 1). Presumably, diatom production continued in these coastal waters, but diatom sedimentation was almost completely masked by the increased clastic sedimentation.

A diatom stratigraphy for the early and middle parts of the Pliocene of California (5.0 to 3.0 Ma) may be pieced together through study of the diatoms of the Harris Grade section, the Centerville Beach section of northern California, and DSDP Site 32 off central California (Fig. 1).

Harris Grade Section

At Harris Grade, a 900-m-thick section of the diatom-rich Sisquoc Formation underlying the Foxen Mudstone overlies the axis of a faulted anticline (Barron and Baldauf, 1986). In contrast to other sections of the Sisquoc Formation, where diatoms disappear a short stratigraphic distance above the Miocene-Pliocene boundary, diatoms persist in abundance to a higher level, corresponding to about 3.8 Ma, in the Harris Grade section. This 3.8-Ma age estimate for the top of the Sisquoc Formation in the Harris Grade section is based on a sedimentation rate of 540 m/m.y., which was determined in the lower part of the Sisquoc Formation of the area based on paleomagnetically calibrated diatom datum levels (Barron and Ramirez, 1991).

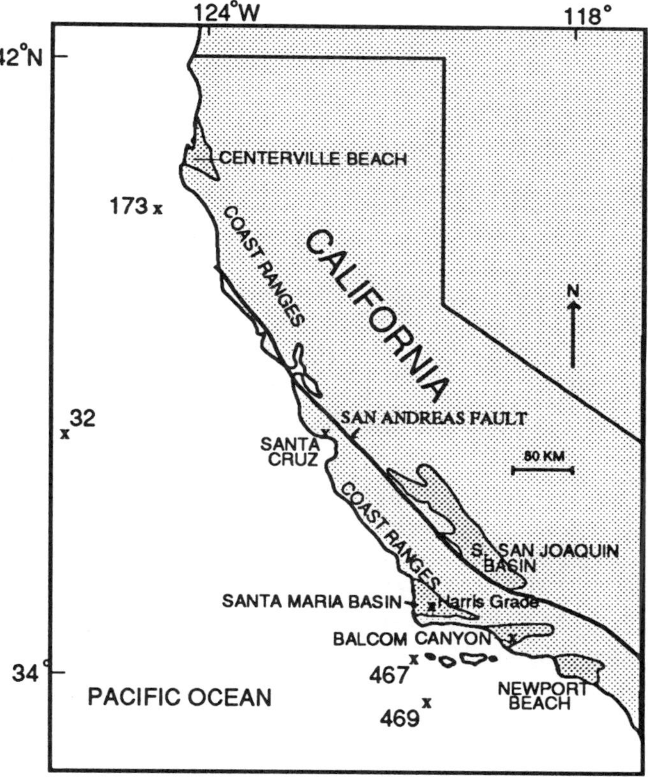

Fig. 1. Location of important Pliocene sections along the margin of California. Outline of selected Neogene onshore basins shown (darker shade).

Study of the diatoms of the Harris Grade section (Barron and Baldauf, 1986) has allowed the assignment of interpolated ages to a number of biostratigraphically useful diatom datum levels for the early Pliocene interval between 5.1 and 3.8 Ma (Barron and Ramirez, 1991). The interpolated ages of these earliest Pliocene diatom datum levels are shown in Table 1 along with the ages of latest Miocene diatom datum levels, which have been calibrated to paleomagnetic stratigraphy in the lower part of the Purisima Formation at Santa Cruz by Dumont et al. (1986).

Centerville Beach

Because published literature (Burckle et al., 1980; McCrory, 1990) suggested that a complete section of diatom-bearing sediments may be present at Centerville Beach in the Eel River basin of northern California (Fig. 1), 33 samples were obtained from P.A. McCrory for diatom study. At Centerville Beach a 300-m-thick sequence of diatomaceous mudstone and siltstone (Pullen and Eel River Formations) is overlain by 800 m of turbidites (lower and middle parts of Rio Dell Formation), which are, in turn, overlain by 700 m of sandy siltstone (upper part of Rio Dell Formation). The samples studied for diatoms (Fig. 2) are from the upper part of the Pliocene (ca. 3.6 to 1.8 Ma) according to the chronology of Fig. 2, which was established based on tephrochronology (Sarna-Wojcicki et al., 1987) and limited paleomagnetic stratigraphy (Dodd et al., 1977). They correspond to the Eel River Formation and basal 400 m of the overlying Dio Dell Formation (Fig. 2).

Diatom preservation is variable in the interval studied at Centerville Beach (Fig. 2), presumably as a result of dilution by terrigenous debris and redeposition of diatom assemblages through downslope transport. Of the 33 samples studied for diatoms, only 14 are moderately well preserved, as indicated by the presence of relatively delicate diatoms such as *Neodenticula, Delphineis,* and *Nitzschia.* An additional 7 samples contain poorly preserved diatoms, while 12 samples, mostly from the middle part of the studied section, are barren of diatoms (Fig. 2).

The first occurrence of *Stephanopyxis dimorpha*, a dissolution-resistant taxon, lies between 274.6 and 291.4 m in McCrory's (1990) section within the basal part Eel River Formation and very close to the Eel River Ash (283.2 m), which is assigned an age of 3.4 Ma by Sarna-Wojcicki et al. (1987, 1991). Thus, the lower Pliocene–upper Pliocene boundary falls within the lower part of the Eel River Formation.

The first consistent occurrence of *Neodenticula koizumii*, within those samples containing moderately-well preserved diatoms, falls between 378.5 and 494.8 m. Although Koizumi and Tanimura (1985) place the first occurrence of *N. koizumii* between 3.6 and 3.26 Ma in the northwest Pacific, based on paleomagnetic calibrations of DSDP Leg 86 material, the chronology of Fig. 2 suggests that this taxon first occurs between 3.1 and 2.8 Ma in the Centerville Beach section.

The last consistent *Neodenticula kamtschatica*, within samples containing moderately well preserved diatoms, lies between 494.8 and 636.6 m in the Centerville Beach section. This assignment agrees well with Koizumi and Tanimura's (1985) 2.58– to 2.5–Ma age for this datum level and is consistent with Dodd et al.'s (1977) approxi-

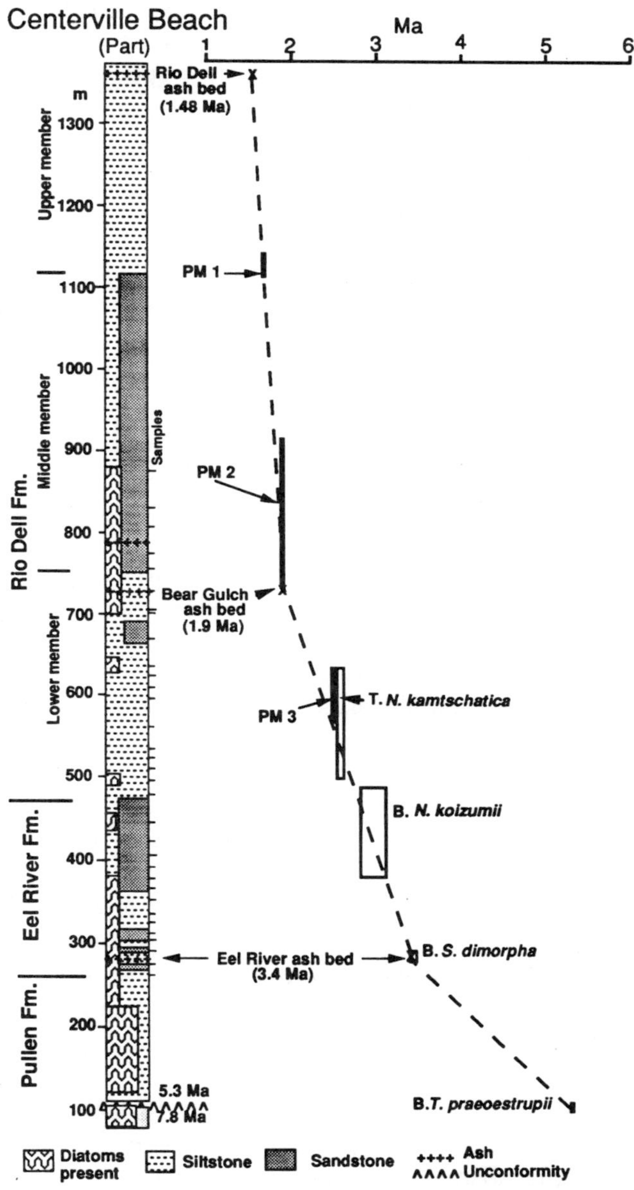

Fig. 2. Age vs. stratigraphic height plot for the middle portion of the Centerville Beach section based on teprochronology, diatom biostratigraphy, and magnetostratigraphy. Ash bed identification and ages from Sarna-Wojcicki et al. (1991). Diatom biostratigraphy from this report with ages from Table 1. Paleomagnetic stratigraphy from Dodd et al. (1977): PM 1 = top of Olduvai Normal-Polarity Subchronozone; PM 2 = base of Olduvai; PM 3 = top of Gauss Normal-Polarity Subchronozone. Generalized lithology from P.A. McCrory (written comm., 1991). Empty boxes = estimated uncertainties of diatom datum levels. Samples studied for diatoms shown by tick marks on the right side of the stratigraphic column.

mate location for the top of the Gauss Normal-Polarity Chronozone (Fig. 2). This occurrence, however, is somewhat higher than Burckle et al.'s (1980) placement in the Centerville Beach section. An isolated and rare specimen of *N. kamtschatica* recorded at 701 m is considered to be reworked.

DSDP Site 32

DSDP Site 173 is often cited as an offshore equivalent of the Centerville Beach section (McCrory, 1990). However, biostratigraphic studies (Sarna -Wojcicki et al., 1987; Barron, 1989) reveal that an interval corresponding to 4.8 to 2.8 Ma (early Pliocene to the early part of the late Pliocene) is missing at a hiatus about 122.5 m below sea floor (mbsf) at Site 173. Because the first occurrences of *S. dimorpha* and *N. koizumii* both fall at this hiatus (Schrader, 1973; Barron, unpublished data), it is difficult to confirm the 3.4–Ma and 3.1– to 2.8–Ma ages assigned respectively to these datums at Centerville Beach (Fig. 2). In an effort to sample this missing 4.8– to 2.8–Ma interval, samples were obtained from diatom from DSDP Site 32 (Fig. 1) off central California (Fig. 1). Diatom stratigraphy combined with planktonic foraminiferal (McManus et al., 1970) and calcareous nannofossil stratigraphy (Bukry and Bramlette, 1970; Gartner, 1970) reveals that a nearly complete interval ranging from 3.0 to 5.0 Ma was recovered in Cores 3 through 6 of Hole 32 (Fig. 3). Based on the stratigraphic positions of the first occurrence of *Globorotalia inflata* (planktonic foraminifer), the last occurrences of *Reticulofenestra pseudoumbilica* and *Amaurolithus* spp. (calcareous nannofossils), and the first occurrence of *Thalassiosira oestrupii* (diatom), a linear post-compaction sediment accumulation rate of 18 m/m.y. is predicted for the early Pliocene of Site 32 (Fig. 3).

The presence of *Neodenticula kamtschatica* without *N. koizumii* at 82.3 mbsf in upper Core 32–3 suggests that this level is older than the 3.1 to 2.8 Ma based on the diatom stratigraphy at Centerville Beach (Fig. 2). The first occurrence of *Stephanopyxis dimorpha* between 92.29 and 88.42 mbsf plots at 3.5 to 3.4 Ma, supporting the same interpolated age for this datum level at Centerville Beach (Fig. 2). Similarly, the last occurrence of *Lithodesmium cornigerum*, the last occurrence of *Thalassiosira hyalinopsis*, and the last occurrence of *T. nativa* sensu Schrader (1973) all plot on a linear age vs. depth plot of Site 32 (Fig. 3), if the ages interpolated from the Harris Grade section (Table 1) are used.

Of the 25 samples studied for diatoms at Site 32, 19 contain moderately-well to well-preserved diatoms based on the presence of specimens of *Neodenticula* and *Nitzschia*. The six samples displaying poor diatom preservation come from the upper part of Core 32–4 and from Core 32–3, corresponding to an interval dated between 3.7 and 3.0 Ma by the microfossil stratigraphy of Fig. 3. Presumably, this interval of poor diatom preservation reflects low diatom numbers in the overlying water column, because Site 32 lies over 400 km offshore near the present-day western limit of the fertile waters of the California Current (Sverdrup et al., 1942).

Ages of stratigraphically important diatom events interpolated from the Centerville Beach and Site 32 sections have been added to Table 1, which summarizes the esti-

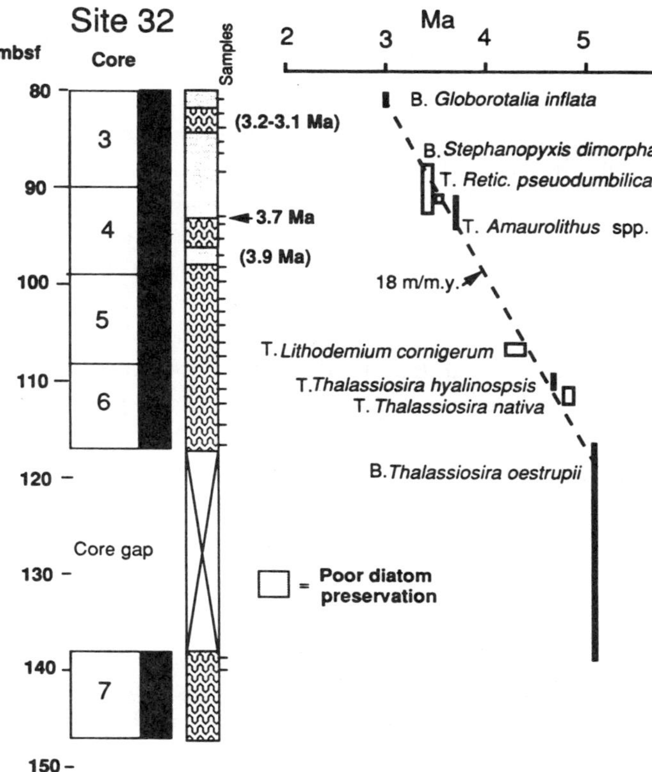

Fig. 3. Age vs. depth plot for the uppermost Miocene through lower Pliocene section recovered at DSDP Site 32. Planktonic foraminiferal and calcareous nannofossil stratigraphy after McManus et al. (1970), Bukry and Bramlette (1970), and Gartner (1970); datum levels after Backman and Shackleton (1983) and Beggren et al. (1985). Diatom stratigraphy after Barron (this paper) and datum levels after Table. 1

mated ages of diatom datum levels for the latest Miocene to early late Pliocene of California ranging between 6.0 and 2.8 Ma.

Onshore Pliocene Record

Figure 4 shows the latest Miocene to late Pliocene sediment records of an important California stratigraphic section in a north to south transect from the Centerville Beach section (40.5° N) to the Newport Beach section of the Los Angeles basin (32.6° N). Diatom stratigraphy (Table 1) has been combined with tephrostratigraphy (Sarna-Wojcicki et al., 1987, 1991) and limited paleomagnetic stratigraphy (Dodd et al., 1977; Madrid et al., 1986) in producing this figure. In the absence of these stratigraphies, benthic foraminiferal stratigraphy has been used (Ingle, 1972, 1980).

The relative abundance of diatoms as well as generalized lithologies also are shown in Fig. 4. Comparisons with the eustatic curve of Haq et al. (1987), times of major change in the motion of the Pacific plate, and the onset of uplift of the Coast Ranges of California have been included on the right side of the figure.

As revealed by Fig. 4, the clastic sediment component (sandstone and siltstone) increased abruptly in the Santa Cruz and southern San Joaquin sections during the latest Miocene (ca. 5.5 to 5.1 Ma). In the Newport Beach section of the deeper Los Angeles basin, this increased terrigenous influx was relatively fine-grained, representing the mudstones of the Capistrano Formation. After the Miocene, diatoms persisted as a major sediment component only in the central Santa Maria basin (e.g., the Harris Grade section) and in the Centerville Beach section of the Eel River basin; but even in these more distal onshore basins, diatom-rich sediments were deposited only until about 4 Ma. The middle part of the Pliocene (4 to 3 Ma) brought further coarsening of terrigenous sediments in the Eel River (Centerville Beach section), Santa Maria, Ventura (Balcom Canyon section), and Los Angeles (Newport Beach section) basins (Ingle, 1980). Diatoms are, therefore, extremely rare in upper Pliocene and Quaternary sediments of onshore California.

Table 1. Ages of selected diatom datum levels in California for the latest Miocene and Pliocene interval ranging between 6.0 and 3.0 Ma. B = first occurrence; T = last occurrence; C = interpolation at Centerville Beach (Fig. 2); H = extrapolation in Harris Grade section (Barron and Ramirez, 1991); S = interpolation in Sweeney Road section (Barron and Ramirez, 1991); NP = North Pacific (Barron, in press); SC* = paleomagnetic calibration at Santa Cruz (Dumont et al., 1986).

Datum	Age (Ma)	Source
B. *Neodenticula koizumii*	2.8–3.1	C
B. *Stephanopyxis dimorpha*	3.4–3.5	C
T. *Lithodesmium cornigerum*	4.2–4.4	H
T. *L. minusculum*	4.2–4.4	H
T. *Thalassiosira hyalinopsis*	4.65	H
T. *T. nativa* sensu Schrader (1973)	4.8–4.9	H
B. *Hemidiscus ovalis*	4.9–5.0	H,S
B. *Raphoneis fatula*	4.9–5.0	H,S
B. *Thalassiosira oestrupii*	5.1	NP
B. *T. praeoestrupii*	5.35	SC*
B. *T. hyalinopsis*	5.6	SC*
L. *T. miocenica*	5.6	SC*
B. *T. miocenica*	5.8	SC*
T. common *Rouxia californica*	6.0	NP

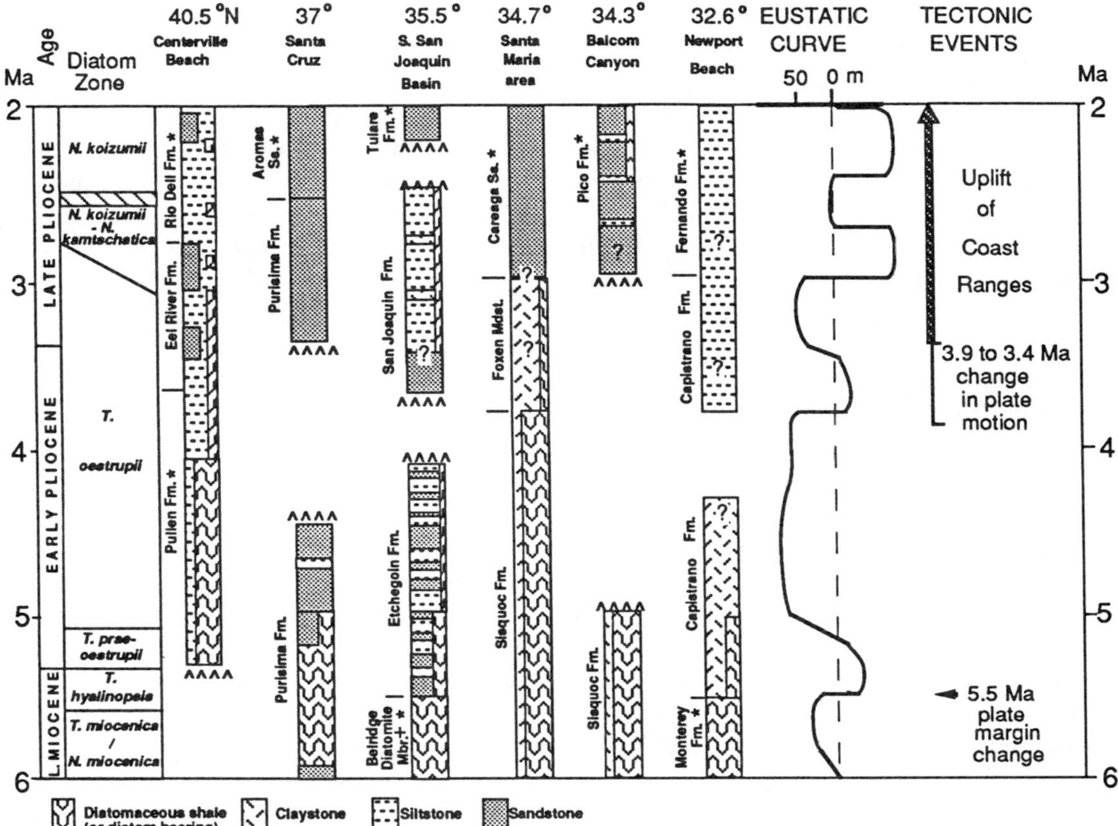

Fig. 4. Comparison of selected onshore California sections for the uppermost Miocene through upper Pliocene interval of 6 to 2 Ma with major tectonic events and the eustatic curve of Haq et al. (1987). Geologic time scale from Berggren et al. (1985). Generalized lithology and geochonology of sections as follows: Centerville Beach (Fig. 2); Santa Cruz (Madrid et al., 1986); southern San Joaquin basin (Woodring et al., 1940; Sarna-Wojcicki et al., 1991; M.P. Dumont, oral comm., 1991); Santa Maria area (Woodring et al., 1943; Barron and Ramirez, 1991); Balcom Canyon (A. Sarna-Wojcicki, oral comm., 1991; Barron, unpublished data); Newport Beach (Ingle, 1972; Barron, 1986b) * = Age of lithologic unit ranges younger than 2.0 Ma or older than 6.0 Ma; + = member of Monterey Formation.

Tectonic and Eustatic Implications

The times of onset of increased influx and coarsening of terrigenous sediments coincide with major falls in global sea level at 5.5, 3.8, and 3.0 Ma as identified by the Haq et al. (1987) eustatic curve (Fig. 4). Similarly, the diatomaceous (or fine-grained) component in the Centerville Beach and Santa Maria sections seems to correspond to an interval of relatively higher global sea level prior to about 3.8 Ma, when one would expect less terrigenous input. On the other hand, this same early Pliocene (5.0 to 3.8 Ma) interval of relatively high sea level also coincides with coarse clastic sediments in the southern San Joaquin and Santa Cruz sections and with a major hiatus in the Balcom Canyon section, suggesting a tectonic overprint.

Ingle (1980) compared the paleobathymetric and sediment histories of southern California continental borderland basins and concluded that tectonic reorganization of the region began in the latest Miocene and was intensified during the Pliocene. Miocene basins began to fill rapidly with terrigenous sediments. Some areas experienced uplift (e.g., the Coast Ranges), while other areas were deepened (e.g., the Ventura and Los Angeles basins) (Ingle, 1980; Crouch et al., 1984). Seismic facies analyses by Teng and Gorsline (1989) reveal that the Pliocene and Quaternary sediments of the borderland basins were deposited as flat-lying ponded sequences that filled the contemporary basins, whereas the underlying upper Miocene diatomaceous sediments follow (and, therefore, predate) much of the topographic structure of the basins. The rapid increase in clastic sedimentation during the Pliocene and Quaternary filled such onshore basins as the San Joaquin, Santa Maria, Ventura, and Los Angeles basins (Ingle, 1980; Teng and Gorsline, 1989).

Sedlock and Hamilton (1991) argue that a major reorganization of the boundary zone between the Pacific and North American plates occurred at about 5.5 Ma due to an eastward jump of the southern triple junction (with the Rivera plate off central Mexico) to the mouth of the Gulf of California. The Gulf of California began to open and the central part of the inboard fault system of California jumped eastward, forming the modern-day San Andreas fault system. Clockwise rotation of the western Transverse Ranges (the area immediately south of the Santa Maria basin in Fig. 1) commenced and inner borderland basins underwent extension.

The late Pliocene and Quaternary brought uplift and folding of the Coast Ranges and the formation of a number of more northerly-trending faults in coastal California, suggesting compression of the California margin according to Harbert and Cox (1989) and Sedlock and Hamilton (1991). Seismic studies by Crouch et al. (1984) and Namson and Davis (1990) suggest that this compressive period began in the middle part of the Pliocene. Harbert and Cox (1989) argue that a major change in motion of the Pacific plate between 3.9 and 3.4 Ma caused a change from strike-slip to transgressive motion along the (coastal California) Pacific-North American plate boundary. They also cite increased tectonism in northern Japan and New Zealand at this time as further evidence for this change in plate motion. Similar to Bachman and Crouch (1987) and Graham (1987), who correlate major unconformities in California basins with this important tectonic event, Fig. 4 shows that this important tectonic event coincides with hiatuses in the Santa Cruz, Balcom Canyon, and southern San Joaquin sections. A major increase in the sediment accumulation rate in the Centerville Beach section also occurs at about 3.4 Ma (Fig. 2) and is further evidence for uplift of the Coast Ranges.

Offshore Pliocene Record

Offshore Pliocene sections are farther removed from coastal areas of uplift and sub-aerial erosion, less diluted by terrigenous debris, and should be less affected by eustatic and tectonic events than onshore Pliocene sections. The offshore sediment record, therefore, should more closely reflect Pliocene paleoceanographic conditions than the onshore record. Uppermost Miocene to upper Pliocene (6.0 to 2.0 Ma) sediment records at offshore DSDP Sites 173, 32, 467, and 469 are compared with that in the Centerville Beach section in Fig. 5. In compiling this figure, stratigraphy from published references such as Sarna-Wojcicki et al. (1987) and Barron et al. (1981) have been supplemented by application of updated diatom stratigraphy (Table 1; Fig. 3). Intervals that either are barren of diatoms or contain only sparse, dissolution-

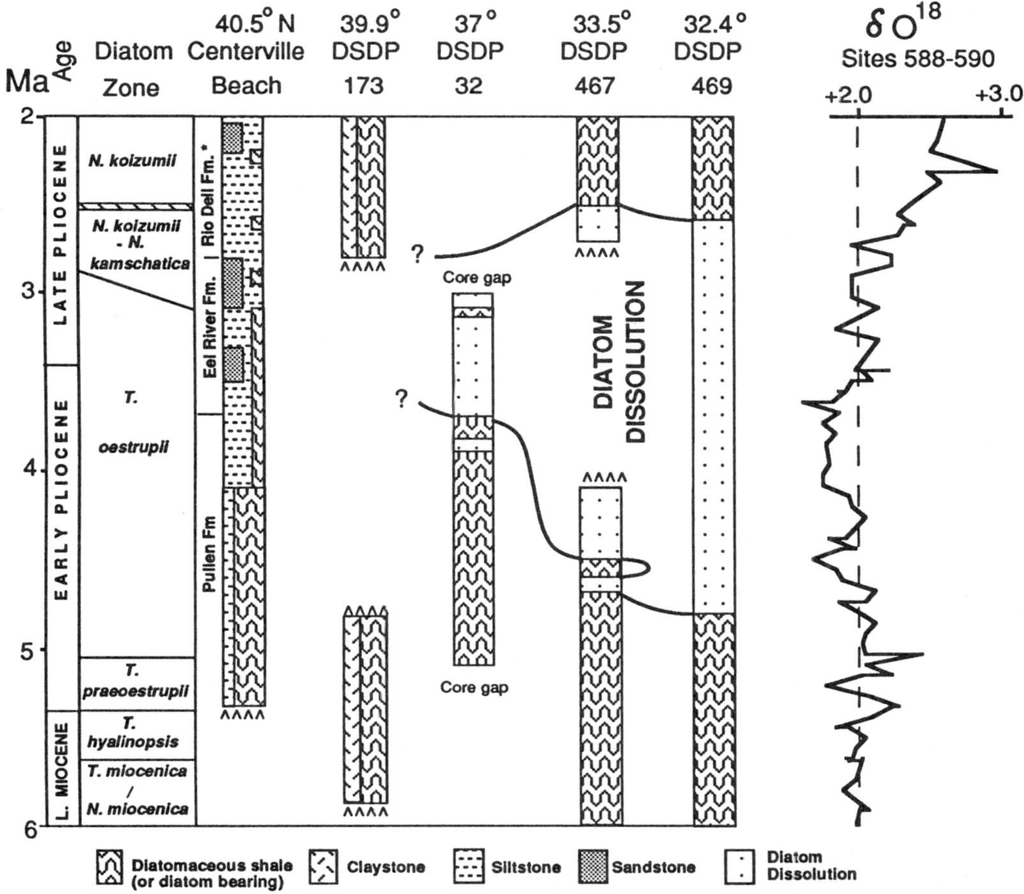

Fig. 5. Comparison of diatom dissolution intervals in offshore California DSDP sections during the latest Miocene through early Pliocene interval from 6 to 2 Ma with a composite benthic foraminiferal oxygen isotope curve from DSDP Sites 588 and 590 (Kennett, 1986). Geologic time scale from Berggren et al. (1985). Geochronology of sections as follows: Centerville Beach(Fig. 2), Site 173 (Sarna-Wojcicki et al., 1987; Barron, 1986b), Site 32 (Fig. 3), Sites 467 and 469 (Barron et al., 1981; updated using Table 1 of this report). * = age of Eel River Formation ranges younger than 2.0 Ma.

resistant forms such as fragments of *Coscinodiscus* and *Stephanopyxis* are also indicated in Fig. 5. The offshore records are compared to the composite benthic foraminiferal oxygen isotope curve from DSDP Sites 588 and 590 in the southwest Pacific (Kennett, 1986) on the right side of the figure.

Paleoceanographic Implications

In present-day waters off California, diatoms owe their abundance to coastal upwelling of nutrient-rich waters, which is driven by the persistent southward winds that are associated with the southward-flowing California Current (Sverdrup et al., 1942). Slackening of the California Current during E1 Nino periods results in a diminished southerly current, decreased upwelling, and diminished diatom fluxes (Lange et al., 1990). Because the preservation and abundance of diatoms in sediments is directly related to their abundance in the overlying surface waters (Calvert, 1974), a decline in spring diatom blooms caused by a slackening of the California Current would result in a decreased diatom sedimentation rate and in a decline in diatom preservation in seafloor sediments. Consequently, one would expect that intervals of poor diatom preservation in otherwise pelagic sediments off California would correspond to high-latitude warming and a slackening of the southward flow of the California Current.

An interval of poor diatom preservation is present throughout much of the Pliocene prior to 2.6 Ma at DSDP sites off California. The records at Site 173 and 467, however, are interrupted by hiatuses which span the intervals from about 4.8 to 2.8 Ma and 4.1 to 2.7 Ma, respectively. In general, this interval of poor diatom preservation at Sites 467, 469, and 32 coincides with relatively lighter oxygen isotope values, which indicate warming of high-latitude surface waters and/or deglaciation. At DSDP Sites 467 and 469 off southern California, poor diatom preservation begins between 4.8 and 4.7 Ma, coincident with the beginning of an early Pliocene interval (ca. 4.8 to 3.6 Ma) characterized by relatively lighter values in the Site 588-590 oxygen isotope curve (Fig. 5). A lightening of planktonic oxygen isotope values is also documented by Hodell et al. (1991) between 4.8 Ma and 3.76 Ma at ODP Site 704 in the South Atlantic, where it is interpreted by them to signal a period of major warming of high-latitude surface waters. Presumably, high-latitude warming and a slackening of the southward flow of the California Current at about 4.8 Ma caused a decline in diatom abundance in the surface waters above Sites 467 and 469 and resulted in poor diatom preservation. Abundant diatoms persisted in more inshore areas (e.g., the Harris Grade and Centerville Beach sections) as well as at the more northerly Site 32 (Fig. 4).

The onset of poor diatom preservation at Site 32 at 3.7 Ma may signal a generalized narrowing of the entire California Current system rather than further high-latitude warming. Diatom preservation also declines abruptly at this time at Sites 471 (23°28.93'N, 112°29.78'W, water depth 3101 m) and 476 (23°02.43'N, 109°05.35'W, water depth 2403 m) off southern Baja California (Barron, unpublished data). Similarly, the onset of this interval of poor preservation 3.7 Ma occurs near the end of the period of warmest Pliocene high-latitude paleotemperatures according to isotope studies (Hodell et al., 1991).

At Sites 467 and 469 off southern California abundant and well-preserved diatoms returned at about 2.6 Ma, coincident with major cooling of high-latitude surface waters and increased amplitudes of the glacial/interglacial oxygen isotope signal (Raymo et al., 1989). Unfortunately, a coring gap above Core 32-3 prevents the identification of the resumption of diatom sedimentation at Site 32, but well-preserved diatoms are present in sediments estimated to be 2.8 Ma in age, immediately above an unconformity at the more northerly Site 173 (Schrader, 1973; Barron, 1989) (Fig. 5).

Diatom abundance and preservation in the more inshore Centerville Beach section is probably more determined by the amount of terrigenous debris masking diatom sedimentation than by productivity in the overlying waters. Whereas offshore sections show increased diatom abundance and preservation after 2.6 Ma, poor diatom preservation characterizes the Centerville Beach diatom record between 3.1 and 2.0 Ma. Presumably, terrigenous input increased during the late Pliocene in the Centerville Beach section, corresponding to both increased uplift of the Coast Ranges and lowered global sea level (Fig. 4).

Comparison of a map showing the preservation of diatoms in North Pacific surface sediments (L.H. Burckle, written comm., 1991) with a map of August North Pacific sea surface temperatures (Sverdrup et al., 1942) suggests that off the coasts of Oregon and northern California diatoms are found only in surface sediments underlying regions that experience August sea surface temperatures of 17°C or less. Such waters correspond to the more inshore regions of the California Current where strong offshore winds cause spring and summer upwelling and result in relatively high seasonal diatom production. If modern conditions can be used as a model for Pliocene time, the intervals of poor diatom preservation at Sites 467, 469 and 32 would be indicative of August sea surface temperatures in excess of 17°C. Presumably, lower paleotemperatures and diatom productivity are confined to more inshore areas, where tectonic uplift of the adjacent borderland has resulted in increased hemipelagic sedimentation, masking the diatom record.

Summary

The early Pliocene diatom record of California is very poorly known due to increased terrigenous dilution in onshore sections, the widespread presence of hiatuses or condensed intervals in both onshore and offshore sections, the provincial nature of diatom assemblages, and the pervasive presence of dissolution in sections south of about 40°N.

A diatom stratigraphy for this 5.0– to 3.0–Ma interval is proposed based on study of diatoms from the Harris Grade section of southern California, DSDP Site 32 off central California, and the Centerville Beach section of northern California. This stratigraphy is combined with tephrostratigraphy and limited paleomagnetic stratigraphy in order to correlate both onshore and offshore California sections and to access the effects of regional tectonic, eustatic, and paleoceanographic events on the sediment record.

Clastic sediment input increased abruptly in onshore California sections during the latest Miocene following an eastward jump in the San Andreas fault system and a global fall in sea level at 5.5 Ma. Diatom sedimentation was effectively masked by terrigenous sedimentation by 5.0 Ma in most sections, although diatom-rich sediments persisted to about 4 Ma in the central part of the Santa Maria basin and in the relatively distal Centerville Beach section of the Eel River basin. Eustatic sea level falls at 3.8 and 3.0 Ma coincide in part with an increased input of terrigenous debris into onshore California basins. However, compression of the California margin and uplift of the Coast Ranges, which followed a change in motion of the Pacific plate between 3.9 and 3.4 Ma, were probably most responsible for locally increased terrigenous sedimentation rates along the Pacific borderland of the western United States during the late Pliocene and Quaternary.

Although terrigenous dilution is less of a problem in the offshore sediment record of California, dissolution of diatom assemblages is commonplace in the early Pliocene and early part of the late Pliocene, indicating a marked decline in the production of diatoms in surface waters. The onset of diatom dissolution at Sites 467 and 469 off southern California at 4.8 Ma coincides with the start of a major period of high-latitude warmth or deglaciation (ca. 4.8 to 3.6 Ma) according to oxygen isotope studies. The southward-flowing California Current presumably slackened at this time resulting in reduced coastal upwelling and diatom blooms in offshore areas. Diatoms, however, persisted in more inshore regions, such as the Santa Maria and Eel River basins, where they were subject to dilution by terrigenous debris. Well-preserved diatoms only return to Sites 467 and 469 after 2.6 Ma, when isotope studies give evidence of significant high-latitude cooling (increased amplitudes in the isotopic glacial-interglacial signals). At Site 32 off central California, the onset of diatom dissolution at 3.7 Ma may indicate a generalized narrowing of the California Current rather than further high-latitude warming, because diatoms also disappear from Sites 471 and 476 off southern Baja California at this same time and do not reappear in younger sediments. Based on the modern distribution of diatoms, it appears that the offshore intervals of diatom dissolution coincided with August sea surface temperatures in excess of 17°C.

Acknowledgements

This manuscript benefitted from the reviews of James C. Ingle, Jr., of Stanford University and Andrei Sarna-Wojcicki of the U.S. Geological Survey. Michael Dumont of ARCO Oil and Gas Company and James Crouch of Crouch and Bachman and Associates also provided very useful discussion, and Scott Starratt of the U.S. Geological Survey edited the manuscript. Patricia McCrory is thanked for supplying samples and data from the Centerville Beach section. DSDP samples were provided by the Deep Sea Drilling Project and Ocean Drilling Program.

References

Bachman, S.B. and Crouch, J.K., 1987. Geology and Cenozoic history of the northern California margin: Point Arena to Eel River. *In:* R.V. Ingersoll and W.G. Ernst (Eds.), Cenozoic Basin Development of Coastal California, Rubey Volume VI. Englewood Cliffs, NJ: Prentice-Hall, pp. 125–145.

Backman, J. and Shackleton, N.J., 1983. Quantitative biochronology of Pliocene and early Pleistocene calcareous nannoplankton from the Atlantic, Indian, and Pacific oceans. *Mar. Micropaleontol.*, 8: 141–170.

Barron, J.A., 1981. Late Cenozoic diatom biostratigraphy and paleoceanography of the middle latitude eastern North Pacifc. *In:* Init. Repts., Deep Sea Drilling Project, 63. U.S. Govt. Printing Office, Washington, D.C., pp. 507–538.

Barron, J.A., 1986a. Paleoceanographic and tectonic controls on deposition of the Monterey Formation and related siliceous rocks in California. *Palaeoceanogr., Palaeoclimatol., Palaeoecol.*, 53: 27–45.

Barron, J.A., 1986b. Updated diatom biostratigraphy for the Monterey Formation of California. *In:* R.E. Casey and J.A. Barron (Eds.), Siliceous Microfossil and Microplankton of the Monterey Formation and Modern Analogs, Pacific Section. *Soc. Econ. Paleontol. Mineral.*, 45: 105–119.

Barron J.A., 1989. The late Cenozoic stratigraphic record and hiatuses of the northeast Pacific; Results from the Deep Sea Drilling Project. *In:* E.L. Winterer, D.M. Hussong, and R.W. Decker (Eds.), The Eastern Pacific Ocean and Hawaii, Volume N, The Geology of North America. Denver, CO.: Geol. Soc. Am., pp. 311–322.

Barron, J.A., in press. Neogene diatom datum levels in the equatorial and North Pacific. *In:* T. Saito and K. Ishizaki (Eds.), The Centenary of Japanese Micropaleontology. Tokyo: Tokyo Univ. Press.

Barron, J.A. and Baldauf, J.G., 1986. Diatom stratigraphy of the lower Pliocene part of the Sisquoc Formation, Harris Grade section, California. *Micropaleo.*, 32(4): 357–371.

Barron, J.A., Poore, R.Z., and Wolfart, R., 1981. Biostratigraphic summary, DSDP Leg 63. Initial Reports, Deep Sea Drilling Project, 63. U.S. Govt. Printing Office, Washington, D.C., pp. 927–940.

Barron, J.A. and Ramirez, P.C., 1991. Diatom stratigraphy of selected Sisquoc Formation sections, Santa Maria basin, California. U.S. Geol. Surv. Open-File Rep.

Berggren, W.A., Kent, D.V., and Van Couvering, J.A., 1985. The Neogene: Part 2. Neogene geochronology and chronostratigraphy. *In:* N.J. Snelling (Ed.), The Chronology of the Geological Record. Mem. Geol. Soc., No. 10 (London), pp. 211–260.

Bukry, D. and Bramlette, M.N., 1970., Coccolith age determinations Leg 5, Deep Sea Drilling Project. Init. Repts., Deep Sea Drilling Project, 5. U.S. Govt. Printing Office, Washington, D.C., pp. 487–494.

Burckle, L.H., Dodd, J.R., and Stanton, R.J., Jr., 1980. Diatom biostratigraphy and its relationship to paleomagnetic stratigraphy and molluscan distribution in the Neogene Centerville Beach section, California. *J. Paleontol.*, 54(4): 664–674.

Calvert, S.E., 1974. Deposition and diagenesis of silica in marine sediments. *Spec. Publ. Inter. Assoc. Sediment.*, 1: 273–299.

Compton, R.R., 1966. Analysis of Plio-Pleistocene deformation and stresses in the northern Santa Lucia Range. *Geol. Soc. Am. Bull.*, 77: 1361–1380.

Crouch, J.K., Bachman, S.B., and Shay, J.T., 1984. Post-Miocene compressional tectonics along the central California margin. *In:* J.K. Crouch and S.B. Bachman (Eds.), Sedimentation along the California Margin. Pacific Section, Soc. Econ. Paleontol. Mineral., Bakersfield, CA, pp. 37–54.

Dodd, J.R., Mead, J., and Stanton, R.J., Jr., 1977. Paleomagnetic stratigraphy of the Pliocene Centerville Beach section, northern California. *Earth Planet. Sci. Lett.,* 34(3): 381–386.

Dumont, M.D., Baldauf, J.G., and Barron, J.A., 1986. *Thalassiosira praeoestrupii* – a new diatom species for recognizing the Miocene/Pliocene Epoch boundary in coastal California. *Micropaleontology*, 32(4): 372–377.

Gartner, S., Jr., 1970. Coccolith age determinations Leg 5, Deep Sea Drilling Project. Init. Repts., Deep Sea Drilling Project, 5. U.S. Govt. Printing Office, Washington, D.C., pp. 495–500.

Graham, S.A., 1987. Tectonic controls on petroleum occurrence in central California. *In:* R.V. Ingersoll and W.G. Ernst (Eds.), Cenozoic Basin Development of Coastal California, Rubey Volume VI. Prentice-Hall, Edgewood Cliffs, NJ, pp. 48–63.

Harbert, W. and Cox, A., 1989. Late Neogene motion of the Pacific plate. *J. Geophys. Res.*, 94(B3): 3052–3064.

Haq, B.U., Hardenbol, J., and Vail, P.R., 1987. Chronology of fluctuating sea level since the Triassic. *Science*, 235: 1156–1167.

Hodell, D.A., Müller, D.W., Ciesielski, P.E., and Mead, G.A., 1991. Synthesis of oxygen and carbon isotopic results from Site 704: implications for major climatic-geochemical transitions during the late Neogene. *In:* Proc. ODP, Sci., Results, 114. College Station, TX: Ocean Drilling Program, College Station, TX, pp. 475–480.

Ingle, J.C., Jr., 1972. Biostratigraphy and paleoecology of early Miocene through Pleistocene benthonic and planktonic foraminifera, San Joaquin Hills-Newport Bay, Orange County, California. *In:* E. Stinemeyer (Ed.), the Pacific Coast Miocene biostratigraphic Symposium. Pacific Section, Soc. Econ. Paleontol. Mineral., Bakersfield, CA, pp. 255–283.

Ingle, J.C., Jr., 1980. Cenozoic paleobathymetry and depositional history of selected sequences within the southern California continental borderland. Memorial to Orville 1. Bandy., *Cushman Foundation Spec. Pub.*, 19: 163–195.

Kennett, J.P., 1986. Miocene to early Pliocene oxygen and carbon isotope stratigraphy in the southwest Pacific, Deep Sea Drilling Project Leg 90. *In:* Init. Repts., DSDP, 90. U.S. Govt. Printing Office, Washington, DC, pp. 1383–1411.

Koizumi, I. and Tanimura, Y., 1985. Neogene diatom biostratigraphy of the middle latitude western North Pacific, Deep Sea Drilling Project Leg 86. *In:* Init. Repts., DSDP, 86. U.S. Govt. Printing Office, Washington, D.C., pp. 269–300.

Lange, C.B., Burke, S.K., and Berger, W.H., 1990. Biological production off southern California is linked to climatic change. *Climatic Change,* 16: 319–329.

Madrid, V.M., Stuart, R.M., and Verosub, K.L., 1986. Magnetostratigraphy of the late Neogene Purisima Formation, Santa Cruz County, California. *Earth Planet. Sci. Lett.*, 79: 431–440.

McCrory, P.A., 1990. Neogene paleoceanographic events recorded in an active margin setting: Humboldt basin, California. *Mar. Micropaleo.*, 80: 267–282.

McManus, D.A. et al., 1970. Site 32. *In:* Init. Repts., DSDP, 5. U.S. Govt. Printing Office, Washingtorn, D.C., pp. 15–56.

Namson, J. and Davis, T.L., 1990. Late Cenozoic fold and thrust belt of the southern Coast ranges and Santa Maria basin, California. *Am. Assoc. Petrol. Geol. Bull.,* 74(4): 467–492.

Raymo, M.E., Ruddiman, W.F., Backman, J., Clement, B.M., and Martinson, D.G., 1989. Late Pliocene variation in Northern Hemisphere ice sheets and North Atlantic deep water circulation. *Paleoceanogr.,* 4: 413–446.

Sarna-Wojcicki, A.M., Morrison, S.D., Meyer, C.E., and Hillhouse, J.W., 1987. Correlation of upper Cenozoic tephra layers between sediments of the western United States and eastern pacific Ocean and comparison with biostratigraphic and magnetostratigraphic age data. *Geol. Soc. Am. Bull.,* 98: 207–223.

Sarna-Wojcicki, A.M., Lajoie, K.R., Meyer, C.E., Adam, D.P., and Rieck, H.J., 1991. Tephrochologic correlation of upper Neogene sediments along the Pacific margin, conterminous United States. *In:* R. Morrison (Ed.), Quaternary Nonglacial Geology: Conterminous United States, Volume K–2, The Geology of North America. *Geol. Soc. Am.,* Dever, CO, pp. 117-140.

Schrader, H.J., 1973. Cenozoic diatoms from the northeast Pacific, Leg 18. *In:* Init. Repts., DSDP, 18. U.S. Govt. Printing Office, Washington, D.C., pp. 673–797.

Sedlock, R.L. and Hamilton, D.H., 1991. Late Cenozoic tectonic evolution of southwestern California. *J. Geophys. Res.,* 94(B2): 2325–2351.

Sverdrup, H.U., Johnson, M.W., and Fleming, R.H., 1942. The Oceans, Their Physics, Chemistry, and General Biology. Prentice Hall, Edgewood Cliffs, NJ, 1087 p., 7 charts.

Teng, L.S. and Gorsline, D.S., 1989. Late Cenozoic sedimentation in California Continental Borderland basins as revealed by seismic facies analysis. *Geol. Soc. Am. Bull.,* 101(1): 27–41.

Woodring, W.P., Bramlette, M.N., and Lohman, K.E., 1943. Stratigraphy and paleontology of the Santa Maria district, California. *Am. Assoc. Petrol. Geol. Bull.,* 27(10): 1335–1360.

Woodring, W.P., Stewart, R., and Richards, R.W., 1940. Geology of the Kettleman Hills oil field, California. *U.S. Geol. Surv. Prof. Pap.,* 195, 170 p., 2 oversized plates.

Neogene Lithofacies and Depositional Sequences Associated with Upwelling Regions along the Eastern Margin of the Pacific

R. E. GARRISON

Earth Sciences Department, University of California, Santa Cruz, California, U.S.A.

Abstract

Sedimentary successions associated with Neogene upwelling systems along the eastern margin of the Pacific can be subdivided into about 15 key lithofacies. Sedimentological interpretations of these lithofacies allow preliminary systems tracts and sequence boundary interpretations of them within a sequence stratigraphic context. Of particular utility in this regard are distinctive early diagenetic carbonates and phosphates which may mark sequence and parasequence boundaries.

Introduction

Neogene sediments deposited beneath paleo-upwelling zones occur around much of the Pacific rim (Ingle, 1981; Dunbar et al., 1990). They are particularly well developed in onshore and offshore basins along the eastern margin of the Pacific, where they appear to be reflections of upwelling associated with the Neogene California and Peru-Chile currents (Fig. 1). These mainly siliceous deposits, which were originally diatom oozes and muds, were deposited in a variety of tectonic settings including forearc basins along the western, convergent margin of South America and transform-margin settings, including "pull-apart" basins, in California. In addition to their scientific importance, these deposits are economically significant because they contain large amounts of petroleum in California and phosphorites in California and Peru.

The Neogene sequences noted above contain a variety of siliceous lithofacies as well as phosphatic and carbonate lithofacies. This suggests that the sediment response to Neogene upwelling varied, either because the upwelling itself varied during the Neogene or because local tectonic or oceanographic factors produced local variations in sedimentation. This paper examines some of the more important of these lithofacies variations along with the stratal stacking patterns of facies and introduces speculations on the causes of these variations.

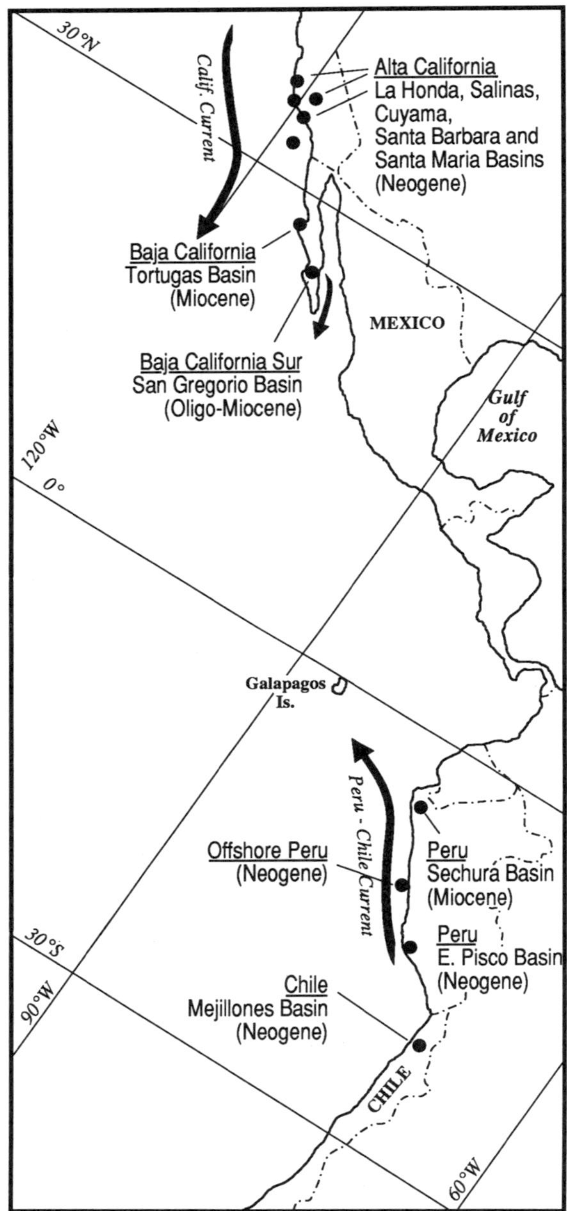

Fig. 1. Important Neogene sedimentary basins along the eastern margin of the Pacific and their geographic relationship to the present California and Peru-Chile boundary currents.

Lithofacies

Biosiliceous Lithofacies

Diatomites, Porcelanites, and Cherts

These rocks typically consist of more than 50–60% siliceous biogenic components, mainly diatoms, with the remainder comprising biogenic calcite, authigenic carbonate and phosphate minerals, organic matter, and fine-grained terrigenous components (clay minerals, etc.). This lithofacies is the major component of the Miocene Monterey Formation in California, the Miocene Tortuga Formation in Baja California, and the Mio-Pliocene Pisco Formation in Peru.

Originally biosiliceous sediments become profoundly modified by diagenesis. Porcelanites and cherts are burial diagenetic transformations of originally diatomaceous sediments, brought about by dissolution and reprecipitation of silica in heated pore waters. The terms porcelanite and chert are field terms which reflect original compositions; porcelanites are impure biosiliceous rocks with the luster of unglazed porcelan, while cherts are relatively pure biosiliceous rocks with a glassy luster. Opal-CT porcelanites and cherts record intermediate stages of alteration of opal-A diatomaceous sediments, at moderate burial depths and pore water temperatures generally below about 70° C; quartz cherts and porcelanites are formed mostly by dissolution of opal-CT and reprecipitation of silica as quartz at temperatures usually in excess of 70 degrees C (Pisciotto, 1978). Whereas porcelanites are diagenetic alterations of relatively impure diatomaceous sediments with admixtures of clay minerals, etc., most cherts record transformations of diatom oozes with no or only very small amounts of such admixtures and generally indicate deposition in distal basins far removed from sources of terrigenous detritus.

This suite of siliceous lithofacies appears to record the most intense phases of upwelling with attendant high fertility and productivity along with phytoplankton populations dominated by diatoms. The predominant sediment component was thus diatom frustules. Variations of this lithofacies are outlined below.

Lithofacies 1a: Thinnly Bedded Porcelanites and Cherts

This lithofacies consists of regularly spaced layers of porcelanite or chert, mostly in the range of 5 to 10 cm thick, separated by paper-thin shale layers (Figs. 2 and 12). Outcrops of this facies closely resemble those of radiolarian-bearing "ribbon cherts" of eugeoclinal terrains, and the rhythmic bedding of such cherts has been variously attributed to cyclic sedimentation, to fine-grained turbidity currents, and to diagenetic "unmixing" (McBride and Folk, 1979; Decker, 1991; Murray et al., 1992). Among Eastern Pacific occurrences, this facies is most abundant and widespread in the Miocene Monterey Formation of Alta California. One origin for this type of bedding can be discerned from observations in the upper part of the Monterey Formation, in the transition between the opal-A and opal-CT zones. Within the opal-A diatomites are alternating layers, each about 5 to 20 cm thick, of laminated, hemipelagic dia-

Fig. 2. Lithofacies 1a. Thinly bedded dark-colored quartz chert and light-colored opal-CT porcelanite, upper part of the Miocene Monterey Formation, Santa Maria Basin, Alta California, USA. Ruler is about one meter long.

Fig. 3. Lithofacies 1a. Thinly interbedded light-colored opal-A diatomite and dark-colored opal-CT chert, Tortugas Formation, Miocene, Tortugas Basin, Baja California, Mexico (cf. Helenes, 1984).

tomites and massive diatomaceous mudrocks; in places the latter contain rip-up clasts, have erosional basal contacts, and appear to be fine-grained turbidites. Selective replacement of the more purely siliceous, laminated, hemipelagic diatomites by opal-CT imparts what appears to be an embryonic rhythmicity to the bedding pattern (Fig. 3). This raises the possibility that one origin for this thinly bedded facies is silicification and diagenetic enhancement of a primary rhymicity caused by alternations of hemipelagic and fine-grained turbidite deposition.

In other parts of the Monterey, thinly bedded porcelanites are interbedded with thin phosphatic marlstone or shale layers without evidence of gravity flow deposition. Therefore, parts of this lithofacies may record cyclicity in sediment supply or preservation, e.g. variations in plankton productivity or preservation. Tada (1991) proposed such an origin to account for rhythmically bedded porcelanites in the Miocene Onnagawa Formation of Japan.

In the Salinas Basin of Alta California, thinly bedded Monterey porcelanites are very widespread across areas shown by Graham (1976) to have been slope regions of a proximal basin. In contrast, thinly bedded cherts and porcelanites in the Santa Maria Basin of Alta California appear to occur in a basinal part of a distal basin. Thus this lithofacies may record either a slope environment dominated by gravity flow redeposition of proximal diatom muds, producing thinly bedded porcelanites, or else a distal, basinal setting where oceanographically controlled cycles of productivity and/ or preservation yielded similar bedding patterns; these latter sequences, however, contain distinctive thinly bedded cherts derived from pure diatom oozes, a lithology not present in the slope sequences of more proximal basins.

Lithofacies 1b: Massive Diatomites, Porcelanites and Siliceous Mudrocks

This lithofacies consists of massive to thickly bedded, impure siliceous rocks that appear to have been completely bioturbated, although individual burrows are seldom visible (Fig. 4). It is present in Neogene siliceous units lying above the Monterey Formation in Alta California (e.g. the Santa Cruz Mudstone and the lower part of the Purisima Formation of the La Honda Basin (Clark, 1981; El-Sabbagh and Garrison,

Fig. 4. Lithofacies 1b. Massive, bioturbated diatomaceous mudrock, Mio-Pliocene Sisquoc Formation, Santa Maria Basin, Alta California, USA. Note blocky fracture pattern.

1990); parts of the Sisquoc Formation in the Santa Maria Basin (Ramirez, 1990)), as well as in the Pisco Formation in the East Pisco Basin of Peru (Dunbar et al., 1990). In all of these settings, the massive diatom muds appear to have been deposited in oxygenated shelfal environments of proximal basins. In some places within this lithofacies, external molds of infaunal and epifaunal bivalves are present, but the shells themselves are rarely preserved.

Lithofacies 1c: Alternations of Massive and Laminated Biosiliceous Rocks

In several onland basins in Alta California, in Baja California, and in Peru, these alternations (Fig. 5) occur in Neogene biosiliceous rocks on a decimeter to meter scale (Govean, 1980; Govean and Garrison, 1981; Pisciotto and Garrison, 1981; Hieshima, 1987; Ramirez, 1990). They are also well developed in Quaternary diatom muds in shelfal and upper slope settings along the present-day margin of Peru (Suess et al., 1988) and in diatom oozes of similar age from the slope of the Guyamas Basin, Gulf of California (Schrader et al., 1980; Calvert, 1966). Govean and Garrison (1981) suggested that cycles of this kind developed in upper slope and shelf sediments deposited near the edges of the oxygen minimum zone (OMZ) by fluctuations in the extent and intensity of the OMZ. During times of expanded and strong OMZ, low oxygen conditions inhibited infaunal burrowing so that primary laminations (including varves, cf. Soutar et al., 1981) were preserved, whereas contraction of the OMZ allowed oxygenated bottom waters, infaunal colonization, and bioturbation. Studies by Savrda and Bottjer (1986, 1987), Hieshima (1987), and Ramirez (1990) have

Fig. 5. Lithofacies 1c. Alternation of laminated (L) and massive, bioturbated (M) diatomaceous mudrocks, Mio-Pliocene Sisquoc Formation, Santa Maria Basin, Alta California, USA. The laminations imparted a bedding fabric to the laminated intervals whereas the massive intervals lack bedding and display a blocky fracture pattern.

shown how the type, degree, and distribution of bioturbated intervals in Neogene biosiliceous sediments were controlled by varying oxygen levels in bottom waters. Because the biosiliceous sediments are compositionally so homogeneous, individual burrows within the massive intervals are seldom visible in outcrop or hand samples, although they may become conspicuous in slabbed samples through use of X-ray radiography or through application of a thin coating of light machine oil to the slabbed surface (Govean, 1980; Savarda et al., 1985).

The sum of the evidence in both Neogene and Quaternary sequences of this kind is that this lithofacies records shelf to upper slope deposition during times of a fluctuating OMZ. Suess et al. (1988), in commenting on Quaternary alternations of this kind recovered from the Peru margin during Leg 112 of the Ocean Drilling Program, suggested that predominantly laminated intervals were formed during interglacial sea level highstands whereas massive, burrowed intervals developed during glacial lowstands. The evidence for this interpretation, however, is not compelling, and the problem requires further investigation.

Lithofacies 1d: Cherts

True cherts — that is, dense siliceous rocks with a glassy luster and chonchoidal fracture — are rare in Neogene biosiliceous sequences of the eastern Pacific margin. Although scattered chert nodules and layers are present in most of these sequences, the only abundant cherts occur within the Monterey Formation in distal basins where they locally serve as important fractured petroleum reservoirs, as in the Point

Fig. 6. Lithofacies 1d. Dark quartz cherts showing distinctive intrastratal folding are interbedded with planar bedded, thick dolomite beds (D) and thinly bedded opal-CT porcelanites. Middle part of the Monterey Formation, Miocene, Santa Maria Basin, Alta California, USA. Photograph provided by R. Behl.

Arguello oilfield in the offshore Santa Barbara Basin of Alta California (Crain et al., 1987). Many cherty intervals exhibit distinctive intrastratal folding (Figs. 6 and 8), shown by Behl (1990) to be produced by differential ductile shortening during tectonic deformation. Thinly bedded chert layers commonly occur interbedded with F-phosphatic marlstone, porcelanite, and authigenic dolomite layers (i.e. lithofacies 1a, 2a, and 3a) in complex cycles a few meters thick (Figs. 6, 7, and 8); interpretation of these cycles remains uncertain (Pisciotto, 1978), but some preliminary diatom age dating of such cyclic sequences suggests they may record fluctuations in diatom productivity on a scale of around 100,000 to 130,000 years, perhaps corresponding to one of the eccentricity cycles within the Milankovitch band of orbital parameters.

If this is confirmed by subsequent work, one possible interpretation is that the chert-porcelanite parts of the cycles record cold intervals of intense upwelling and dominantly biosiliceous sedimentation whereas the phosphatic marlstone-dolomite parts indicate warm, less fertile periods of the cycles. In any event, the cherts themselves indicate relatively undiluted biosiliceous sedimentation in distal basins. Such cherts are nearly 100% silica, and nearly all of this silica appears to have been of biosiliceous origin, although burial diagenesis has profoundly altered the original sediment. Among the alterations were silica phase changes, embrittlement (leading to pervasive fracturing), brecciation, and the development of distinctive intrastral folding (Behl, 1990).

Fig. 7. Lithofacies 1d. Aerial view of chert-bearing cycles in the middle part of the Miocene Monterey Formation along the coastline of the Santa Maria Basin, Alta California, USA (cf. Figs. 6 and 8). The cherty upper parts of the cycles form the resistant ledges, and about a dozen individual cycles (each a few meters thick) are visible in this view.

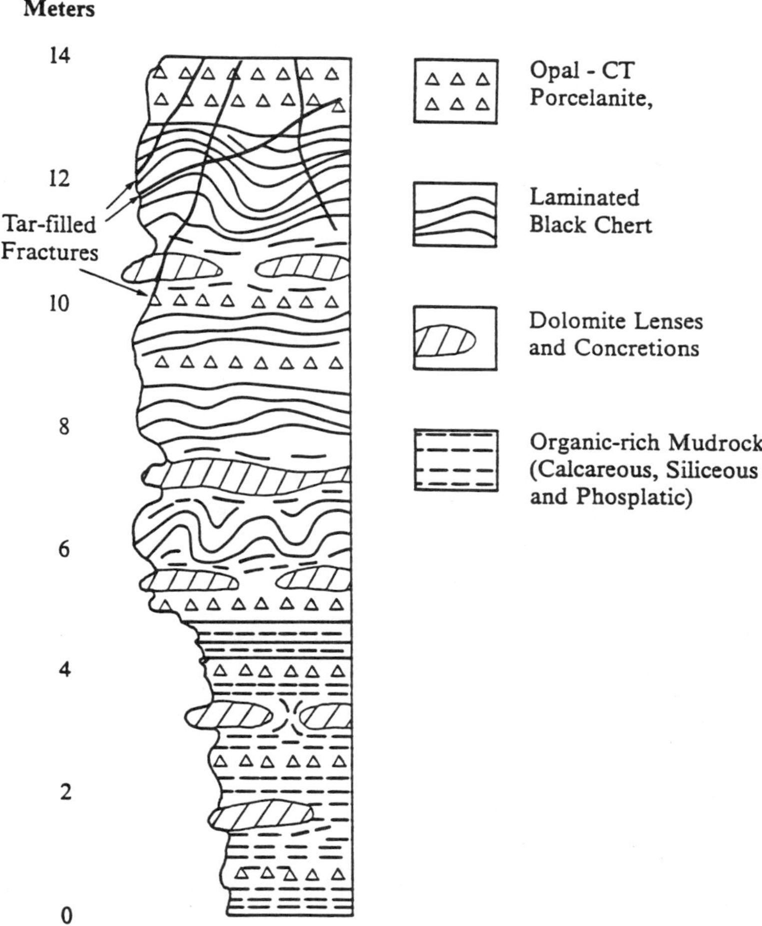

Fig. 8. Lithofacies 1a, 1d, 2a, and 3a are packaged into distinctive cycles in the middle part of the Miocene Monterey Formation of the Santa Maria Basin, Alta California, USA. Two such cycles are illustrated in this drawing. The top parts of each cycle contain intrastratally folded quartz cherts of lithofacies 1d. (cf. Fig. 6). These occur interbedded with opal-CT porcelanites (lithofacies 1a), phosphatic marlstones with F-phosphates (lithofacies 2a), and with layers of authigenic dolomite concretions (lithofacies 3a). The brittle cherts are intensively fractured, and fractures commonly are filled with tar; these cherts form fractured petroleum reservoirs in some oil fields of Alta California.

Phosphatic Lithofacies

Phosphatic rocks occur in three main lithofacies within Neogene upwelling units of the Eastern Pacific (Garrison, 1992). The discussion below utilizes the descriptive classification scheme proposed by Garrison and Kastner (1990) for phosphatic rocks recovered from the Peru margin during Leg 112 of the Ocean Drilling Program. Although the three phosphatic lithofacies described are volumetrically minor in the Neogene units under discussion, all three apparently formed during very early diagenesis on or just beneath the sea floor, and hence are important as records of depositional conditions. We believe these facies may also be important in sequence stratigraphic interpretations, as discussed in a later section.

Lithofacies 2a: F-phosphates

F-phosphates are friable, generally light-colored peloids, micronodules, and laminae of authigenic carbonate fluorapatite (CFA) which occur within a pelagic or hemipelagic host rock (Fig. 9). Composition of the host rocks, which typically contain only very small amounts of detritus, varies from biosiliceous to calcareous. Host rocks are commonly organic-rich (up to 30% TOC), laminated to banded, and some laminated host rocks contain microbial mat structures (Reimers et al., 1990). This type of phosphate, called pristine phosphate by some workers (Föllmi and Garrison, 1991; Föllmi et al., 1991), appears to have formed by CFA precipitation just below the sea floor in organic-rich sediments deposited in low oxygen settings. They are common in the middle part of the Monterey Formation in distal basins of Alta California (Garrison et al., 1987), in the Oligo-Miocene San Gregorio Formation of Baja California (Grimm et al., 1991), and in Neogene sections from shelfal forearc and upper slope basins along the Peru margin (Garrison and Kastner, 1990). Reimers et al. (1990) suggested that part of the phosphate for CFA precipitation is derived from degradation of sulfur-oxidizing bacterial communities which tend to be enriched in phosphate. Rhythmically bedded F-phosphate sections in Alta California were interpreted by Föllmi and Garrison (1991) as products of early diagenesis of fine-grained, organic-rich turbidites. For Quaternary Peru margin successions, Garrison and Kastner (1990) speculated that F-phosphates and associated laminated diatom

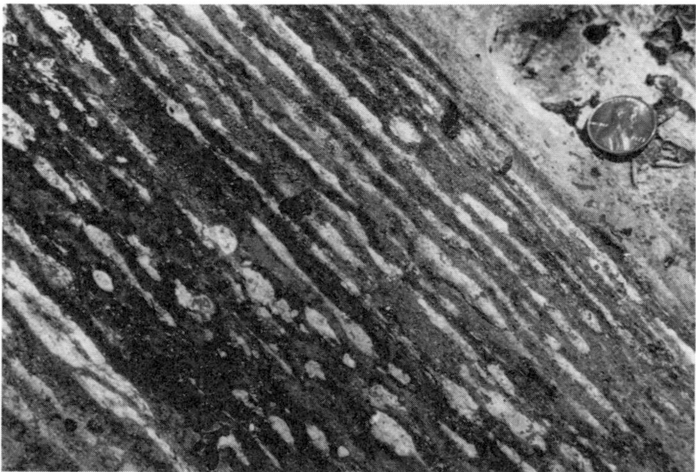

Fig. 9. Lithofacies 2a. Light-colored small lenses and layers of F-phosphates within dark, organic-rich marlstone, middle part of the Miocene Monterey Formation, Santa Barbara Basin, Alta California, USA. Coin is about 1.7 cm in diameter.

muds formed preferentially during interglacial periods of high sea level and an expanded OMZ.

Lithofacies 2b: P-phosphates

P-phosphates consist of sand-size peloids of CFA, which vary from structureless ovoids and intraclasts to coated grains which resemble ooids (Fig. 10). Electron microscopy reveals that many of the coatings are phosphatized microbial laminae surrounding a nucleus, the latter varying from diatom frustules to fish remains to siliciclastic detrital grains (Garrison et al., 1987; Ledesma, 1992). P-phosphates occur in distinct phosphatic sandstone beds, 10 to 300 cm thick, within shelfal successions that are predominantly biosiliceous hemipelagic deposits. They are thus products of diagenesis and deposition mainly in proximal settings, but some represent redeposition via gravity flows into more distal and deeper parts of basins (Garrison et al., 1987; Föllmi and Grimm, 1990). P-phosphate beds are most abundant, however, in shelf facies in proximal basins, where they constitute economically significant or

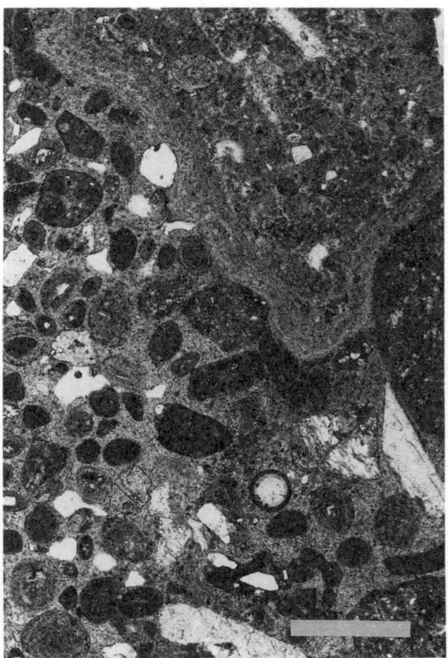

Fig. 10. Lithofacies 2b and 2c. Thin section photomicrograph of a D-phosphate nodule which incorporates two generations of P-phosphates. One generation (lower part of the photo) consists of sand-size phosphatic peloids, coated grains and intraclasts along with light-colored siliciclastic grains. The second generation (upper part of the photo) contains less distinct phosphate peloids and siliciclasic grains. Both generations were cemented by phosphate to form the nodule, and they are separated by an irregular laminated layer which may be a phosphatized microbial mat which coated the nodule during a stage of its growth. Sample is of Pliocene age and is from a core recovered from the Peru shelf during Leg 112 of the Ocean Drilling Program. Scale bar is 500 microns.

potentially significant phosphorites; examples include the Cuyama Basin of Alta California (Roberts and Vercoutere, 1985), the Oligo-Miocene forearc basin of Baja California Sur (Galli-Olivier et al., 1990), and the Sechura and Pisco (offshore and onshore) basins of Peru (Cheney et al., 1979; McClellan, 1989; Marty, 1989; Garrison and Kastner, 1990; Dunbar et al., 1990). Most such beds have erosional basal contacts, which suggest the beds are either event deposits (e.g. tempestites) or else deposits associated with erosional surfaces developed during a transgression. Because we remain uncertain about the origin of P-phosphates, uncertainties remain also about their sedimentological significance, as discussed in a later section.

Lithofacies 2c: D-phosphates

D-phosphates are dense and hard nodules, pebbles, and hardgrounds of CFA which are generally a dark brown to black color, at least on exterior surfaces (Figs. 10 and 11). Quaternary deposits of this type occur at the edges of the OMZ along the modern Peru margin (Burnett, 1977, 1980, 1990). A very minor lithofacies in terms of volume, D-phosphates occur in thin layers which commonly appear to mark unconformities or condensed sections. Two subfacies are recognized. Subfacies D1 consists of phosphate nodules and nodule-bearing gravels that occur at submarine unconformities or hiatuses. Some D1 phosphates appear to record exhumation of early-formed F-phosphate mircronodules, winnowing and reworking by bottom currents, reburial, and additional phosphatization which densifies and hardens the normally soft and friable F-phosphates; at some localities, this cycle of processes appears to have been repeated several times to produce compound D1 phosphate nodules marked by concentric, laminated growth bands. Allen and Dunbar (1988) described

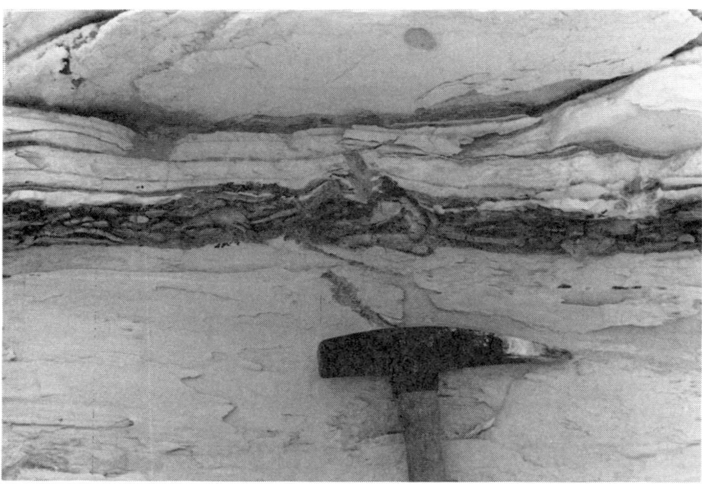

Fig. 11. Lithofacies 2c, D-phosphate. Thin layer in the center of the photograph, interbedded with light-colored diatomite, consists of flat clasts and nodules of D-phosphate. These clasts and nodules appear to have been reworked along a hiatal surface which may be a sequence or parasequence boundary. Miocene Pisco Formation, East Pisco Basin, Peru.

"ball-bearing" nodules of this type in the onshore Pisco Basin of Peru. Subfacies D2 phosphates are mainly hardgrounds formed by CFA cementation of D1 nodules and gravels as well as of P-phosphates. D2 hardgrounds from the Peru shelf commonly contain organic borings and serpulid worm tube encrustations, whereas those in the deeper water successions of Alta California are devoid of organic colonization except for some probable microbial mat coatings (now phosphatized) on hardground surfaces (Föllmi et al., 1991). D2 phosphates occur in condensed sections where sedimentation was substantially retarded (Föllmi and Garrison, 1991). Both D1 and D2 phosphates appear to mark important surfaces within a sequence stratigraphic context, as described below.

Lithofacies 2d: Banktop Glaucophosphatic Rocks

Glauconitic rocks are relatively rare in Neogene successions of the eastern Pacific, and this lithofacies, like the D-phosphates, is volumetrically a minor one, occurring mainly in the Monterey Formation of Alta California. It consists of sand-size peloidal glauconite and phosphate grains, in places with an admixture of phosphate nodules and siliciclastic sand grains. It occurs in thin units, typically a meter or less thick, whose stratigraphic context suggests they were condensed deposits formed on sediment-starved submarine banktops, probably near the edges of the OMZ (Garrison et al., 1987).

Carbonate Lithofacies

Lithofacies 3a: Authigenic Dolomites and Limestones

Authigenic dolomite and calcite occur as isolated crystals in all of the upwelling units herein discussed, and, for example, according to Isaacs (1984), dolomite crystals disseminated in biosiliceous and calcareous-siliceous host rocks constitute, volumetrically, the most abundant form of dolomite in the Monterey Formation. More evident in these units, however, are distinct beds or concretionary horizons of authigenic carbonates which cemented and commonly replaced a host sediment (Figs. 6, 8, and 12). Most are composed of dolomite, some of calcite. Isotopic work by Murata et al. (1969) and Pisciotto (1978) has shown wide ranges in carbon isotopes in Monterey dolomites, indicating precipitation in varied zones of organic matter degradation, including the zone of sulfate reduction. Following the experimental work of Baker and Kastner (1981), a plausible explanation for the origin of dolomite in these organic-rich units is precipitation within and beneath the zone of sulfate reduction. In low oxygen settings with organic-rich sediments, this zone may occur at or just beneath the sea floor, a fact consistent with the occurrence of early diagenetic dolomite crystals at the sediment-water interface disseminated in diatomaceous sediments on the Peru margin and fully lithified dolomite layers only 1.5 meters below this interface (Suess et al., 1988). We believe, therefore, that most dolomite layers formed near the sea floor during early diagenesis, although their isotopic compositions may have become modified by recrystallization or additional components of dolomitization during later burial diagenesis (Kushnir and Kastner, 1984; Baker and

Fig. 12. Lithofacies 1a and 3a. Thinly bedded opal-CT porcelanites (lithofacies 1a) contain two layers of authigenic dolomite (lithofacies 3a) marked by the letter "D". Note that the dolomite layers are somewhat thicker and more massive than the porcelanite beds. Miocene Monterey Formation, Santa Maria Basin, Alta California, USA.

Fig. 13. Lithofacies 3a. Upper bedding plane surface of an authigenic dolomite layer is perforated by organic borings indicating that this was a hardground exposed on the sea floor. Chilcatay Formation, Oligo-Miocene, East Pisco Basin, Peru. Diameter of coin is about 2.5 cm.

Burns, 1985). Sea floor dolomitization is clearly evident in dolomitic hardground surfaces perforated by organic borings (Fig. 13) and in places colonized by hard-surface encrusting organisms; dolomites of this kind occur in shelfal biosiliceous deposits of the Peruvian Pisco Formation (Baker and Martin, 1988) and of the Oligo-Miocene successions of Baja California Sur (Grimm et al., 1991).

Dolomite layers in the Monterey and Pisco Formations occur at seemingly regularly spaced intervals (Fig. 12). White (1989), using diatom age dating of selected Monterey sections (Khan et al., 1989) showed that the dolomites occur at a probable frequency of about 20,000 to 50,000 years, suggesting possible correlations with the precession and obliquity cycles of the Milankovitch parameters. We offer the following speculative scenario for the origin and significance of the dolomite layers. Compton and Siever (1986) and Compton (1988) demonstrated that early dolomitization in Monterey sediments could be explained by diffusion of ions from the overlying seawater and that the residence time of the dolomite-precursor sediment near the sediment-water interface is crucial when a seawater source of magnesium or calcium limits dolomite precipitation. Shimmield and Price (1984) showed that modern shelfal dolomite forms in sea floor sediments off Baja California in areas of low net sedimentation. These observations suggest that the most intense episodes of dolomitization (and formation of dolomite beds or concretions) occurred during periods of slow sedimentation when diffusion from seawater supplied maxium quantities of magnesium and calcium to layers undergoing dolomitization just beneath the seafloor. (Such early formed dolomites could be easily uncovered and exposed on the sea floor as hardgrounds in shelfal areas like the Neogene hardgrounds in Peru and Baja California.) If the dolomite layers (1) occur at the precession and obliquity frequencies and (2) mark slowdowns in sedimentation, a likely underlying cause is an Milankovitch cycle-related rise in sea level, trapping detrital sediment in nearshore settings and thereby reducing sedimentation rates in more distal settings. If this interpretation is correct, the dolomites may be regarded as flooding surfaces at parasequence boundaries (Van Wagoner et al., 1988; Vail et al., 1991), as will be discussed later.

Lithofacies 3b: Pelagic Hemipelagic Calcareous Siliceous Rocks

Micritic siliceous limestones comprise this facies, which is best developed within the early to early middle Miocene, lower part of the Monterey Formation, in distal basins like the Santa Barbara Basin of Alta California (Isaacs, 1980). Though burial diagenesis and silica phase changes have altered many of the primary sediment components, the most likely precursor sediment was a coccolith-foraminiferal-diatom ooze, probably a product of deposition beneath relatively low fertility zones during a warm highstand interval with attenuated coastal upwelling. Many calcareous, coccolith-bearing rocks in the Monterey Formation are laminated, organic-rich, and contain low-oxygen-tolerant benthic foraminiferal assemblages (e.g. buliminids), all suggesting low oxygen depositional settings possibly associated with an expanded OMZ. Of uncertain significance in determining the final sediment composition are variations in preservation of calcareous and siliceous microplankton shells in low-oxygen environments. There is some evidence for preferential dissolution of siliceous tests in low-oxygen environments (Boucher, 1984), hence it is possible that the calcareous nature of this facies is at least partly a matter of preferential preservation.

Lithofacies 3c: Bioclastic Shelfal Limestones

Limestones composed of shallow-water skeletal debris occur associated with shelfal biosiliceous rocks in the Oligo-Miocene of Baja California Sur, where they show prominent cross bedding (Grimm et al., 1991), and with in Mio-Pliocene successions in Chile (Tsuchi et al., 1988). Biota represented include barnacles, bivalves, and bryozoa, indicating that these are temperate water carbonates in the sense of Nelson (1978). We suggest these are nearshore facies in regions where shallow coastal waters remained cool due to vigorous upwelling, but we note that this facies has been little studied.

Siliciclastic Lithofacies

Three general types of siliciclastic lithofacies are associated with Neogene biosiliceous facies within proximal basins in upwelling regions of the eastern Pacific. These are shelfal sandstones, shelfal mudrocks, and turbidite fans. In Alta California, they occur interbedded with or overlying the commonly organic-rich biosiliceous, petroleum source rocks of the Monterey Formation, and the sandstone lithofacies serve, in some basins (e.g. the San Joaquin Basin), as petroleum reservoirs.

Lithofacies 4a: Shelfal Sandstones

Shelfal sandstones occur interbedded with biosiliceous rocks in a variety of subfacies. In clastic-dominated basins or in the proximal portions of shelf-dominated basins, this lithofacies includes alluvial fan, estuarine, deltaic, and littoral sandstones as well as offshore sandstones (DeVries, 1988; Callaway, 1990). Phillips (1983) described sandstones with large-scale cross bedding formed in a tidally dominated shelfal region of the La Honda Basin in northern Alta California. These rocks, assigned to the Santa Margarita Sandstone which lie between biosiliceous facies of the Monterey Formation, correlate with a sea level lowstand at ca. 10–11 Ma, but they also appear to reflect a period of basin uplift (Phillips, 1983). Similar siliciclastic lithofacies of about the same age are widespread in the Neogene basins of Alta California and likewise appear to coincide with a middle Miocene lowstand and a period of uplift.

Upper Miocene shelfal sandstones interbedded with siliceous mudrocks, claystones, and P-phosphate sandstones occur in the Cuyama Basin of Alta California (Lagoe, 1987a, 1987b; Roberts and Vercoutere, 1985), but they have not been studied in detail. Most are thoroughly burrowed, and hence lack traction structures such as cross bedding. Some thinner beds (ca. 50–200 cm) have erosional basal contacts with the underlying hemiplelagic rocks, and some contain ripup clasts of the latter rocks as well as abraded mollusk shells. They resemble storm deposits (tempestites), but diagnostic structures such as hummocky cross stratification have not been identified. Coring on the Peru shelf during ODP Leg 112 recovered Quaternary, shell-bearing sands which are possible analogs (Schneider and Wefer, 1990).

Lithofacies 4b: Shelfal Mudrocks

These fine-grained rocks occur in proximal basins where they occur interbedded with sandstones (Lithofacies 4a) and biosiliceous mudrocks (Lithofacies 1b). Examples include those in the Neogene of the Cuyama Basin, Alta California (Roberts and Vercoutere, 1985). This lithofacies has not been studied in detail, and its relationship to a sequence stratigraphic framework remains unresolved; our preliminary interpretation is that it records deposition on a shelf during a sea level highstand (Fig. 16).

Lithofacies 4c: Tubidite Fans

Tubidite fans and associated channel facies occur interbedded with and overlying hemipelagic facies of the Monterey Formation in proximal basins of Alta California. Among the best documented examples are those in the San Joaquin Basin (Webb, 1981). According to Callaway (1990), most such fans are parts of low-stand systems tracts formed during sea level falls, but a few also formed when high-stand shelfal clastic deposits reached the shelf-slope break and spilled over through gravity flows to form high-stand "slope bypass" fans. Ward (1984) and Garrison and Ramirez (1989) documented sandstone- and conglomerate-filled channels which appear to be the upper parts of submarine canyons incised near the shelf-slope break.

Slump Structure and Breccia Lithofacies

This is volumetrically a quite minor lithofacies but is included herein because (1) one variety (5a) serves to identify paleoslopes, and (2) it may mark sequence boundaries.

Lithofacies 5a: Slop Breccias and Slump Structures

Hieshima (1987) and Garrison and Ramirez (1989) described a variety of breccias and slump structures in the Monterey Formation of Alta California which are indicative of slope failure. These include small slump folds, slide packets, and intraformational breccias consisting of clasts of plastically to brittlely deformed hemipelagic rocks and early-formed dolomite concretions (Fig. 14). Thicknesses of the slope breccias vary from 0.5 to over 50 meters. Garrison and Ramirez (1989) noted that these phenomena appear to have developed most commonly during sea level lowstands, when increased hemipelagic sedimentation rates led to oversteepened slopes and slope failure. According to Bohacs (1990), the occurrence of gravity flow deposits and slide and slump structures in the Monterey Formation serves to identify the hemipelagic facies in this unit.

Fig. 14. Lithofacies 5a. Intraformational slope breccia composed of resedimented authigence dolomite concretions (light-colored clasts) and slabs of laminated F-phosphate shales, encased in a fine-grained matrix of hemipelagic sediment. Monterey Formation, Miocene, Santa Barbara Basin, Alta California, USA. Scale is indicated by the rock hammer just below the letter "H".

Lithofacies 5b: Intraformational Shelf Breccias and Slump Structures
Phenomena similar to those in lithofacies 5a occur also in shelfal settings, but they are typically less well developed. Intraformational folds and breccias, for example, are present within shelfal diatomites of the Pisco Formation in the onshore East Pisco Basin, and they also are evident in cores of diatomaceous sediment recovered from the modern Peru shelf (Suess et al., 1988). They typically occur in intervals no thicker than about 40–50 cm, and this small scale may serve to differentiate them from their better developed and thicker counterparts from slope areas. Their origin is uncertain; they may, however, record sediment loading and failure, perhaps during periods of increased sedimentation during lowstands.

Distribution of Lithofacies and Depositional Sequences

Figure 15 and Table 1 summarize the depositional environments of the lithofacies discussed in previous sections, and a sequence stratigraphic interpretation of these lithofacies is provided in Fig. 16 and Table 1. Compared to passive margin settings, where eustatic sea level fluctuations are superimposed upon predictable subsidence patterns, the sequence stratigraphic interpretations of the active margin settings of the eastern Pacific margin are much more complex due to pronounced and rapid variations in patterns of uplift and subsidence. Many uncertainties remain, and the interpretations portrayed in Fig. 16 and Table 1 must be regarded as preliminary and subject to considerable refinement.

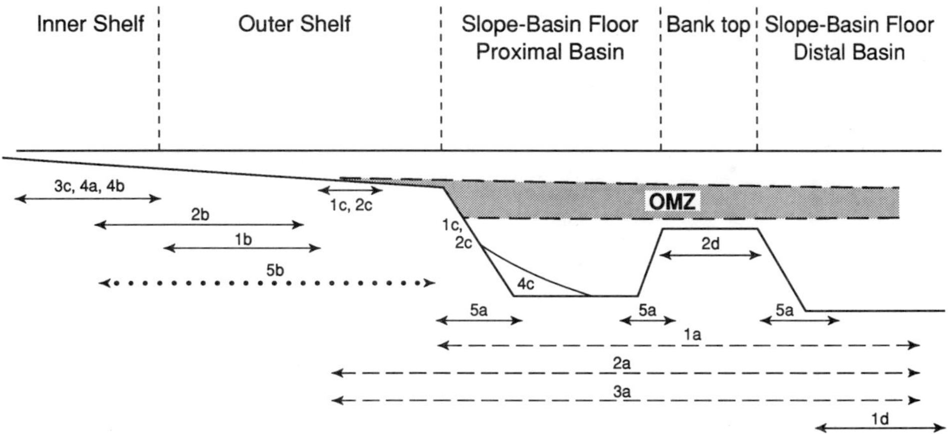

Fig. 15. Schematic protrayal of the distribution of Neogene lithofacies in different depositional settings of the eastern margin of the Pacific. Numbers correspond to the lithofacies discussed in the text and summarized in Table 1. OMZ refers to the oxygen minimum zone.

Table 1. Major Neogene lithofacies and systems tracts associated with paleo-upwelling regions in the eastern Pacific.

1. Biosiliceous Lithofacies
    1a. Thinly bedded porcelanites – HST & LST (?)
    1b. Massive diatomites, siliceous mudrocks – HST
    1c. Alternating massive-laminated biosiliceous rocks – ps in HST
    1d. Cherts – distal basin facies

2. Phosphatic Lithofacies
    2a. F-phosphates – HST
    2b. P-phosphates – TST
    2c. D-phosphates – sb & cs
    2d. Banktop glaucophosphorites – cs on banktops

3. Carbonate Lithofacies
    3a. Authigenic dolomites and limestones – fs & psb
    3b. Pelagic calcareous-siliceous rocks – HST
    3c. Bioclastic shelfal limestones – HST

4. Silicilcastic Lithofacies
    4a. Shelfal sandstones – HST, TST, & LST
    4b. Shelfal mudrocks – HST (?)
    4c. Turbidite fans – LST

5. Slump structures and slump breccias
    5a. Slope breccias and slump structures – sb in LST
    5b. Shelfal slump structures – psb in LST (?)

Systems tract designations: HST = highstand systems tract; LST = lowstand systems tract; TST = transgressive systems tract; fs = flooding surface; cs =condensed section; sb = sequence boundary; ps = parasequences; psb = parasequence boundary.

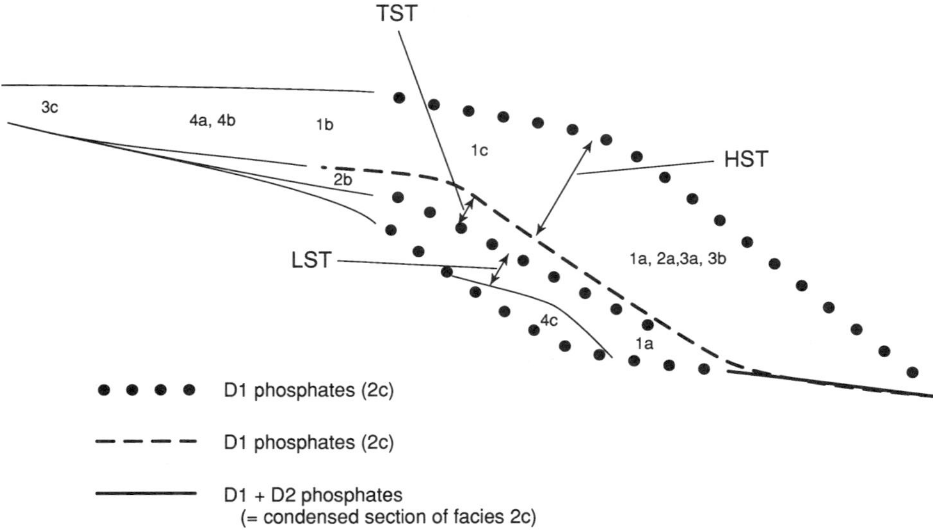

Fig. 16. Schematic portrayal of an idealized Neogene upwelling-related sequence and its component systems tracts along a proximal to distal transect. Numbers refer to lithofacies discussed in the text, summarized in Table 1, and related to depositional setting in Fig. 15. LST = lowstand systems tract, TST = transgressive systems tract, HST = highstand systems tract. Note the interpreted significance of D-phosphates (lithofacies 2c) in defining the doundaries of systems tracts and sequences.

In Fig. 16, we follow the classification of Van Wagoner et al. (1988) in recognizing a sequence as a relatively conformable succession of geologically related strata bounded by unconformities and their correlative conformities. Sequences consist of combinations of systems tracts, of which three major types are recognized: Lowstand Systems Tract (LST), Transgressive Systems Tract (TST), and Highstand Systems Tract (HST). This system of nomenclature was developed chiefly in passive margin successions and, as implied above, may not be wholly applicable to tectonically active margins like the western margins of North and South America. Nevertheless, along with Callaway (1990) and Bohacs (1990), we believe a sequence stratigraphic classification can be partially applied.

Among the key lithofacies in these successions of upwelling regions are early diagenetic phosphates and carbonates. As noted previously, regularly spaced dolomites appear to mark flooding surfaces at parasequence boundaries within all three types of systems tracts. D-phosphates are most common at sequence boundaries. As suggested in Fig. 16, D1 phosphate gravels and nodules are concentrated along the boundaries between HST's and the overlying LST's, and also between LST's and the overlying TST's (= the "first flooding surface" of Vail et al., 1991). These concentrations probably reflect sediment reworking and erosion along such boundaries, processes which produce slow sedimentation (and hence favor phosphate formation) and exhumation (hence concentration of phosphatic diaclasts). At the "maximum flooding surface" (= boundary betwen TST's and HST's), phosphate concentrations tend to be of the D2 type, i.e., phosphatic hardgrounds in which cemented D1 nodules and peloidal phosphates (facies 2c) record seafloor lithification coincident with high sea level stands and very slow sedimentation. In very distal, sediment-starved locations, the D1 and D2 systems tract boundaries may merge to form highly condensed sections consisting of both D1 and D2 phosphate layers.

Sequence stratigraphic interpretation of the remaining phosphatic lithofacies is somewhat less certain, though F-phosphates (facies 2a) appear to occur in the distal parts of HST's. We tentatively assign the P-phosphates (facies 2b) to the shelfal portion of TST's and HST's because of their association with siliceous mudrocks of these tracts; though tenuous, similar interpretations were made for P-phosphates in other areas (Glenn and Arthur, 1990; Southgate and Shergold, 1991). We surmise also that banktop glaucophosphorites (2d) were formed in distal settings during prolonged sea-level highstands and sediment starvation, and hence are the offshore equivalents of HST's.

We believe that most of the pelagic and hemipelagic facies (e.g. biosiliceous lithofacies 1a-d and carbonate lithofacies 3b) can be assigned to HST's, though we surmise that thinly bedded porcelanites (1a) generated by dilute turbidity currents formed also in LST's. In the Monterey Formation, Bohacs (1990) distinguished between pelagic and hemipelagic facies in distal basins and proposed that their stratal stacking patterns could be important in recognizing depositional settings. Coccolith-rich pelagic calcareous siliceous rocks (facies 3b) in Alta California appear to be characteristic of middle Miocene waters of moderate fertility, whereas the biosiliceous facies mark times of higher fertility and probable strong coastal upwelling (Pisciotto and Garrison, 1981).

Turbidite fans of proximal basins (facies 4c) formed mostly in LST's, although HST submarine fans are also reported in the San Joaquin Basin of Alta California (Callaway, 1990). Shelfal sandstones (facies 4a), including deltaic facies, appear to be largely parts of HST's in proximal basins, though low-stand deltaic, tidally-dominated, and incised channel fill sandstone are described by Callaway (1990). We tentatively interpret the shelfal mudrocks of facies 4b as parts of HST's, but this facies requires substantial additional investigation. Slope breccias and slump structures in the Neogene of Alta California were shown by Garrison and Ramirez (1989) to have formed most commonly during times of eustatic sea level lowering and/or tectonic uplift due to probable increased sedimentation rates on slopes; hence some but probably not all of these features mark the lower boundaries of LST's, i.e., some are sequence boundaries. We are uncertain how to interpret shelfal slump structures, but hypothesize that they may mark parasequence boundaries within LST's.

Conclusions

The above preliminary interpretations rely largely on the sedimentological analysis of some key Neogene lithofacies. Possibilities for extensions and refinements of these interpretations are manifold, and might include more precise characterization of stratal stacking patterns and sequence boundaries using integrated well-log/seismic grid and outcrop studies along with measurement of geochemical and paleontological parameters in the manner proposed by Bohacs (1990). Modern cyclostratigraphic studies are likewise much to be desired in Neogene successions all along the eastern margin of the Pacific. Perhaps the most critical need, however, is enhanced ability to date and correlate the various lithofacies produced as a consequence of Neogene upwelling and tectonics in the eastern Pacific. Also needed is increased understanding of the sequence stratigraphic significance of early diagenetic carbonates and phosphates. These thin layers appear to record geochemical and physical processes associated with abrupt changes in sedimentation rates, and hence may mark sequence and parasequence boundaries within succession of rocks which tend to be lithologically rather uniform.

Acknowledgements

Work reported herein has been supported over a number of years by grants from the National Science Foundation, the Petroleum Research Fund administered by the American Chemical Society, the Pacific Rim Research and UC MEXUS programs of the Univ. of California, and the Committee on Research of the Univ. of California, Santa Cruz. For discussions and advice on questions herein addressed, I am grateful to P. Baker, J. Barron, R. Behl, R. Dunbar, K. Grimm, J.C. Ingle, and R. Tsuchi. I thank Eugenio Gonzales for thin section preperation and Evelyn Portillo-Hegemier for drafting.

References

Allen, M.R. and Dunbar, R.B., 1988. Phosphatic sediments of the Pisco Basin. *In:* Cenozoic Geology of the Pisco Basin. Guidebook, IGCP 156 Field Workshop: Genesis of Cenozoic Phosphorites and Associated Organic-rich Sediments: Peruvian Continental Margin (R.B. Dunbar, and P.A. Baker; eds.), pp. 109–126.

Baker, P.A. and Burns, S.J., 1985. The occurrence and formation of dolomite in organic-rich continental margin sediments. *AAPG Bull.,* 69: 1917–1930.

Baker, P.A. and Kastner, M., 1981. Constraints on the formation of sedimentary dolomites. *Science,* 213: 214–216.

Baker, P.A. and Martin, J.B., 1988. Diagenesis of sediments of the Pisco Basin. *In:* Cenozoic Geology of the Pisco Basin. Guidebook, IGCP 156 Field Workshop: Genesis of Cenozoic Phosphorites and Associated Organic-rich Sediments: Peruvian Continental Margin (R.B. Dunbar and P.A. Baker; eds.), pp. 99–108.

Behl, R.J., 1990. Certification in the Monterey Formation of California (abs.). *Amer. Assoc. Petroleum Geologists Bull.,* 74: 608.

Bohacs, K.M., 1990. Sequence stratigraphy of the Monterey Formation, Santa Barbara County: Integration of physical, chemical, and biofacies data from outcrop and subsurface. *In:* Miocene and Oligocene Petroleum Reservoirs of the Santa Maria and Santa Barbara-Ventura Basins, California (M.A. Keller and M.K. McGowen; eds.), SEPM Core Workshop No. 14, pp. 139–199.

Boucher, J.M., 1984. Silica dissolution and reaction kinetics in southern California border-land sediments. M.S. Thesis, Univ. of Southern California, Los Angeles, 149 p.

Burnett, W.C., 1977. Geochemistry and origin of phosphorite deposits from off Peru and Chile. *Geol. Soc. Am. Bull.*, 88: 813–823.

Burnett, W.C., 1980. Apatite-glauconite associations off Peru and Chile. *J. Geol. soc.* (London), 137: 764–767.

Burnett, W.C., 1990. Phosphorite growth and sediment dynamics in the Modern Peru shelf upwelling system. *In:* Phosphate Deposits of the World, v. 3, Neogene to Modern Phosphorites (W.C. Burnett and S.R. Riggs; eds.), Cambridge Univ. Press, pp. 62–72.

Callaway, D.C., 1990. Organization of stratigraphic nomenclature for the San Joaquin Basin, California. *In:* Structure, Stratigraphy and Hydrocarbon Occurrences of the San Joaquin Basin, California (J.G. Kuespert and S.A. Reid; eds.), *Pacific Section S.E.P.M.*, 64: 5–22.

Calvert, S.E., 1966. Origin of diatom-rich varved sediments from the Gulf of California. *Jour. Geology*, 74: 546–565.

Cheney, T.M., McClellan, G.H., and Montgomery, E.S., 1979. Sechura phosphate deposits, their stratigraphy, origin, and composition. *Economic Geology*, 74: 232–259.

Clark, J.C., 1981. Stratigraphy, paleontology, and geology of the central Santa Cruz Mountains, California Coast Ranges. U.S. Geological Survey Prof. Paper 1168, 51 p.

Compton, J.S., 1988. Sediment composition and precipitation of dolomite in the Neogene Monterey and Sisquoc formations, Santa Maria basin area, California. *In:* Sedimentology and Geochemistry of Dolostones (V. Shukla and P.A. Baker; eds.), *SEPM Spec. Pub.*, 43: 53–64.

Compton, J.S. and Siever, R., 1986. Diffusion and mass balance of Mg during early dolomite formation, Monterey Formation. *Geochim. Cosmochim. Acta*, 50: 125–135.

Crain, W.E., Mero, W.E., and Patterson, D., 1987. Geology of the Point Arguello Field. *In:* Cenozoic Basin Development of Coastal California (R.V. Ingersoll and W.G. Ernst; eds.), Prentice Hall, New Jersey, pp. 407–426.

Decker, K., 1991. Rhythmic bedding in siliceous sediments – an overview. *In:* Cycles and Events in Stratigraphy (G. Einsele, W. Ricken, and A. Seilacher; eds.), Springer-Verlag, Berlin, pp. 464–479.

DeVries, T.J., 1988. Paleoenvironments of the Pisco Basin. Cenozoic Geology of the Pisco Basin. *In:* Cenozoic Geology of the Pisco Basin. Guidebook, IGCP 156 Field Workshop: Genesis of Cenozoic Phosphorites and Associated Organic-rich sediments: Peruvian Continental Margin (R.B. Dunbar and P.A. Baker; eds.), pp. 41–50.

Dunbar, R.B., Marty, R.C., and Baker, P.A., 1990. Cenozoic marine sedimentation in the Sechura and Pisco Basins, Peru. *Paleogeography, -climatology, -ecology*, 77: 235–262.

El-Sabbagh, D. and Garrison, R.E., 1990. Silica diagenesis in the Santa Cruz Mudstone (Upper Miocene), La Honda Basin, California. *In:* Geology and Tectonics of the Central California Coastal Region, San Francisco to Monterey (R.E. Garrison, et al. eds.), Pacific Section Amer. Asoc. Petrol. Geologist Volume and Guidebook, Book GB 67, pp. 123–132.

Föllmi, K.B. and Garrison, R.E., 1991. Phosphatic sediments, ordinary or extraordinary deposits? The example of the Miocene Monterey Formation (California). *In:* Controversies in Modern Geology (D.W. Müller, J.A. McKenzie, and H. Weissert; eds.), Academic Press, London, pp. 55–84.

Föllmi, K.B., Garrison, R.E., and Grimm, K.A., 1991. Stratification in phosphatic sediments: illustrations from the Neogene of central California. *In:* Cycles and Events in Stratigraphy (G. Einsele, W. Ricken, and A. Seilacher; eds.), Springer, Berlin, pp. 492–507.

Föllmi, K.B. and Grimm, K.A., 1990. Doomed piomeers: gravity-flow deposition and bioturbations in marine oxygen-deficient environments. *Geology*, 18: 1069–1072.

Galli-Olivier, C., Garduño, G., and Gamiño, J., 1990. Phosphorite deposits in the Upper Oligocene San Gregorio Formation at San Juan de la Costa, Baja California Sur, Mexico. *In:* Phosphate Deposits of the World, vol. 3, Neogene to Modern Phosphorites (W.C. Burnett and S.R. Riggs; eds.), Cambridge Univ. Press, pp. 122–126.

Garrison, R.E., 1992. Neogene phosphogenesis along the eastern margin of the Pacific. Revista Geológica de Chile, v. 19, pp. 91–111.

Garrison, R.E. and Kastner, M., 1990. Phosphatic sediments and rocks recovered from the Peru margin during ODP Leg 112. *In:* E. Suess, R. von Huene et al., 1990. Proc. ODP, Sci. Results, 112: College Station, TX (Ocean Drilling Program), pp. 111–134.

Garrison, R.E., Kastner, M., and Kolodny, Y., 1987. Phosphorites and phosphatic rocks in the Monterey Formation and related Miocene units, coastal California. *In:* Cenozoic Basin Development in Coastal California (R.V. Ingersol and W.D. Ernst; eds.), Prentice Hall, Englewood Cliffs, N.J., pp. 349–381.

Garisson, R.E. and Ramirez, P.C., 1989. Conglomerates and breccias in the Monterey Formation and related units as reflections of basin margin hisotry. *In:* Conglomerates in Basin Analysis: A symposium Dedicated to A.O. Woodford (I.P. Colburn, PL. Abbott, and J. Minch; eds.), *Pacific Section S.E.P.M.,* 62: 189– 206.

Glenn, C.R. and Arthur, MA., 1990. Anatomy and origin of a Cretaceous phosphorite-greensand giant, Egypt: *Sedimentology,* 37: 123–154.

Govean, F.M., 1980. Some paleoecologic aspects of the Monterey Formation, California. Ph.D. Thesis, Univ. of California, Santa Cruz, 278 p.

Govean, F.M. and Garrison, R.E., 1981. Significance of laminated and massive diatomites in the upper part of the Monterey Formation. *In:* The Monterey Formation and Related Siliceous Rocks of California (R.E. Garrison, and R.G. Douglas; eds.), Pacific Section S.E.P.M. Pub., pp. 181–198.

Graham, S.A., 1976. Tertiary sedimentary tectonics of the central Salinian block of California. Ph.D. Thesis, Stanford Univ., California, 510 p.

Grimm, K., Garrison, R., Ledesma, M., and Fonseca, C., 1991. The Oligo-Miocene San Gregorio Formation of Baja California Sur: An early record of coastal upwelling along the eastern Pacific margin (abs.). *AAPG Bull.*, 75: 366.

Helenes, J., 1984. Stratigraphy, depositional environments and foraminifera of the Miocene Tortugas Formation, Baja California Sur, Mexico. *Mexicana Geologos Boletin*, 41, nos. 1, 2: 47–67.

Hieshima, G.B., 1987. Sedimentology of Miocene Monterey Formation diatomites, California. M.S. Thesis, Univ. of Wisconsin, Madison, 113 p.

Ingle, J.C., 1981. Origin of Neogene diatomites around the north Pacific rim. *In:* The Monterey Formation and Related Siliceous Rocks of California (R.E. Garrison and R.G. Douglas; eds.), *Pacific Section S.E.P.M.* Pub., pp. 159-–80.

Isaacs, C.M., 1980. Diagenesis in the Monterey Formation examined laterally along the coast near Santa Barbara, California. Ph.D. Thesis, Stanford Univ., California, 329 p.

Isaacs, C.M., 1984. Disseminated dolomite in the Monterey Formation, Santa Maria and Santa Barbara area, California. *In:* Dolomites of the Monterey Formation and Other Organic-rich Units (R.E. Garrison, M. Kastner, and D.H. Zenger; eds.), *Pacific Section S.E.P.M.*, 41: 155–170.

Khan, S.M., Coe, R.S., and Barron, J.A., 1989. High-resolution magnetic polarity stratigraphy of Shell Beach section of the Monterey Formation in Pismo Basin, California. *In:* Japan-U.S. Seminar on Neogene Siliceous Sediments of the Pacific Region, Syllubus and Fieldtrip Guidebook (R.E. Garrison; ed.), Univ. of California, Santa Cruz, pp. 106–124.

Kushnir, J. and Kastner, M., 1984. Two forms of dolomite occurrences in the Monterey Formation, California – a comparative mineralogical, geochemical, and isotopic study. *In:* Dolomites of the Monterey Formation and Other Organic-rich Units (R.E. Garrison, M. Kastner, and D.H. Zenger; eds.), *Pacific Section S.E.P.M.*, 41: 171–184.

Lagoe, M.B., 1987a. Middle Cenozoic basin development, Cuyama Basin, California. *In:* Cenozoic Basin Development of Coastal California (R.V. Ingersoll, and W.G. Ernst; eds.), Rubey Vol. 6, Prentice-Hall, Inc., Englewood Cliffs, N.J., pp. 172–206.

Lagoe, M.B., 1987b. Middle Cenozoic unconformity-bounded stratigraphic units in the Cuyama and southern San Joaquin Basins: Record of eustatic and tectonic events in active margin basins: Cushman Foundation for Foraminiferal Research, Special Publication 24, pp. 15–32.

Ledesma, M.C., 1992. Petrology of phosphatic sedimentary rocks within the Oligo-Miocene San Gregorio Formation of Baja California Sur, Mexico. M.S. Thesis, Univ. of California, Santa Curz, 154 p.

Marty, R.C., 1989. Stratigraphic and chemical sedimentology of Cenozoic biogenic sediments from the Pisco and Sechura Basins, Peru. Ph.D. Thesis, Rice Univ., Houston, Tex., 239 p.

McBride, E.F. and Folk, R.L., 1979. Features and origin of Italian Jurassic radiolarites deposited on continental crust. *Jour. of Sed. Petrology*, 49: 837–868.

Mcclellan, G.H., 1989. Geology of the phosphate deposits at Sechura, Peru. *In:* Phosphate Deposits of the World, Vol. 2, Phosphate Rock Resources (A.J.G. Notholt, R.P. Sheldon, and D.F. Davidson; eds.), Cambridge Univ. Press, pp. 123–130.

Murray, R.W., Jones, D.L., and Buchhooltzten Brink, M.R., 1992. Diagenetic formation of bedded chert: evidence from chemistry of chert-shale couplet. *Geology,* 20: 271–274.

Murata, K.J., Friedman, I., and Madsen, B.H., 1969. Isotopic composition of diagenetic carbonates in marine Miocene formations of California and Oregon. U.S. Geological Survey Prof. Paper 614B, 24 p.

Nelson, C.S., 1978. Temperate Shelf carbonate sediments in the Cenozoic of New Zealand. *Sedimentology,* 25: 737–771.

Pisciotto, K.A., 1978. Basinal sedimentary facies and diagenetic aspects of the Monterey Shale, California. Ph.D. Thesis, Univ. of California, Santa Cruz, 450 p.

Pisciotto, K.A. and Garrison, R.E., 1981. Lithofacies and depositional environments of the Monterey Formation, California. *In:* The Monterey Formation and Related Siliceous Rocks of California (R.E. Garrison and R.G. Douglas; eds.). *Pacific Section S.E.P.M.*, pp. 97–122.

Phillips, R.L., 1983. Late Miocene tidal shelf sedimentation, Santa Cruz Mountains, California. Marine Sedimentation, Pacific Margin, U.S.A. (D.K. Larue and R.J. Steel; eds.), *Pacific Section S.E.P.M.* Pub., pp. 73–88.

Ramirez, P.C., 1990. Stratigraphy, sedimentology, and paleoceanographic implications of the Late Miocene to Early Poliocene Sisquoc Formation, Santa Maria area, California. Ph.D. Thesis, Univ. of California, Santa Cruz, 388 p.

Reimers, C.E., Kastner, M., and Garrison, R.E., 1990. The role of bacterial mats in phosphate mineralization with particular reference to the Monterey Formation. *In:* Phosphate Deposits of the World, vol. 3, Neogene to Modern Phosphorites (W.C. Burnett, and S.R. Riggs; eds.), Cambridge Univ. Press, pp. 300–311.

Roberts, A.E, and Vercoutere, T.L., 1985. Geology and petrology of the upper Miocene phosphate deposit near New Cuyama, Santa Barbara County, California. U.S. Geological Survery Bull. B-1635, 61 p.

Savrda, C.E. and Bottjer, D.M., 1986. Trace-fossil model for reconstruction of paleo-oxygenation in bottom waters. *Geology,* 14: 3–6.

Savrda, C.E., Bottjer, D.J., and Gorsline, D.S., 1985. An image-enhancing oil technique for friable, diatomaceous rocks. *Jour. Sed. Petrology*, 55: 604–605.

Savrda, C.E. and Bottjer, D.J., 1987. Trace fossils as indicators of bottom-water redox conditions in ancient marine environments. *In:* New Concepts in the Use of Biogenic Sedimentary Structures for Paleoenvironmental Interpretation (D.J. Bottjer; ed.), *Pacific Section S.E.P.M.* Book 52, pp. 3–26.

Schnider, R. and Wefer, G., 1990. Shell horizons in Cenozoic upwelling-facies off Peru: distribution and mollusk fauna in cores from Leg 112: Proc. Ocean Drilling Program. *Sci. Results,* 112: 335–352.

Schrader, H.J. et al., 1980. Laminated diatomaceous sediments from Guayamas Basin slope (central Gulf of California): 250,000 year climate record. *Science*, 207: 1207–1209.

Shimmield, G.B. and Price, N.B., 1984. Recent dolomite in hemipelagic sediments off Baja California. *In:* Dolomites of the Monterey Formation and Other Organic-rich Units (R.E. Garrison, M. Kastner, and D.B. Zenger; eds.), *Pacific Section S.E.P.M.*, 41: 5–18.

Soutar, A., Johnson, S.R., and Baumgartner, T.R., 1981. In search of modern depositional analogs to the Monterey Formation. *In:* The Monterey Formation and Related Siliceous Rocks of California (R.E. Garrison and R.G. Douglas; eds.), *Pacific Section S.E.P.M.* Pub., pp. 123–148.

Southgate, P.N. and Shergold, J.H., 1991. Application of sequence stratigraphic concepts to Middle Cambrian phosphogenesis, Georgina Basin, Australia. *BMR Jour. Austral. Geology & Geophysics*, 12: 119–144.

Suess, E. et al., 1988. Ocean Drilling Program Leg 112, Peru continental margin: Part 2, Sedimentary history and diagenesis in a coastal upwelling environment. *Geology*, 16: 939–943.

Tada, R., 1991. Origin of rhythmical bedding in Middle Miocene siliceous rocks of the Onngagawa Formation, northern Japan. *Jour. Sed. Petrology*, 61: 1123–1145.

Tsuchi, R., Shuto, T., Takayama, T., Fujiyoshi, A., Koizumi, I., Ibaraki, M., and Martinez-Pardo, R., 1988. Fundamental data on Cenozoic biostratigraphy of Chile. Reports of Andean Studies, Shizuoka Univ., Special Vol. 2, pp. 71–95.

Vail, P.R., Audemard, F., Bowman, S.A., Eisner, P.N., and Perezcruz, C., 1991. The stratigraphic signatures of tectonics, eustacy and sedimentology – an overview. *In:* Cycles and Events in Stratigraphy (G. Einsele, W. Ricken, and A. Seilacher; eds.): Springer-Verlag, Berlin, pp. 617–659.

Van Wagoner, J.C. et al., 1988. An overview of the fundamentals of sequence stratigraphy and key definitions. *In:* Sea-Level Changes: An Integrated Approach (C.K. Wilgus et al.; eds.), SEPM Spec. Pub., 42, pp. 39–46.

Ward, R.B., 1984. Sedimentology, diagenesis and provenance of the slope channel deposits exposed within the upper siliceous member of the Monterey Formation exposed west of Gaviota Beach, California. M.S. Thesis, Stanford Univ., California, 101 p.

Webb, G.W., 1981. Stevens and earlier Miocene turbidite sandstones, southern San Joaquin Valley. *Amer. Assoc. Petroleum Geologists Bull.*, 65: 438–465.

White, L.D., 1989. Chronostratigraphic and paleoceanographic aspects of selected chert intervals in the Miocene Monterey Formation, California. Ph.D. Thesis, Univ. of California, Santa Cruz, 236 p.

Neogene Planktonic Foraminiferal Biostratigraphy on the Coast of Peru and Its Paleoceanographic Implications

Masako IBARAKI

Geoscience Institute, Faculty of Science, Shizuoka University, Shizuoka 422, Japan

Abstract

Bio- and chronostratigraphy of planktonic foraminifera is examined in selected marine Neogene sequences on the coast of Peru. The Pampa Las Salinas section includes horizons of Zone P21a of ealy Oligocene age; the Cerro Las Salinas section, Zone N4 to Zones N6–7 of early Miocene; the Camana section, Zone N6 to Subzone N8b of early Miocene to earliest middle Miocene; and Loma Cuesta Chilcatay section, Zones N16-17 of late Miocene. Diatomaceous facies suggesting coastal upwelling are intercalated in early Miocene horizons, but they are predominant from Middle Miocene time up to the present. Two warm marine events in the earliest Middle Miocene and Late Miocene ages are presumed on the basis of dominant ratios of warm-water planktonic foraminifera.

Introduction

Marine Neogene sequences are scattered on the coastal area of Peru. Studies on the planktonic foraminiferal biostratigraphy of the area and on the coast of Chile were carried out as part of the Andean Studies Program of Shizuoka University in 1990–1992. Representative sections of the areas studied are Pampa Las Salinas, Cerro Las Salinas and Loma Cuesta Chilcatay in central Peru, and Camana in southern Peru. In this paper, correlations and geologic age assignments of these sections are made by utilizing age-diagnostic planktonic foraminifera. An attempt is also made to consider paleoceanographic implications based on planktonic foraminifera with reference to bio- and chronostratigraphic results of the ODP 112–Site 682 off Peru and those of the Caleta Herradura de Mejillones section in northern Chile.

Working Method

Field work on the Pacific coast of South America was conducted during 1985–1990. In 1985, 1986 and 1988, field examinations were carried out in Colombia, Ecuador, Peru and Chile, and in 1990, additional work was done in Peru and Chile in cooperation with collaborators from those countries. Summaries of the studies are presented in a series of the Reports of Andean Studies of Shizuoka University (Tsuchi, 1988, 1990 and 1992).

A total of 137 rock samples for planktonic microfossil examination were obtained from selected horizons of Neogene sequences in Peru. Planktonic foraminiferal analyses of all of the samples were made in the laboratory of Shizuoka University. Dried rock samples of 100g each were crushed and treated by the Naphtha method or Glauber's salt method. Then samples were washed through a 200-mesh screen and dried in an oven. Species of planktonic foraminifera were picked up from washed residues, and the frequency of occurrence of each species was calculated. The planktonic foraminiferal zonation used here is according to Blow (1969) and Ibaraki (1986), with age-assignments in accordance with Berggren et al. (1985).

Planktonic Foraminiferal Bio- and Chronostratigraphy on the Coast of Peru

Pampa Las Salinas

The Pampa Las Salinas section is located in central Peru, south of Paracas, 260 km south of Lima. Limited exposures of marine sequences were found a little north of a salt lake basin of the Cerro Las Salinas. The section is at least 30 meters in thickness and consists mainly of siltstone with intercalations of sandstone layers. Planktonic foraminifera were obtained from three horizons. The locality Pe–88–7–11 contains *Chilogumbelina cubensis, Globorotalia opima opima* and *Globorotalia insolita*, which indicate Zone P21a of early Oligocene age (Fig. 1). Locality Pe–90–7–3 is assignable to Zones P18–21a of early Oligocene age, including *Chilogunbelina cubensis* and *Cassigerinella chipolensis*.

Cerro Las Salinas

The Cerro Las Salinas section, near the Pampa Las Salinas, consists of marine sequences exposed on the east side of a salt lake basin. The section has a thickness of 200 meters or more and is composed mainly of sandstone in the lower part and siltstone in the upper part, with some intercalations of limestone, tuff, and diatomaceous layers.

Planktonic foraminifera were obtained from four horizons in the middle part of the sequence (Fig. 1). *Globorotalia kugleri* and *Globorotalia pseudokugleri*, known from Zone N4 of the earliest Miocene, were found in Pe–88–7–8. Locality Pe–88–7–2, from a horizon about 60 m above the Pe–88–7–8, contains *Globigerinoides immatulus* and *Globoquadrina dehiscens*, which indicates an early Miocene age. Locality Pe–86–17–0, a horizon 50 m above the Pe–88–7–2, is assignable to Zones N6–7 of early Miocene age, including *Globigerina praebulloides pseudociperoensis, Globigerinoides sacullifer, Globoquadrina dehiscens, Globoquadrina altispira,* and *Globorotalia peripheroronda*. Planktonic foraminifera obtained from another locality, Pe–86–17–3, are considered to be Zones N6–7 of early Miocene age based on the occurrence of *Globorotalia scitula praescitula* and *Globigerinoides sacculifer*. Thus, the section includes horizons from Zone N4 to Zones N6–7 of early Miocene age. The contact of this sequence with that of the Pampa Las Salinas must be a fault, as the structures of both sequences are different, but this could not be ascertained due to the thick cover of a sand dune.

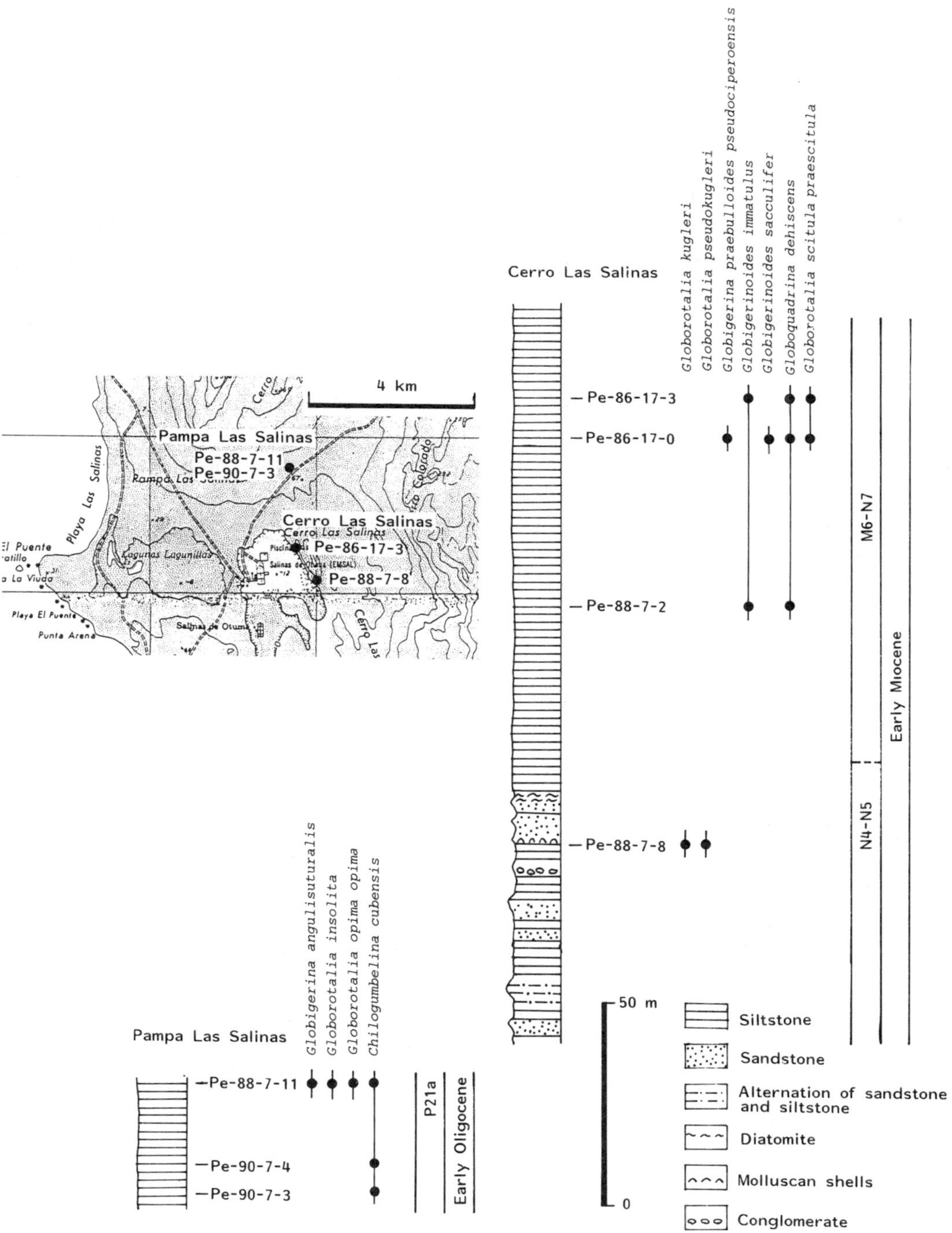

Fig. 1. Locations, stratigraphic profiles, and age-assignments of the Pampa Las Salinas and Cerro Las Salinas sections by means of planktonic foraminifera.

Camana

The Camana area is located 840 km south of Lima in southern Peru. Marine Neogene sequences, exposed in a basin along the coast, attain a thickness of 400–500 meters; they consisting of calcareous sandstone in the lower part and siltstone in the upper part. Pecho and Morales (1969) have reported occurrences of miogypsinid larger foraminifera in the lower part of the sequence. Well-preserved planktonic foraminifera are obtained from many localities, with rich occurrences of miogypsinid larger foraminifera (Fig. 2).

The locality Pe–86–21, located in the eastern end of the basin, is assigned to Zone N6 of the early Miocene by the occurrences of *Globigerinatella insueta* and *Globoquadrina dehiscens praedehiscens*. The locality Pe–90–9–1, located at the western end of the basin, yielded *Catapsydrax stainforthi* and *Globorotalia minutissima*. The horizon can, therefore, be assigned to Zones N6–7 of the early Miocene age. In localities Pe–90–9–7 and Pe–90–9–8, which are located in the middle part of the basin, *Globigerinoides sicanus* and *Globorotalia birnageae* were found without *Praeorbulina glomerosa*. These horizons can, therefore, be assigned to Subzone N8a of the latest early Miocene.

In locality Pe–88–3, located in an eastern part of the basin, *Globigerinoides sicanus* and *Praeorbulina glomerosa* were obtained, which indicate Subzone N8b of the earliest middle Miocene. As specimens of *Praeorbulina glomerosa* were obtained also

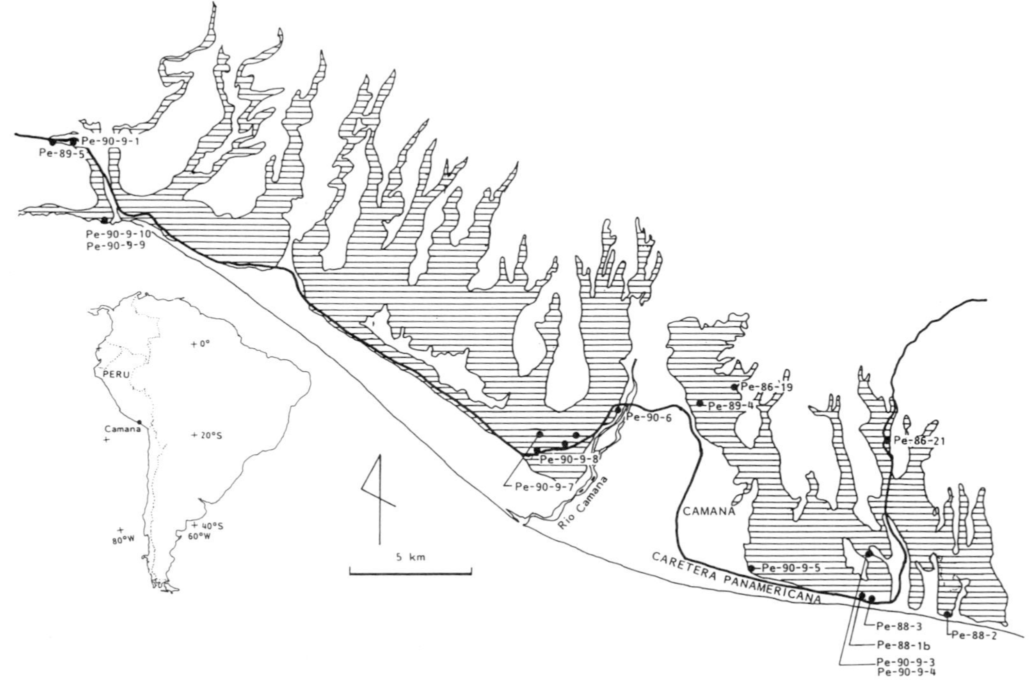

Fig. 2. Sample locality map of the Camana Formation at Camana, on the coast of southern Peru.

from Pe–90–9–5, in an eastern part of the basin, the horizon is assignable to Subzone N8b of the earliest middle Miocene in age.

In locality Pe–90–9–3, located a little north of Pe–88–3, *Globigerinoides sicanus, Praeorbulina glomerosa glomerosa* and *Praeorbulina glomerosa circularis* were found with none of *Orbulina* spp. The horizon is, therefore, assinged to the top of Subzone N8b. As mentioned above, the Camana Formation is assignable to range from Zone 6 to Subzone N8b of the early Miocene to earliest middle Miocene.

Loma Cuesta Chilcatay

The Loma Cuesta Chilcatay section is located in central Peru, west of Ica. Neogene sequences are well exposed in a desert. The sequences consist mainly of siltstone, with interbedded diatomaceous layers in the lower part and sandstone and limestone layers in the upper part.

Dunbar et al. (1990) named the Chilcatay Formation for well-exposed sections of uppermost Oligocene to lower middle Miocene rocks within the Pisco Basin in the area of Pampa Chilcatay. The geologic age of the Chilcatay Formation in the Quebrada Huaracangana area, southeast of Pampa Chilcatay, has been assigned to planktonic foraminiferal Subzone N8b, and diatom Subzone A of the *Cestodiscus peplum* Zone, both of earliest middle Miocene age (16.5–15 Ma) (Dunbar et al., 1990). The section studied here is probably the same as the Pampa Chilcatay section of Dunbar et al. (1990), according to the location map in their text. Dunbar and colleagues utilize diatoms for the geologic age-assignment of the Chilcatay Formation, but the occurrence list of diatoms from the Pampa Chilcatay has not yet been reported.

Planktonic foraminifera obtained from two horizons in the middle part of the Loma Cuesta Chilcatay section, including *Neogloboquadrina acostaensis* and *Globoquadrina dehiscens advena*, are assignable to N16–17 of late Miocene age.

Bio- and Chronostratigraphy of the ODP Leg 112--Site 682 off Peru

During the ODP Leg 112, 10 sites were drilled on the continental shelf and slope off Peru. The drill sites are located within the present coastal upwelling center and neighboring areas strongly influenced by upwelling. The commencement of diatomaceous sedimentation in the area has been recorded on Hole 112–682A, which is situated off Lima on the landward lower slope of the Peruvian Trench at a water depth of 3788 meters. The sediments recovered from the top through Core–34X–CC (311 m sub-bottom depth) consist of diatomaceous mudstone, having diatom contents of 20–40% and becoming negligible rapidly below this horizon. Sediments of Core–34X–CC are assignable to Zone NN4 of calcareous nannoplankton, the *Denticula nicobarica* Zone of diatoms, and the basal *Calocycletta costata* Zone to *Dorcadospyris alata* Zone of radiolaria, respectively, and are dated at about 17 Ma of the early Miocene (Suess et al., 1988). The index planktonic foraminifera *Globigerina falconensis* and *Catapsydrax stainforthi* were found in Core–35X–CC, a little lower than the above-mentioned core, indicating an assignment to Zones N6–7 at 17–18 Ma

(Ibaraki, 1990a). Horizons just above the horizon of N6–7 in Hole 682A were poorly recovered by drilling (Fig. 3).

Paleoceanographic Implications of Planktonic Foraminifera

A bio- and chronostratigraphic correlation chart of Neogene marine sequences on the coastal area of Peru is presented in Fig. 3. The correlation and geologic age-assignment of these sequences are made by utilizing age-diagnostic planktonic foraminifera. In the Rio Grande and the Rio Pisco sections, planktonic foraminifera were not obtained. In both sections, age-assignments were made by age-diagnostic diatoms (Tsuchi et al., 1988, 1990).

In the ODP Hole 682A, the dominant lithology is organic rich diatomaceous mudstone from the top through 311 m subbottom depth, which has been dated at about 17 Ma of the early Miocene. Index planktonic foraminifera were found in a horizon 320 m deep, being assignable to Zones N6–7 at 17–18 Ma. From this hole, some horizons a little above those of early Miocene were poorly recovered, which might suggest earliest middle Miocene sediments.

On the coastal area of Peru, middle Miocene and later sequences are composed mostly of marine diatomaceous mudstone similar to that in sections of Rio Grande, the lower part of Chilcatay, and Rio Pisco (except for the Camana section) (Fig. 3). The Cerro Las Salinas section, being assignable to N4 to N6–7 of early Miocene, consists of sandstone and siltstone with several intercalations of diatomaceous layers in the middle part. Based on planktonic foraminiferal biostratigraphy of off-shore and on-shore sections in Peru, the beginning of the coastal upwelling off Peru is considered to be early Miocene in age, although the diatomaceous facies have been predominant in the middle Miocene.

The Camana section of early Miocene to earliest middle Miocene age consists mostly of calcareous sandstone and siltstone, containing just a few diatomaceous facies. Planktonic foraminifera of Subzone N8a of the Camana section are dominantly warm water dwellers, such as *Globigerinoides* spp., *Globoquadrina* spp., and *Globorotalia siakensis*. Rich assemblages of miogypsinid larger foraminifera occur in the Camana section in association with warm-water planktonic foraminifera. The Subzone N8b horizons of the earliest middle Miocene Camana horizons are characterized mostly by warm-water taxa with few eurythermal elements. This suggests a warm event during the earliest middle Miocene.

Middle Miocene and later sequences on the coast of Peru are composed mostly of marine diatomaceous mudstone, containing few planktonic foraminifera and no warm-water taxa. The Loma Cuesta Chilcatay section of late Miocene age consists mostly of siltstone, with intercalated diatomaceous and limestone layers. Planktonic foraminifera of the N16–17 horizons of the Loma Cuesta Chilcatay section consist of warm-water and eurythermal dwellers, with a few cold-water elements such as *Globigerina bulloides*, *G. decoraperta*, *G. falconensis*, *Globigerinoides ruber*, *G. sacculifer*, *Globoquadrina dehiscens deshiscens*, *G. venezuelana*, *Globigerinita glutinata*, *G.*

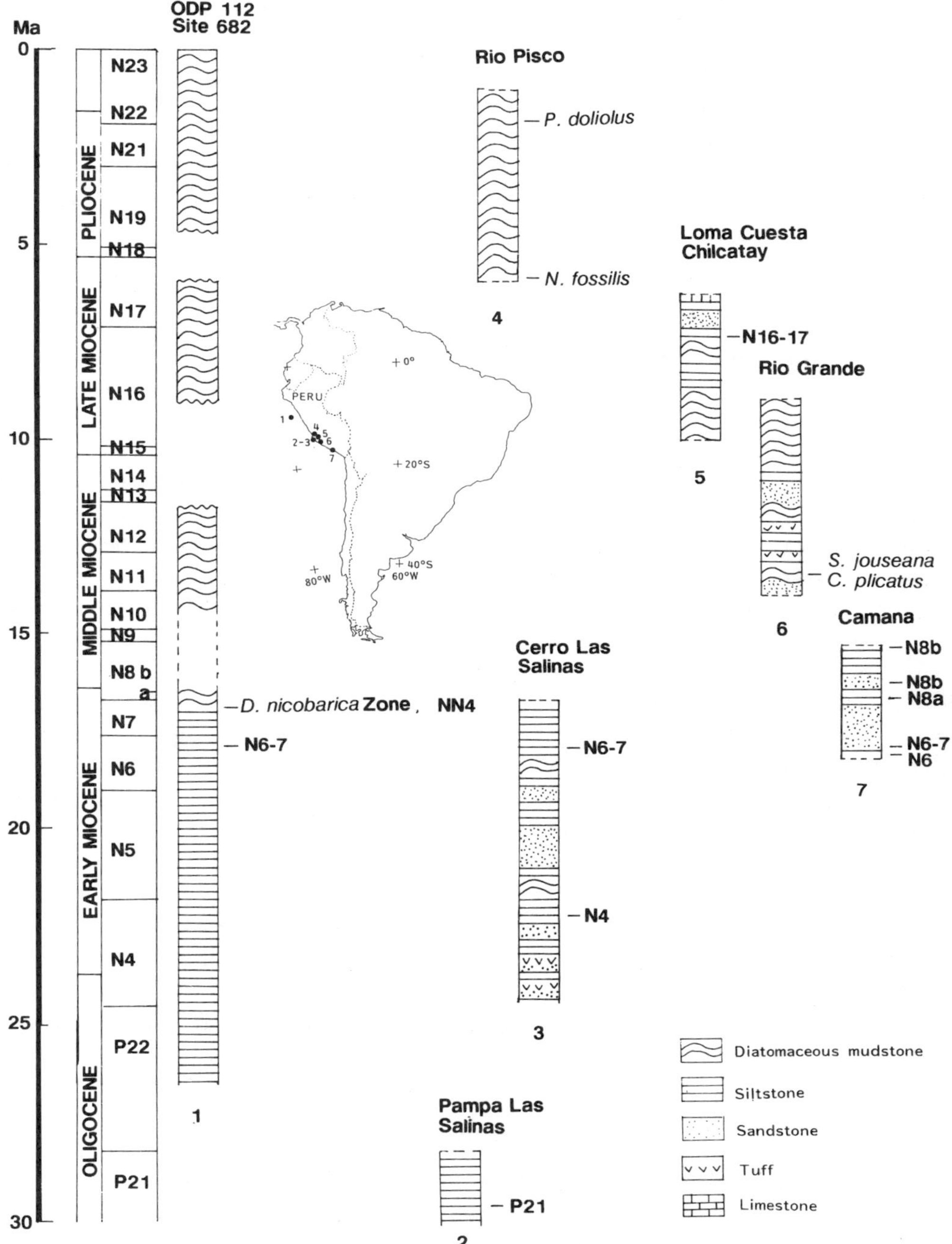

Fig. 3. Chronologic correlation of selected Neogene sequences on the coast of Peru. Planktonic foraminiferal zonal assignments at the respective horizons are indicated.

uvula, Neogloboquadrina acostaensis, and *Globorotalia obesa.*

The warm-water elements account for 43.3% of the total planktonic foraminiferal fauna showing no upwelling feature. This high ratio of warm-water planktonic foraminifera suggests another warm event during the late Miocene.

As mentioned above, two marine warm events of earliest middle Miocene and late Miocene times are suggested on the coast of Peru. Two such warm events are also recognized in the section of Caleta Herradula de Mejillones, north of Antofagasta in northern Chile (Ibaraki, 1990b).

The Neogene sequence of the Caleta Herradura de Mejillones includes horizons of the interval from N7 through N16–17 of early Miocene to late Miocene in age. The planktonic foraminiferal fauna of the Caleta Herradura de Mejillones secton is mostly occupied by warm-water and eurythermal elements. The percentages of tropical-subtropical planktonic foraminifera to the total fauna in successive horizons of the section are shown in Fig. 4. The warm-water species defined here are: *Globigerinoides* spp., *Globoquadrina* spp., *Globigerina druryi, Globigerina decoraperta, Globigerina nepenthes, Globorotalia archeomenardii, Globorotalia birnageae, Globorotalia menardii, Globorotalia praemenardii, Globorotalia siakensis, Globorotaloides hexagona, Catapsydrax stainforthi,* and *Sphaeroidienllopsis* spp. These species have been regarded as characteristic in tropical and subtropical areas (Kennett and Srinivasan, 1983). The percentage of warm-water elements to the total assemblage increases through early Miocene time and attains a maximum interval including Subzone N8a, N8b and the base of Zone N9 of latest early Miocene to earliest middle Miocene age. Then, the percentage suddenly drops in Zones N10–N12 of middle Miocene age. The ratio increases again in N16–N17 of late Miocene age, but the nature of the ratio during the late middle Miocene age is obscure because of the paucity of planktonic foraminifera. It is thought that these different values of the percentage of warm-water taxa are related to changes in surface sea-water temperature. Thus, two warm events in the early middle Miocene and late Miocene can be presumed on the coastal areas of Peru and Chile.

During the Neogene of the Pacific side of central Japan, the maximum seawater temperature in Subzone N8b of the earliest middle Miocene is followed by an abrupt cooling in the early middle Miocene and a general warming toward the latest Miocene. These events are known from the fluctuation in the ratio of tropical-subtropical planktonic foraminifera to the total assemblage (Ibaraki, 1990c).

Conclusion

Biostratigraphic studies are made in representative Neogene sequences on the coast of Peru by means of planktonic foraminifera. Age-diagnostic planktonic foraminifera were obtained from sections of Pampa Las Salinas, Cerro Las Salinas, Loma Cuesta Chilcatay, and Camana. The Pampa Las Salinas section includes horizons of Subzone P21a of early Oligocene age, the Cerro Las Salinas section ranges from Zone N4 to Zones N6–7 of early Miocene age, the Camana section ranges from Zone N6 to

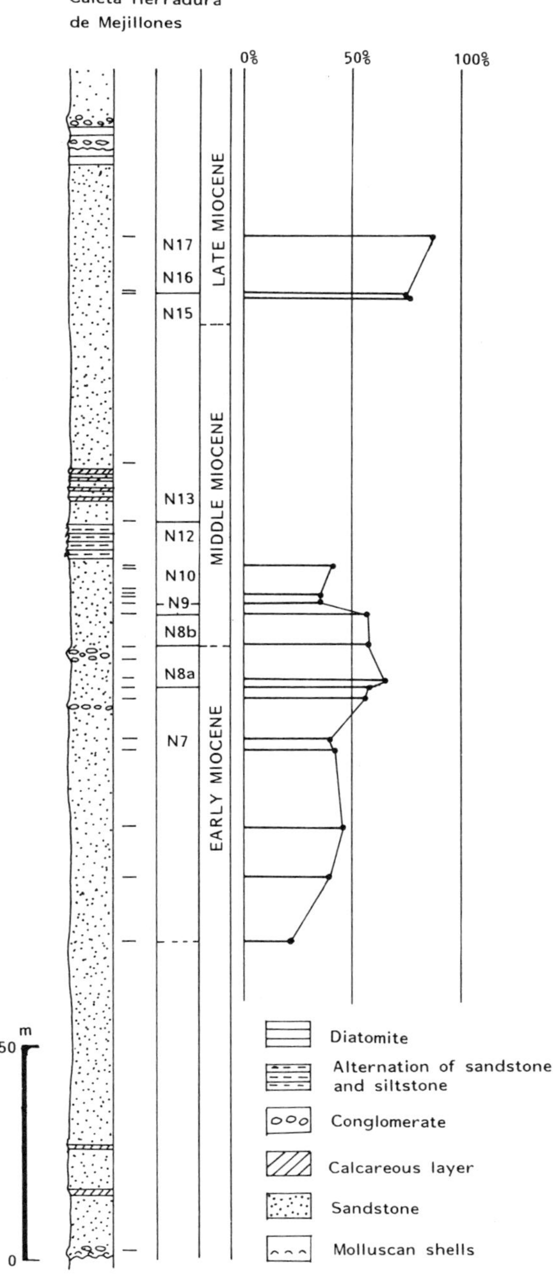

Fig. 4. Stratigraphic profile of the Caleta Herradula de Mejillones section in northern Chile and ratio of warm-water taxa to the total planktonic foraminifera.

Subzone N8b of late early Miocene to earliest middle Miocene age, and the Loma Cuesta Chilcatay section is assigned to the zonal interval N16–17 of late Miocene age.

Based on the planktonic foraminiferal biostratigraphy of off-shore and on-shore sections in Peru, the beginning of the coastal upwelling off Peru is considered to be early Miocene in age. Diatomaceous facies become predominant in middle Miocene age.

Two marine warm events in the earliest middle Miocene and late Miocene ages can be presumed based on planktonic foraminiferal faunas. Planktonic foraminifera from horizons of Subzone N8b in the Camana section and those of Zones N16–17 in the Loma Cuesta Chilcatay section consist dominantly of warm-water elements. These warm events can be correlated with those recognized in the Caleta Herradula de Mejillones section in Chile and those on the Pacific coast of Japan.

Acknowledgements

Doctor J. Barron of the U.S. Geological Survey at Menlo Park, California, kindly reviewed the manuscript. The author's cordial thanks are due to him for his critical suggestions. The author is grateful to Dr. R. Tsuchi, Professor Emeritus of Shizuoka University, for his kind reading of the manuscript.

References

Berggren, W.A., Kent, D.V., Flynn, J.J., and Van Couvering, J., 1985. Cenozoic geochronology. *Geol. Soc. Am. Bull.*, 96: 1407–1418.

Blow, W.H., 1969. Late middle Eocene to Recent planktonic foraminiferal biostratigraphy. *Proc. 1st Internatl. Conf. Planktonic Microfossils. Geneva*, 1967, v.1, pp. 199–422.

Dunbar, R.B., Marty, R.C., and Baker, P.A., 1990. Cenozoic marine sedimentation in the Sechura and Pisco basins, Peru. *Palaeogeogr., Palaeoclimatol., Palaeoecol.*, 77: 235–261.

Ibaraki, M., 1986. Neogene planktonic foraminiferal biostratigraphy of the Kakegawa area on the Pacific coast of central Japan. *Rep. Shizuoka Univ., Fac. Sci.*, 20: 39–173.

Ibaraki, M., 1990a. Eocene through Pleistocene planktonic foraminifera off Peru. Leg 112-biostratigraphy and paleoceanography. *In:* E. Suess, R. von Huene, et al., *Proceedings of the Ocean Drilling Program. College Station, TX.*, 112: 239–262.

Ibaraki, M., 1990b. Planktonic foraminifera biostratigraphy of the Neogene of Caleta Herradura de Mejillones, north Chile. *In:* R. Tsuchi; ed., *Reports of Andean studies Shizuoka University, Shizuoka University*, Special Volume, no.3, 9–16.

Ibaraki, M., 1990c. Neogene planktonic foraminifera and events in Japan. *Palaeogeogr. Palaeoclimatol., Palaeoecol.*, 77: 335–343.

Kennett, J.P. and Srinivasan, S., 1983. Neogene planktonic foraminifera. *A Phylogenetic Atlas*. Hutchinson and Ross. Strondsburg, Pa., 265pp.

Pecho, V.G. and Morales, G.S., 1969. Geologia de los Cuadrangulos de Camana y la Yesera. *Ministerio de Energia y Minas, Servicio de Geologia Mineria, Boletin*, No. 21, pp.1–71, Lima.

Suess, E., von Huene, R. et al., 1988. Proceedings of the Ocean Drilling Program. *Initial Reports. Ocean Drilling Program, College Station, TX.*, 112: 1015p.

Tsuchi, R. (ed.), 1988. Reports of Andean studies, Shizuoka University, Special Volume 2, Trans-Pacific correlation of Cenozoic geohistory, 108p.

Tsuchi, R., 1990. Reports of Andean studies, Shizuoka University, Special Volume 3, Trans-Pacific correlation of Neogen geologic events. 77p.

Tsuchi, R., 1992. Reports of Andean studies, Shizuoka University, Special Volume 4, Pacific Neogen events in Japan and South America: Their timing and nature (in press).

New Zealand Cenozoic Marine Paleoclimates:
A Review Based on the Distribution of
Some Shallow Water and Terrestrial Biota

N. de B. Hornibrook

Research Associate, DSIR Geology and Geophysics, Lower Hutt, New Zealand

Abstract

Isolated occurrences of warm-water molluscan genera in the Paleocene indicate that sea temperatures were warmer than at present, rising in the Late Paleocene to a sudden peak of warmth in the Paleocene/Eocene boundary zone when the first *Asterocyclina* appeared in the Southern Hemisphere. Marine climates remained within the warm subtropical zone (20°C–25°C) throughout most of the Paleogene, when many Indo-Pacific molluscan genera, some large foraminifera, the genus *Lingula*, mangroves and, occasionally, *Cocos* ranged as far south as the southern part of the South Island. Climate cooled somewhat in the Eocene/Oligocene boundary zone although the marked cooling event shown in the isotope record is not evident in the fossil record of shallow-water biota.

Rising temperature in the Late Oligocene continued into the Early Miocene, reaching a peak of warmth in the upper Altonian and Clifdenian stages in the Early/Middle Miocene boundary zone when larger foraminiferal genera reached their southernmost recorded limit of distribution and reef-building coral genera were well established in the northern North Island. In the Middle Miocene, the Lillburnian climate cooled somewhat, then warmed in the late Lillburnian and early Waiauan when most of the larger foraminiferal genera made a final brief reappearance in the North Island. In the Late Miocene, the climate cooled and probably fluctuated during the Tongaporutuan but remained warm subtropical in the northern North Island. The Late Miocene, Kapitean Stage, was a time of an unsettled oceanic environment under the influence of increased Antarctic glaciation. The distribution of warm-water molluscs and also *Amphistegina* indicate a climate at times no cooler than that of the present day in early Kapitean time. Yet markedly decreasing molluscan diversity throughout the stage seems to be at variance with the presence of stenothermal warm-water biota and also with the strongly fluctuating isotope signal indicating initially cooler temperatures followed by lighter ^{18}O values and warming in the late Kapitean. Early Pliocene, Opoitian, climate was probably marginally warmer than the present climate, but temperatures declined and fluctuated in the Late Pliocene from the Waipipian Stage onwards, cooling rapidly in the latest Pliocene Nukumaruan Stage, ca 2.4 Ma, when elements of the subantarctic marine fauna migrated northwards.

The complex history of Pleistocene climates, most fully recorded in the stratigraphic record of glacio-eustatic marine cyclothems, correlated with oxygen isotope stages, and in the record of terrestrial glaciations, is beyond the scope of this review.

Introduction

Throughout the Cenozoic, New Zealand, with its very full record of marine animal and terrestrial plant life, has been well situated to record evidence of changing climatic trends in the southwest Pacific over the past 65 My. Various kinds of evidence can be used. Fluctuating proportions of the stable isotopes of oxygen and carbon have shown marine climatic trends in much greater detail than has the fossil record, and it is, perhaps, fair to question whether a parallel exercise based on the more haphazard occurrences of stenothermal warm-water marine organisms would have sufficient validity to support independent assessments of past climates. The view taken here is that paleontological and palaeobotanical evidence is as valid as any other in spite of its admitted limitations due to the incompleteness of the fossil record. To obtain a complete picture of paleoclimates all the available evidence from both isotopes and the fossil record needs to be evaluated.

In this review two main complementary categories of evidence are relied upon. The principal one is that of fossil occurrences of warm-water stenothermal benthic marine life. Nearly all the taxa considered lived in shallow inner to mid-shelf depths and comprise very distinctive assemblages of genera and species that are mostly characteristic of warm subtropical or tropical seas to the north of New Zealand. Their descendants progressively became extinct in high southern latitudes along with cooling seas during the Late Miocene and Pliocene.

The second category comprises particular plants that grew at the water's edge on the coast and lived under the immediate influence of the marine climates. The relevant fossils include the mangrove palm *Nipa* and the coconut palm *Cocos zeylandica*.

The distribution patterns of planktic microfossils including foraminifera and calcareous nannofossils, in continuous stratigraphic sequences, also show both broad climatic trends and brief events in the open ocean (Jenkins, 1968, 1973; Edwards, 1968a) and are used where appropriate as supporting evidence.

Marine Fossils: Foraminifera, *Lingula*, Corals, Molluscs

The "larger foraminifera" which include *Asterocyclina, Heterostegina, Cycloclypeus, Marginopora, Fabiania, Lepidocyclina, Miogypsina, Operculina* and *Amphistegina* were distributed most abundantly in low equatorial latitudes in the Cenozoic, and their living relatives do not live in water temperatures below 18° to 20°C (Adams et al., 1990).

Amphistegina is an especially important genus because it is so widely and consistently present in New Zealand from Late Eocene to Early Pliocene. Although a single specimen of *Amphistegina* has been reported from New Zealand waters in bottom sediments near the Cavalli Islands off Northland, Lat 35°S, there are no reports of living populations established in the area (Hayward, 1980). *Amphistegina* is abundant in beach sands at Lord Howe and Norfolk Islands, Lat 30°S, in the warm subtropical climatic zone. Off eastern Australia it is abundant in Hervey Bay, north of Brisbane, Lat 25°S, but it is not definitely reported living further south (A.N. Carter, A.D. Albani, pers. comm.). Specimens in the collections of DSIR Geology and Geophysics from a dredging in Watsons Bay, near the entrance of Sydney Harbour, may be part of a reworked Holocene fauna as *Amphistegina* was not found by Albani (1968) in his survey of the Recent foraminifera. If this is a genuine record of a Recent population, it was living in water with a mean annual temperature of 20°C (Wisely, 1959). The record of *Amphistegina* off the southeastern coast of Australia at Lat 37°C in 925 m depth (Todd, 1976) is of dubious significance as the genus typically lives in the photic zone at inner shelf depths and the specimen has obviously been reworked for a long distance from a shallower locus of possible Holocene age. *Amphistegina* is similarly confined to the warmer water "West Indian Province" of northern South America extending from Lat 0° to 30°S in the Western Atlantic Ocean and also the "Panamanian Province" on the Pacific Side from Lat 0°–20°N (Boltovskoy and Wright, 1976).

Discocyclinidae

The reason for isolated occurrences of *Asterocyclina* and *Lepidocyclina*, often in distinct beds, in the New Zealand Cenozoic, has long been open to two alternative interpretations. One explanation is that these larger foraminifera, typical of warm shoal environments, were well established here throughout the Eocene and in the Miocene but are found only where beds of appropriate facies are exposed. The other explanation is that they were isolated populations of a typically subtropical or tropical fauna able to advance and retreat in response to warming and cooling climatic events. The occurrence of a bed of abundant *Asterocyclina speighti* in basal Eocene limestone of the Chatham Is, which has been isolated from mainland New Zealand throughout the Cenozoic, connected only by the submarine Chatham Rise (Campbell et al., in press) shows that Discocyclinidae were able to migrate quickly over at least 860 km of ocean during a warming event. The same ability to cross ocean barriers is probably shared by other larger foraminifera such as the Lepidocyclinids, *Fabiania* and *Amphistegina* and also by corals. This raises a question when depicting climatic trends diagramatically, as to whether beds of *Asterocyclina* and *Lepidocyclina* should be shown as part of an overall warming trend or as brief warming events.

Lingula

The brachiopod genus *Lingula* has been shown by Lee and Campbell (1987) to have a living distribution in shallow near-shore habitats, and its distribution as fossils in the Cenozoic is consistent with very shallow paleoenvironments. Its present and past distribution is very similar to that of *Amphistegina* as it is associated with subtropical and warmer climatic zones.

Corals

The presence of warm subtropical reef coral genera in the Lower Miocene of northern New Zealand has long been known (Squires, 1962, 1968; Keyes, 1968). They are found, mostly not in situ, but in the vicinity of submarine volcanoes that would have provided a good substrate.

Molluscs

While abundant and diverse throughout much of the New Zealand Cenozoic, molluscs tend to be concentrated in beds of particular facies over a range of predominantly shelf depths. The presence of "cowries" (*Cypraea*) and other typically warm-water Indo-Pacific genera is undeniable evidence of climates much warmer than at present throughout most of the Paleogene and the Miocene.

Land Plants: *Cocos* and Mangroves

The mangrove palm, *Nipa*, represented by the fossil pollen genus *Spinizonocolpites*, is found in Eocene and Oligocene brackish water sediments. Its presence is one of the strongest pieces of evidence of very warm climates throughout most of the Early and Middle Eocene and is the principal basis for accepting a warmer climate in New Zealand at this time than was accepted in some earlier models (Hornibrook, 1968, 1978; Pocknall, 1990). *Cocos zeylandica* Berry is represented by small nuts at several levels in the Cenozoic; they are most abundant in the upper Miocene where they occur in beds in Northland. Even isolated occurrences must be considered to be good evidence of a warm climate as drift was probably minimal, as attested by the fact that coconuts from the mid Pacific islands are not presently being washed ashore in Northland. Terrestrial paleoclimates would have been affected by some different factors than those which affected marine climates. The composition of inland floras would have been influenced by topography, rock and soil type, and patterns of rainfall and prevailing winds in addition to temperature. Land climates derived from palynological evidence could be expected to differ somewhat from marine climates derived from marine biota.

Considerations in Constructing a Diagram Showing Paleoclimatic Trends

Prior to the onset of glaciation in Antarctica and the initiation of the circum-Antarctic current system in the Early Oligocene, southern seas were more uniformly warm (Barker and Kennett et al., 1990), and the latitudinal climatic gradient was probably less than it was in post-Eocene time.

The history of sea-floor spreading movement during the Paleogene, when New Zealand was drifting northwards, adds uncertainty as to what allowance should be made in any comparison that is made with present climates, which exist under very different meteorological and oceanographic regimes. Increasing dislocation and rotation of the South Island and Chatham Is., (Walcott, 1987) further back in time, adds a further complication. Crook and Belbin (1978), in their paleolatitude maps of the south-west Pacific, show New Zealand passing through 15° of paleolatitude during the Paleogene to reach close to its present position by the Early Miocene. How the paleoclimatic trend (Fig. 1) should be adjusted to allow for paleolatitude is not necessarily a straightforward matter. However, the paleoclimatic trend of Shackelton and Kennett (1975) for the Cenozoic of the Campbell Plateau, based on the isotopic content of planktic foraminifera, implies a surface temperature of 20°C during the Early Eocene, which is only marginally cooler than estimates adopted here from other faunal and floral evidence.

A conservative approach was adopted in the construction of Fig. 1, which shows the southernmost fossil occurrences of some typically warm stenothermal animals and plants within New Zealand's present range of 12°Lat, (see also Fig. 2). Obviously, Neogene fossil occurrences are likely to have more validity as indicators of paleoclimates related to present latitudes than are Paleogene ones.

Paleogene

Paleocene

The Teurian (Dt), although a relatively thin stratigraphic unit, is a long stage, equivalent to the whole Paleocene in the time scale adopted in Fig. 1. Hornibrook and Edwards (1971) proposed a three-fold division based on the LOD of *Globigerina pauciloculata* and *Globoconusa daubjergensis* and the FOD of *Discoaster multiradiatus*.

The well-known diverse molluscan fauna of the Wangaloan Stage (Lower Teurian) at Wangaloa, Lat 46°S, on the southeastern coast of South Island, contains most of the molluscan species known from the Early Paleocene of New Zealand. Beu (1966) considered the Wangaloan taxa *Cucullaea*, Pteriidae, *Miltha, Polinices* and Ficidae to be good evidence of subtropical marine climate. A record of *Lingula* from the Lower Teurian, below the Abbotsford Mudstone, Lat 46°S (Lee and Campbell, 1987), is also consistent with a warmer climate than at present.

At Chatham Is., Lat 44°S, 860 km east of the South Island, a diverse molluscan fauna of about 75 species in the Red Bluff Tuff of late Teurian age (Late Paleocene) contains *Spondylus, Ctenoides, Perotrochus* and four taxa of cowries, all typical members of warm subtropical Indo-Pacific faunas (Beu and Maxwell, 1990). The only southern record, in the Red Bluff Tuff at the Chatham Islands, of the foraminiferal genus *Fabiania* (Campbell et al., in press), found elsewhere in the Caribbean and Indo-Pacific regions, is consistent with the molluscan evidence of warm climate. Tropical elements such as the genus *Didymeles*, represented by the pollen type *Schizocolpus marlinensis*, appeared in the New Zealand flora (Mildenhall, 1980).

Early and Early Middle Eocene (Waipawan, Dw, Mangaorapan, Dm, Heretaungan, Dh, and Porangan, Dp, Stages)

The Waipawan Stage, which is mainly Early Eocene and possibly latest Paleocene in part, has provided little positive evidence of marine climate from outcrops on the New Zealand mainland. A diverse foraminiferal fauna from the Matanganui Limestone at Chatham Is., Lat 44°S, includes a bed of *Asterocyclina speighti* (Chapman), the oldest and southernmost Southern Hemisphere record of the genus (Campbell et al., in press). Despite the higher paleolatitude of the Chathams during the Eocene, a much higher temperature than present-day is accepted.

Palynological identifications of the mangrove palm, *Nipa*, from brackish marine facies in the Taranaki Basin and from coal measures and shallow marine beds in Can-

NEW ZEALAND CENOZOIC MARINE CLIMATES FROM SOUTHERNMOST RECORDS OF WARM SHALLOW WATER INVERTEBRATES, MANGROVES AND COCOS NORMALISED TO LAT. 42° S
MAIN CENOZOIC SEDIMENTARY BASINS
WINTER SURFACE TEMP °C
SUBTROPICAL CONVERGENCE
18° 17° 16° 13° 12° 13° 11°
34° S 36° S 38° S 40° S 42° S 44° S 46° S
PLATE BOUNDARY ZONE 15 M.y. (Walcott, 1984, 1987)
TIME SCALE: N.Z. SERIES AND STAGES (Edwards et al 1988)
DANNEVIRKE ARNOLD LANDON PAREORA SOUTHLAND TARANAKI WANGANUI
TEURIAN Dw Dm Dh Dp Ab Ak Ar Lwh Lr Lwh up Ld Lw Po Pl Sc Sl Sw Tt Tk Wo — Wn Wc
PALEOCENE EOCENE OLIGOCENE MIOCENE PLIOC Q
E L E M L E L E M L E L
65 My 60 55 50 45 40 35 30 25 20 15 10 5 0
°C 25° 20° 15° 10°
TROPICAL
°C 25° 20° 15° 10°
PRESENT MEAN ANNUAL BOUNDARY POSITION 25° S LAT.
WARM SUBTROPICAL
climatic trend
HIGH MOLLUSC DIVERSITY CHATHAM IS 44° S
Lingula 46° S
Asterocyclina ABUNDANT CHATHAM IS 44° S
MANGROVES 40° S – 46° S
Amphistegina Asterocyclina SOUTH ISLAND 44° S
Cocos
Lingula 45° S
(last Morozovella)
Asterocyclina RARE 44° S
Heterostegina 43° S
Lingula 45° S
Amphistegina REAPPEARS ABUNDANT 45° S
Asterocyclina 45° S
Amphistegina Operculina Halkyardia 46° S
MANGROVES 36° S 38° S
(fluctuating isotope values)
Cocos Lingula 44° S
S. & W. SOUTH IS
E COAST S.IS. PENGUINS, WHALES
Eulepidina Miogypsina REEF CORALS 38° S 39° S
Heterostegina 42° S
REEF CORAL Cocos Lepidocyclina 36° S 37° S
Lingula Amphistegina 46° S
Lingula 46° S
Cycloclypeus Heterostegina Nephrolepidina
Amphistegina 46° S
(fluctuating isotope values)
LAST Cycloclypeus Heterostegina Operculina Marginopora Nephrolepidina
Cocos → 35° S
Cocos 40° S
Amphistegina 46° S 41° S
SUBANTARCTIC FAUNA MOVED N. GLACIATIONS
ISOLATED OCCURRENCES OF MOLLUSCS
INDO-PACIFIC MOLLUSCS THROUGHOUT INDICATE WARM SUBTROPICAL SEAS
EXTINCTIONS

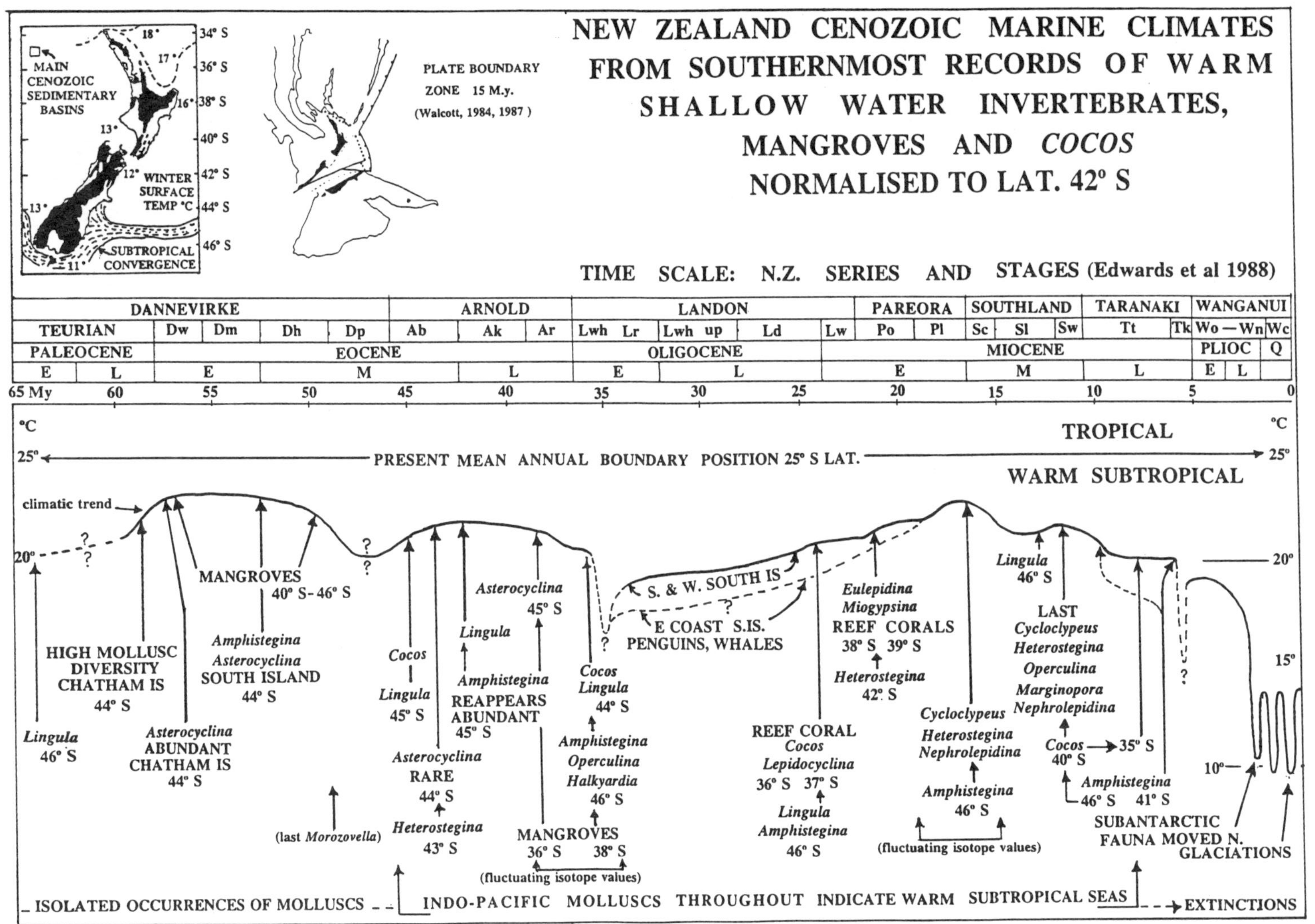

Fig. 1

terbury Basin (Boundary Creek and Abbotsford Mudstone) show that mangroves were established on intertidal mudflats during the whole of this period within a warm climatic zone (20°–24°C, Pocknall, 1990) extending over the whole of New Zealand. The tropical genus *Anacolosa* (Olacaceae), restricted today to the Indomalaysian area, appeared in New Zealand in the Late Paleocene or Early Eocene (Mildenhall, 1980).

The Porangan Stage tends to be patchily distributed, variable in thickness and difficult to correlate by both marine fossils and palynomorphs in shallow facies. The sudden extinction of keeled planktic foraminifera of the genus *Morozovella* at the end of the Heretaungan (Hornibrook et al., 1989) is likely to be the result of a marked environmental event, possibly a sudden cooling episode (Fig. 1). A possible mismatch between a warm Porangan terrestrial climate deduced from the palynology (Pocknall, 1990), and a likely cooling episode deduced from foraminifera may indicate some mis-correlation in this interval.

The only recorded occurrence of larger foraminifera in the Early Eocene of the New Zealand mainland is at Whites Creek, Mid Canterbury, Lat 44°S, where *Asterocyclina speighti* (Chapman), and *Amphistegina eyrei* Larsen, are present in tuffaceous sandstone of the Eyre Sand Group of either Mangaorapan or Heretaungan age (Finlay, 1946; Larsen, 1978a). It is likely that this particular occurrence of larger foraminifera is due to the unusual shallow-water facies as much as it is to a brief climatic warming event. The distinctly warm water assemblage of *Quadrilatera, Septifer, Spondylus, Cypraea* and *Eocithara* from the tuffs at Whites Creek is the only well-dated molluscan fauna from the Early Eocene in New Zealand (Beu and Maxwell, 1990).

Fig. 1

1. Winter sea surface temperatures around New Zealand are taken from Garner and Ridgway (1962).
2. Paleontological evidence for Paleocene paleoclimates is scarce and relies mainly on a few isolated occurrences of diverse molluscan faunules (Beu and Maxwell, 1990).
3. Records of mangroves (*Nipa*) are from Pocknall (1989, 1990).
4. Records of *Lingula* are from Lee and Campbell (1987).
5. The cooling event suggested in the Porangan (Dp) is speculative, based on the extinction of keeled *Morozovella*.
6. The brief cooling event shown in the Early Oligocene is based on oxygen isotope paleotemperatures from Burns and Nelson (1981).
7. The speculative cooler climatic trend shown during the Oligocene and Early Miocene on the east coast of South Island refers to the Canterbury Basin.
8. Fluctuating isotope values shown within the E/M Miocene boundary zone refer to DSDP Site 588 (Kennett, 1986).
9. The time scale adopted is not able to show details of Late Pliocene and Pleistocene climates.
10. Stage symbols, e.g. Dw, Ab, Lwh, Pl, Sw, Tt, Wo, are repeated in the text against the full stage names.
11. First published as an abstract in the programme of the 5th Congress on Pacific Neogene Stratigraphy and IGCP-246, Shizuoka, Oct. 1991, amended and updated Dec. 1991.

Late Middle Eocene

The Bortonian Stage (Ab) is represented in the North Island principally by deep water outer shelf or bathyal facies in which molluscs are scarce. In the South Island it is best represented in glauconitic shelf facies in southern Canterbury Basin. Highly diverse molluscan assemblages occur in the standard coastal section for the stage north of Hampden and in the Waihao Greensand further inland. The Bortonian fauna is recognized as one of the most diverse in the New Zealand Cenozoic with an apparently massive influx of molluscs into the region, possibly exaggerated by comparison with the poor representation of the group earlier in the Eocene. The Bortonian taxa *Pteria*, *Tellina*, *Placamen*, *Costacallista*, *Rimella*, Cypraeidae, Ovulidae, Conidae, *Marshallina*, *Gemmula*, *Fusiaphera* and *Granosolarium* are restricted at present to tropical and subtropical regions (Beu and Maxwell, 1990). *Lingula* occurs in the Tapui Sandstone in the same district (Lee and Campbell, 1987).

Foraminifera are not strongly indicative of a warm climate throughout the whole of the Bortonian, although a single occurrence of a bed of *Asterocyclina* in the Waihao River sequence, correlated with the Waihao Greensand (Riddolls, 1966), is certainly suggestive of a warm subtropical climate, possibly during a short event, and small nuts of *Cocos* have been found in the Waihao Greensand, 45°S (Campbell et al., 1991). Also, *Heterostegina* is recorded by Maxwell (1968) from the Kaiwara district, North Canterbury. *Amphistegina* (*Asterigerina*), common from the Late Eocene onwards, is absent, but the warm water planktic genus *Hantkenina*, occurs in the uppermost part of the stage (Cameron and Waghorn, 1985) possibly during another warm event that preceded rising temperature in the Late Eocene.

Late Eocene

Kaiatan (Ak) molluscan faunas and the more poorly known Runangan (Ar) ones contain *Arca*, *Quadrilatera*, *Spondylus*, *Bolma*, *Cheilea*, *Ficus*, *Colubraria*, *Gemmula*, *Cochlespira*, *Marshallena* and *Cordieria* indicating a continuing warm subtropical climate in southern New Zealand (Beu and Maxwell, 1990).

**Amphistegina* (*Asterigerina*) appears in the Kaiatan of the Greymouth Basin (Hoskins, 1974) and continues to be common in inner and mid shelf facies in the Runangan and basal Oligocene throughout New Zealand. *Hantkenina* reappears in a narrow zone in the upper Kaiatan and lower Runangan (Srinivasan and Vella, 1968). *Asterocyclina* is present in the lower Runangan Totara (Ototara) Limestone in southern Canterbury Basin (Cole, 1967), and rare *Discocyclina* is recorded from the Runangan in Greymouth Basin (Maxwell, 1968).

---

* The generic name *Asterigerina* is retained for asymmetrical species that are otherwise similar to *Amphistegina*. Larsen (1978b), in his review of *Amphistegina*, includes them in that genus.

Summary

In spite of the fact that New Zealand was located at considerably higher latitudes than it is at present, the Early to late Middle Eocene was characterized by consistently warm subtropical or marginally tropical types of marine biota and coastal mangroves and *Cocos*. Apart from a possible brief Middle Eocene cooling event, marine climates do not show evidence of significant cooling until the very latest Eocene.

Land Paleoclimates

During the Early and Middle Eocene, when New Zealand was surrounded by a more uniformly warm Southern Ocean, it was a land of generally low relief; judging from palynological evidence and from the presence of *Cocos*, it had climates in accord with those indicated by the marine fauna. However, in the Late Eocene, an increase in pollen of the *Nothofagus fusca* group is interpreted as an indication of a "cool temperate" land climate cooler than the marine climate (Pocknall, 1989). Both palynology and the marine fauna are in accord in supporting a record of a fluctuating cooling trend across the Eocene/Oligocene boundary zone.

Early Oligocene

The sudden extinction of *Globigerapsis index* at the end of the Runangan Stage, followed soon after by *Pseudohastigerina micra* and the typically Eocene benthic foraminifera *Cibicides parki* and *Uvigerina bortotara bradleyi* (Hornibrook et al., 1989), coincides with an isotopic record consistent with a declining and fluctuating climatic trend. Burns and Nelson (1981) show isotopic paleo-temperatures descending to 12°C followed by a rise to 18°C in a brief event in the very Early Oligocene in the Cape Foulwind section in the Greymouth sub-basin on the west coast of the South Island. Devereux (1968), in his pioneering study of oxygen isotopes in the New Zealand Cenozoic, also showed a marked drop in temperature across the Eocene/Oligocene boundary zone to 7°C in shallow bottom waters. Bryozoan limestone from Kakanui in the southern Canterbury Basin, on which these results were based, is known not to be in isotopic equilibrium with sea water, and Devereux's paleotemperatures were questioned by Burns and Nelson, who recalculated them to produce a figure closer to their own. Both of these studies predate the subsequent demonstration of Early Oligocene glaciation in Antarctica and may be subject to some upward revision to allow for an ice accumulation factor.

Paleontological confirmation of a cooling event of the severity indicated by the isotope record is lacking or at least inconclusive. Srinivasan (1965) records *Amphistegina* (*Asterigerina*) *cyclops* present continuously across this short interval at Cape Foulwind, posing an apparent contradiction to the isotope temperatures which would imply the survival and reproduction of a strongly warm-water stenothermal genus in temperatures well below those known from its living distribution in the South Pacific. Possibly it could have survived in sheltered coastal niches.

Edwards (1968a, 1991) reviewed a range of positive paleontological evidence for warm subtropical or warm temperate climate in the latest Eocene and Early Oligocene in the Oamaru district in southern Canterbury Basin and discussed the disparity between isotopic and paleontologically derived paleotemperatures in some depth. He also questioned the reliability of Devereux's results from bryozoa, concluding that the few isolated samples available to him were insufficient to define a climatic trend. A brief cooling event might easily be missed in shallow-water sequences. However, the writer has found *Amphistegina* (*Asterigerina*) to be abundant in the basal Oligocene McDonald Limestone at Kakanui but generally absent in the Early Oligocene in the Canterbury Basin thereafter.

A decline in Early Oligocene temperature in New Zealand has long been hypothesized because of the absence of most larger foraminifera other than *Amphistegina* and the scarcity of other warm-water taxa in the Canterbury Basin (Hornibrook, 1978). However, discoveries of *Lingula* and *Cocos* in the Early Oligocene lower Wharekuri Greensand in the Canterbury Basin (Campbell et al., 1991) do not support a marked cooling. Records of *Operculina, Amphistegina,* and *Halkyardia* in the Lower Oligocene of the Southland Basin, Lat 46°S (Hornibrook in Wood, 1969) and of *Amphistegina* in the South Island West Coast Basin, and of *Operculina, Asterigerina* and mangroves (*Nipa*) in the Mangakotuku Siltstone, Waikato Basin, North Island, (Hornibrook et al., 1989; Pocknall, 1989) all indicate that the climate was marginally warm subtropical but was probably not warm enough to support larger foraminifera or reef-building corals.

The Early Oligocene molluscan fauna is not as abundant or as diverse as Late Eocene ones, but records of warm-water taxa *Arca, Quadrilatera, Bolma, Cypraea,* and *Conus* in southern Canterbury Basin (Beu and Maxwell, 1990) are in accord with the foraminifera in not supporting a very marked post-Eocene cooling.

A regional Middle Oligocene disconformity in the Canterbury Basin, possibly related to the Middle Oligocene drop in eustatic sea-level (Haq et al., 1987), has removed the uppermost part of the *Globigerina angiporoides* range-zone and any evidence of a possible climatic event associated with the extinction of *G. angiporoides* ca. 32 Ma.

The southern Canterbury Basin is well known for the abundance of fossil cetaceans and penguins in the Oligocene (Fordyce, 1977). There are also records from the Late Eocene and Early Miocene. However, it is uncertain whether the abundance of these typically Southern Ocean animals and birds has any special climatic significance. Eastern South Island was in higher paleolatitudes and was more southwards-facing during that early phase of sea-floor spreading movement than it is at present. There is evidence of marine upwelling around a Late Eocene volcanic high when the Oamaru Diatomite formed (Edwards, 1991) on the outer part of a broad shelf that persisted into the Early Miocene. An environment well suited to whales and penguins may have developed independent of climate. The post-Early Oligocene marine faunas of the Canterbury Basin do not seem to support the continuance of climates as warm as those of the Eocene, and give some grounds for speculation that there were parallels with the present situation when the cool Southland Current follows the east coast northwards.

Late Oligocene

In the Waikato Basin, North Island, where the Upper Whaingaroan (Lwh) is best developed, it is typically represented by outer shelf and upper bathyal facies lacking shallow water assemblages. The overlying Duntroonian (Ld), in contrast, is typically represented by shallower facies in many parts of New Zealand. In the Southland Basin, Lat 46°S, *Amphistegina* is present in the Chatton Marine Beds (Hornibrook in Wood, 1956). Diverse molluscan assemblages of warm-water taxa contain *Maoricardium, Solecurtus, Tapes, Clavagella, Pyrazus, Typhis*, two species of Conidae, and *Scalptia* (Beu and Maxwell, 1990). *Lingula* is recorded from the Gore Lignite Measures (Lee and Campbell, 1987). In the North Canterbury Basin, *Amphistegina* has been identified by the writer, present in the Duntroonian in the Trelissick sub basin. A diverse molluscan assemblage from the upper Wharekuri Greensand, Lat 45°S, contains the warm-water taxa *Arca, Solecurtus, Tapes, Clavagella, Echinophoria, Ficus, Typhis, Morum* (*Oniscidea*) and *Umbraculum* (Maxwell, 1969, 1991). The South Island marine climate during the Late Oligocene was certainly warmer than at present although cooler than during the Eocene.

In the North Island, the writer has found *Amphistegina* common in shallow facies of the Duntroonian mapped as Aotea Sandstone in the Waikato Basin. Recent discoveries of the small coconut, *Cocos zeylandica*, in limestone of probable late Duntroonian age in Northland and a single head of the reef coral *Leptastria* of about the same age in Coromandel Peninsula, Lat 37°S, complement the long-known presence of *Heterostegina* and *Lepidocyclina* in a redeposited bed in the Motatau Limestone in Hokianga Harbour (Hayward et al., 1990). Northland should have been well within the warm subtropical zone in agreement with the temperature range of 22°C–25°C estimated by Hornibrook (1968).

Summary

Sea temperatures in the earliest Oligocene were not much cooler than they were in the latest Eocene. They probably declined somewhat subsequently, although, in contrast to the extinctions of some planktic foraminifera and nannoplankton species, the shallow-water fauna give little positive support for the abrupt drop in isotopic paleotemperatures. Climate warmed in the Late Oligocene, and larger foraminifera and *Cocos* and reef corals began to reappear in the far north.

Land Paleoclimates

It is likely that during the Oligocene, when the New Zealand land area had been much reduced, possibly to an archipelago (Stevens and Suggate, 1978), the terrestrial flora were directly affected by a cooling marine climate. Pocknall (1989, Fig. 11) show a marked increase in *Nothofagus brassii* group pollen across the Eocene/Oligocene boundary followed by a sudden but short-lived decrease in the Early Oligocene. If accurately correlated with the Cape Foulwind sequence, this trend could help to confirm the brief Early Oligocene cooling event shown in the isotope record by Burns and Nelson. *N. brassii* beech pollen once more became dominant with rising humidity in the Late Oligocene. Mildenhall (1980), on the other hand, saw Oligocene forests dominated by *Nothofagus brassii* associated with other taxa, as indicating humid "warm-temperate" conditions over the whole of New Zealand with little evidence of the marked cooling identified from isotopic paleotemperatures.

Neogene

Early Miocene

In the Waitakian (Lw), *Amphistegina* is present in the Chatton Formation in the Southland Basin (Hornibrook in Wood, 1956), and it also occurs in the Middle Canterbury Basin in a calcareous sandy facies in the upper Tengawai River.

Waitakian molluscan faunules from southern Canterbury and Southland Basins, although less diverse, generally resemble those of the Duntroonian in containing the warm-water genera, *Arca, Solecurtus, Guildfordia, Parazus, Ficus, Oniscidea*, and *Gemmula*, all indicative of warmer than present climate (Beu and Maxwell, 1990). There is a record of *Lingula* from White Rock River in the middle Canterbury Basin (Lee and Campbell, 1987).

In the Otaian (Po), and the Altonian (Pl), molluscan evidence for a "temperate fauna" in southern Canterbury Basin is based on the presence of much less diverse assemblages than are found to the north and even in the Southland Basin (Beu and Maxwell, 1990).

In the North Island, isolated heads of reef corals of 17 Indo-Pacific genera including *Porites* and *Turbinaria* appear in the far north, Lat 36°S, and at East Cape, Lat 38°S, in the Otaian. Most live in the Great Barrier Reef today in a range of 19°C to 28°C (Squires, 1962, 1968; Keyes, 1968; Hayward, 1977). *Heterostegina* occurs in the Blue Bottom Formation at Fletcher Creek in the Murchison sub basin, Lat 42°S, western South Island, which represents possibly the southernmost record of larger foraminifera in the Otaian. Additional genera including *Eulepidina, Nephrolepidina, Cycloclypeus* and *Miogypsina* had become well established in the northern half of the North Island (Chaproniere, 1984).

A situation similar to that reported in the Early Oligocene, when a marked brief downward fluctuation in isotopic paleotemperature seems not to have produced a corresponding marine faunal response, is described by Nelson and Burns (1982) in the Otaian Taumutamaire Formation in the Waikato Basin, where a sequence of ^{18}O measurements shows a brief interval of increasing heavier isotopic values interpreted as a 3°C drop in temperature. Such examples, if representative of actual marine temperatures, would indicate that the paleontological record shows a smoothed climatic trend that does not adequately represent the degree of actual climatic variability.

By Altonian (Pl) time all of these warm subtropical foraminiferal genera, except for *Miogypsina*, had spread southward, possibly along the west side of the South Island, as did the molluscs, to reach their southernmost limits of recorded distribution at Clifden in the Southland Basin, Lat 46°S (Hornibrook, 1978; Beu and Maxwell,

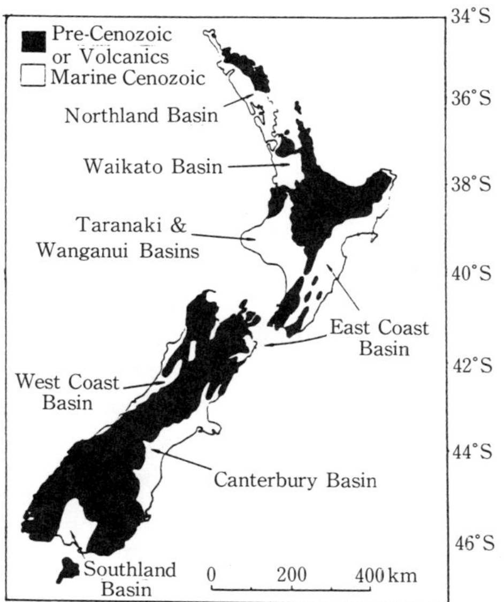

Fig. 2. Main marine Cenozoic sedimentary basins of New Zealand.

Fig. 3

Smoothed benthic foraminiferal delta ^{18}O data for DSDP Leg 90, Holes 588, 588a and 588c and biostratigraphic zonation for Hole 588 plotted against age. Adapted from Kennett 1986, p.1389, Fig. 4, with delta ^{13}C data removed and N.Z. stages (biostratigraphic correlations) shown on right.

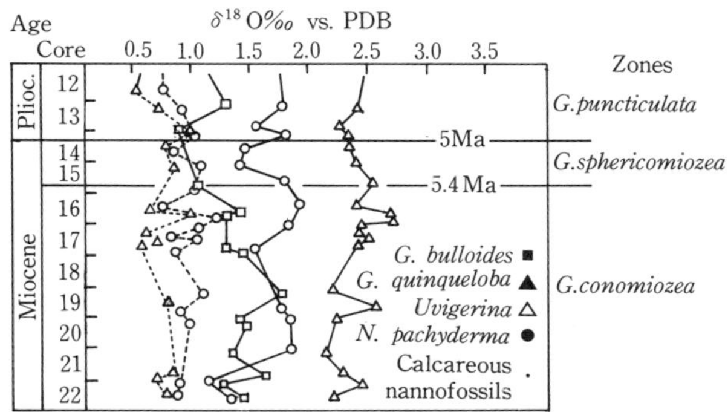

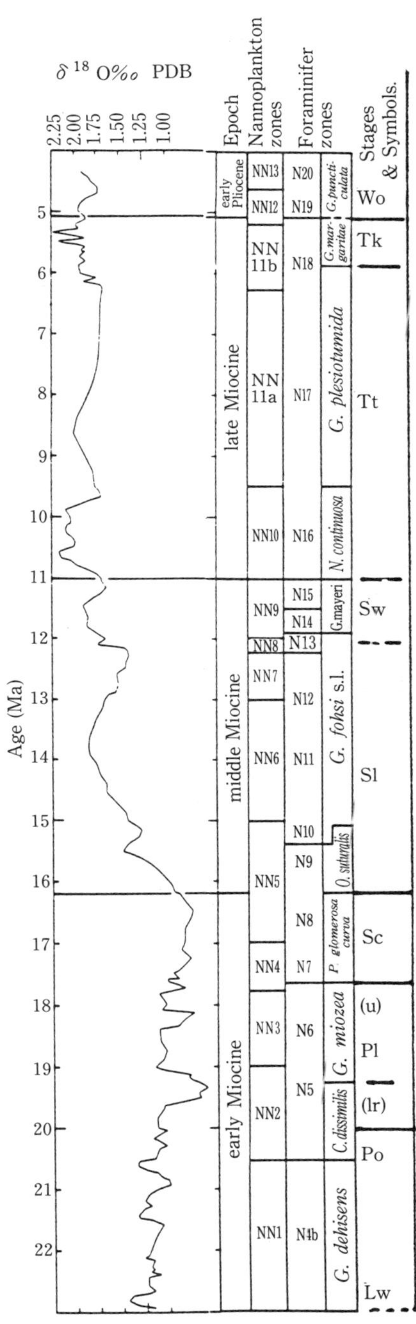

Fig. 4

Oxygen isotope data from DSDP holes 284, 284a, adapted from Kennett et al. (1979).

1990). A warm subtropical or possibly marginally tropical sea, the warmest since the Middle Eocene, surrounded the whole country with the possible exception of the Canterbury Basin, in which most larger foraminifera are rare or absent.

Middle Miocene

The Clifdenian (Sc) and Lillburnian (Sl) are mostly represented by deep-water facies unsuitable for larger foraminifera. *Heterostegina, Cycloclypeus, Lepidocyclina,* and *Amphistegina*, all apparently reworked downslope into deeper water, occur in the Clifdenian Stillwater Mudstone in the Greymouth Basin, Lat 42°S (Hoskins, 1974).

At Clifden, in the Southland Basin, Lat 46°S, the molluscan fauna include *Arca, Spondylus, Septifer, Chama, Trachycardium (Regozara), Cypraea (Notadusta), Ficus, Chicoreus (Siratus), Morum (Oniscidea)*, Conidae, and *Microdrillia*, indicating that sea temperatures in Southland and Westland were at least as warm as they were during Altonian time (Beu and Maxwell, 1990). Planktic foraminifera reached their peak of Miocene diversity in the Clifdenian *Praeorbulina glomerosa*-zone (Jenkins, 1973). Edwards (1968a, Fig. 1), shows a peak of warmth with calcareous nannofossils Braarudosphaerids, *Ericsonia ovalis* and rhabdoliths present in the Southland Basin.

This whole Altonian-Clifdenian interval between the FOD of *Globorotalia praescitula* and the FOD of *Orbulina suturalis* is clearly differentiated in the isotope record of DSDP Holes 588A, 588B and 588C, Lat 26°S, where low ^{18}O values from deep-sea benthic foraminifera show a fluctuating peak of highest Neogene temperatures (Fig. 3). Isotopic paleotemperatures show a marked rise near the FOD of *Globorotalia miozea* in the upper Altonian followed by strongly fluctuating values with very high warm peaks in the *Praeorbulina glomerosa curva*-zone in the Clifdenian (Kennett, 1986; compare also Devereux, 1967, Fig. 1; 1968, Fig. 1).

The increase in ^{18}O values that followed, indicating a substantial cooling of deep bottom waters, may have been responsible for the appearance of *Uvigerina mioschwageri* and *U. notohispida* in the lower Lillburnian, but the shallow-water fauna provides little positive evidence to support a marked cooling of marginal seas around New Zealand. *Lingula* is present in the stage stratotype at Clifden (Lee and Campbell, 1987), and a badly worn reworked individual of *Lepidocyclina*, and a few records of abundant *Amphistegina* in the Southland Basin (Hornibrook, in Wood, 1969) do not suggest particularly cool temperatures. Nevertheless, the lack of records of *Lepidocyclina* and other larger foraminifera in the Lillburnian, although possibly due to the absence of shallow facies, seems to support a cooling of the climate in the Middle Miocene.

In the upper Lillburnian (Sl) and lower Waiauan (Sw), ca. 12 Ma, shallow carbonate facies became widespread (e.g. Kingma, 1971). *Lepidocyclina* s.l., *Operculina, Heterostegina, Cycloclypeus*, and *Marginopora* appeared in the North Island at Pourere, Lat 40°S, in the East Coast Basin, and in Northland, in the Waikuku Limestone, Lat 34°30'S (Leitch et al., 1969; Chaproniere, 1984). Except for *Amphistegina*, the end of the Waiauan, or some time within the stage, marked the final extinction of most larger foraminifera in New Zealand. Nuts of *Cocos* have been found in southern Hawkes Bay, Lat 40°S (Kingma, 1971), and also in turbidites in the Poverty Bay district, Lat 38°30'S (Ballance et al., 1981). *Globorotalia menardii*, seldom found south of the warm subtropical zone, appeared in the North Island in the Waiauan.

Waiauan (Sw) and Tongaporutuan (Tt) molluscs occur mostly in deep-water assemblages; they are difficult to distinguish from each other and rather unspecific as to climate. The Tongaporutuan is a long stage of about 4 My, but the overall numbers of molluscan genera becoming progressively extinct indicate cooling climate (Beu and Maxwell, 1990; Beu, 1990). *Amphistegina* persisted as far south as Lat 41°S (Whakapuni, E. Wellington, loc. T28/F6671, lower Tongaporutuan), indicating that the North Island was still marginally warm subtropical. *Globoquadrina dehiscens*, present since the Early Miocene, also persisted into early Tongaporutuan in the North Island before its extinction in this latitude at 9.4 Ma (Wright and Ashby, 1985). *Globorotalia menardii* appeared at times during the Tongaporutuan in the North Island (Hornibrook et al., 1989). Beds of *Cocos*, of probable Tongaporutuan age, found at Mongonui, in Northland, Lat 35°S (Berry, 1926, as "Pliocene"), show that the genus was well established there in the Late Miocene.

Oxygen isotope values from benthic foraminifera in DSDP Holes 588, 588A, 588B, 588C (Kennett, 1986) show a distinct increase in the heavy isotope at a level equivalent to Lower Tongaporutuan, indicating cooling followed by a longer warmer period (Fig. 3). However, this interpretation seems to be at variance with the planktic foraminiferal and molluscan records.

Latest Miocene

The Kapitean (Tk) stage coincides, in part, with the onset of the terminal Miocene glaciation of West Antarctica and with the Mediterranean "salinity crisis" often called the "Terminal Miocene Event", when there was a marked increase in circulation of cold Antarctic water (Kennett and von der Borch, 1986b). This was also at the end of a stepwise descent in molluscan diversity resulting in a short period of marked faunal impoverishment (Beu, 1990).

The typically cold subantarctic low diversity siliceous and calcareous microplankton assemblages in the Late Miocene of DSDP Leg 90, Hole 594, situated on the present southern boundary of the Subtropical Convergence Zone, Lat 45°31.41'S (Kennett, von der Borch et al., 1986a) would be present in Late Miocene on land sequences if the Convergence Zone had moved far to the north of its present position.

Foraminiferal response to what should have been a marked cooling episode was somewhat varied and does not show a consistent trend. The Kapitean was a time of fairly low planktic foraminiferal diversity with predominantly sinistral populations of and peak of abundance of the cold water planktic foraminifer, *Neogloboquadrina pachyderma* (Vella and Kennett, 1975), yet populations of mature, apparently autochthonous *Amphistegina* are present in the Clay Creek Limestone in South Wairarapa, southern North Island, Lat 41°S. These populations should have been adversely affected by any significant northward shift of the Subtropical Convergence Zone along the east of the North Island bringing cold water to this latitude. Vella's (1973) explanation of the presence of *Amphistegina* in the Clay Creek Limestone as an example of non breeding populations is difficult to reconcile with their occurrence in such abundance. Vella and Kennett (1975) concluded that frequency oscillations of *N. pachyderma* were the most useful criteria for determining paleoclimatic history but that this species did not adopt its present-day coiling pattern until the Early Pliocene or later.

Beu (1974) argued vigorously in support of warmer than present temperatures during the Kapitean because of the presence of the molluscan genera (names updated) *Sassia, Ranella, Charonia, Terebra,* large *Pterynotus, Conus, "Isognomes",* and *Izumonauta* as far south as Lat 38°S in the North Island and the presence of *Chama, Semicasis (Kahua),* large *Lepsiella* and *Aturia* in the Southland Basin Lat 46°S. Beu emphasized that lower Kapitean molluscan faunas show no significant changes in the stenothermal fossils that occur through most of the better-known sections. However, Beu (1990) subsequently has given more weight to the high percentage of extinctions and low generic diversity of upper Kapitean molluscan assemblages to recognize a marked environmental deterioration during the Terminal Miocene Event.

Apparent discrepancies between Kapitean paleotemperatures derived from benthic and planktic faunas and those from stable isotopes have long been a problem (Hornibrook, 1978; Beu, 1974, 1975; Vella and Kennett, 1975). What is not particularly in question is that this was a time of accumulation of ice in Antarctica and that the rather thin, patchy, and interrupted nature of much of the sedimentary record could be interpreted as glacio-eustatically controlled (Kennett 1967, 1968; Kennett and Watkins, 1974; Wright and Vella, 1988).

How is the persistence of these stenothermal warm-water genera to be interpreted? Were they truly stenothermal? Are the north – south correlations of these occurrences valid? Do they represent warm phases of a strongly fluctuating interglacial climate? How nearly are the isotopic values related to temperature and how much to fluctuations in polar ice? How are we to account for an apparent juxtaposition of a warm, shallow-water foraminifera with a cool planktic fauna (Kennett, 1968). A comparable situation possibly existed in the Late Miocene Mohnian Stage of California when very warm shallow-water molluscs co-existed with sinistrally coiled populations of *Neogloboquadrina pachyderma* which are considered to indicate 5°C water temperature, possibly living at greater and colder depths (Addicott, 1969). Antarctic cold bottom water surrounded southernmost New Zealand, possibly creating a steep thermal gradient that insulated the shallow-water fauna from the effects of Antarctic glaciation.

The uppermost Kapitean (*Globorotalia sphericomiozea* range-zone), which correlates most closely with the upper part of the Mediterranean "Salinity Crisis" (Edwards, 1987), is, in most New Zealand sequences, a deepening phase with few molluscs preserved, perhaps giving an exaggerated impression of very low faunal diversity. Furthermore, the ^{18}O record of DSDP Site 284, Lat 40°S (Fig. 4), shows a reversal to lighter values in the *G. sphericomiozea* range-zone, implying a short warming event (Kennett et al., 1979; Hornibrook, 1984).

One or more possible cooling events are allowed for in Fig. 1, principally to accommodate the implications of strongly fluctuating isotopic values, but it is difficult to find much positive evidence from shallow-water faunas that marginal sea temperatures were much below these of the present day even in the southern part of the South Island.

Summary

Climate warmed in the Early Miocene to become warm subtropical or marginally tropical in northern North Island. Both the isotope record and marine fauna closely agree in identifying a Neogene climatic optimum in a narrow Early/Middle Miocene boundary-zone. Climate probably cooled in the Middle Miocene and warmed again with the brief reappearance of some larger foraminiferal genera for the final time in the New Zealand fossil record.

Climate slowly cooled in the Late Miocene and finally entered a phase of strong isotopic fluctuations and apparently contradictory marine faunal signals, which indicate an unsettled oceanic environment.

Land Paleoclimates

During the Early to Middle Miocene *Nothofagus brassii* forest covered New Zealand and climatic conditions were alternatively "cool" and "humid and warm-temperate" (Mildenhall and Pocknall, 1984, 1989). Assemblages in the Gore Lignite Measures in Southland, Lat 46°S, and the Manuherikia Group in Central Otago (Mildenhall, 1989; Mildenhall and Pocknall, 1989, 1990) are rich in variety and probably represent optimum diversity in the Cenozoic.

In the Late Miocene the vegetation pattern changed to one where *N. brassii* was more common in the north and west of New Zealand. Podocarps and other cool temperate taxa show a pattern of gradual northward migration (Mildenhall and Pocknall, 1984). As with the marine fauna, there is evidence of cooling climate and withdrawal northwards of more thermophyllic biota. Marked differences in the regional dominance of tree species and in the relative percentages of warm and cool taxa suggest that climate fluctuated markedly from warm-temperate to cool-temperate (Mildenhall and Pocknall, 1984). Increasing topographic relief (Stevens and Suggate, 1978) associated with orogenic activity, which increased markedly from this time onwards, was probably an additional factor in creating a greater variety of ecological situations for floral elements that were adapted to cooler conditions.

Pliocene and Pleistocene

Early Pliocene marine climate was warm-temperate (or cool-subtropical) over most of New Zealand except in northern North Island, which was probably marginally warm subtropical. Large, apparently autochthonous, thick-walled *Amphistegina* is common in the Whakapunaki Limestone (upper Opoitian) in northern Hawkes Bay in the East Coast Basin, Lat 39°S. Also, *Globorotalia menardii*, from the northern warm subtropical zone, occurs sporadically in the Opoitian (Wo). Several molluscan taxa such as *Proterato, Neocola, Gemmula*, and *Zemacies* reappeared after an apparent Late Miocene absence (Beu and Maxwell, 1990).

The history of the Mollusca and benthic foraminifera in the Late Pliocene is predominantly one of progressive extinction and withdrawal northwards of Miocene taxa as climate cooled and fluctuated and of the appearance in New Zealand of elements of the cool-water Southern Ocean fauna. Beu (1974) considered that the presence of the large bivalve genera *Phialopecten, Maoricardium, Crassostrea*, and *"Isognomon"* in the Waipipi Shell Bed (Waipipian Stage (Wp), early Late Pliocene) in Wanganui Basin, in association with many other taxa now extant only in northern New Zealand, indicates conditions "significantly warmer" than at present.

The cool Southland Current moving northwards along the east coast of the South Island would have brought cooler water to the southern half of the North Island, and any northward shift of the Subtropical Convergence Zone would have intensified cooling of offshore waters (Beu et al., 1977). Devereux et al. (1970) concluded, from oxygen isotope ratios and coiling direction ratios of *Neogloboquadrina pachyderma*, that Pliocene and Pleistocene paleotemperatures off eastern New Zealand fluctuated strongly from moderately warm to very cold. These effects probably reached as far north as Hawkes Bay, Lat 39°S, where fluctuating reversals of dominantly sinistral and dextral populations of *N. pachyderma* and intervals with warmer-water planktic species are recorded in the Wairoa Drillhole (Hornibrook, 1976).

Amphistegina, present in the shallow-water Tahaenui Limestone, northern Hawkes Bay, Lat 39°S, indicates that there was a warm interval in the early Waipipian (Wp), possibly warmer than at present, in the East Coast Basin, whereas the cool water genus *Hyalinea* is confined to deeper muddy facies (Collen, 1974) in both the East Coast and Wanganui Basins.

The first dramatic cooling event was marked by the sudden northward migration of elements of the subantarctic fauna, including *Chlamys patagonica delicatula* and the spider crab *Jacquinotia edwardsii*, which appeared in the southern half of the North Island in the early Nukumaruan (Wn) (Beu et al., 1977) in the Matuyama Chron, ca. 2.4 Ma, probably coinciding with the earliest known (Ross) glaciation.

The subsequent history of the interplay between successive Pleistocene glacial and interglacial events and their influence upon both the marine and land fauna and flora is too complex to be dealt with within the timescale adopted in Fig. 1 and is beyond the scope of this review of Neogene climates. The most complete record of Late Pliocene and Pleistocene marine climates is believed to be that inferred from the stratigraphic record of glacio-eustatic marine cyclothems in Wanganui and East Coast Basins and in DSDP Site 594 correlated with oxygen isotope stages (Nelson et al., 1986; Haywick et al., 1991; Beu et al., 1987).

Land Paleoclimates

During the Pliocene climatic changes increased in magnitude and frequency. Many subtropical and humid warm-temperate floral elements became extinct or were substantially reduced in distribution. A number of warm-temperate or subtropical taxa persisted in the Early Pliocene, indicating a climate similar to that of the Late Miocene. As with the marine fauna, cooling climates in the Late Pliocene caused the extinctions of a large number of warm-temperate taxa, which were replaced by cool-temperate taxa. A marked latitudinal temperature gradient existed in Late Pliocene time, with more warm-temperate taxa in the north and more cool temperate taxa in the south (Mildenhall and Pocknall, 1984).

A rapid rise of the main axial mountain chain during the Late Pliocene and Pleistocene exaggerated the climatic effects of four glaciations in the South Island, producing an extensive complex of moraines and outwash gravels (Suggate, 1990). At times, temperatures may have dropped by as much as 4.5°C. In the central North Island the record is complicated by the deposits from huge volcanic ash eruptions. Recurrent glaciations markedly changed the patterns of vegetation, and the last remnants of tropical and subtropical taxa disappeared. Floral assemblages varied widely in space and time, ranging from arid in northern coastal sites to grassland in the south (Mildenhall, 1980).

Directions of Future Research in Neogene Paleoclimates

The past 5 My of the Pliocene and Pleistocene geological history of New Zealand have, at times, been characterized by dramatic fluctuations of climate and associated rises and falls of sea-level. The effects on the distribution and survival of previously long-established plants and animals has been profound. The composition of the present-day land flora and of the marine life, is the result of a long-term trend of cooling climate.

New Zealand, with its excellent sedimentary sequences, has the most complete record of late Neogene history in the southwest Pacific area. It is a record that holds many answers to the kinds of environment that existed during the Pliocene when climates were closer to the likely results of a possible "greenhouse climate". The rapidity of sea-level rises during the Pleistocene is a question whose answer is to be found in the sedimentary sequences and is closely related to rates of climatic change.

New Zealand is a fertile field of research on these questions, which are of direct relevance to the concerns over the future environmental results of the present climatic warming trend. They will require coordinated attention, more particularly in the fields of biostratigraphy, magnetostratigraphy, and isotope geochemistry.

Acknowledgements

The writer is grateful to the following colleagues who read this manuscript and contributed useful and much-appreciated comments: A.G. Beu, G.R. Stevens, G.H. Scott, I.W. Keyes, D.C. Mildenhall, and D.T. Pocknall. The Ms was typed by Irene Galuszka.

References

Adams, C.G., Lee, D.E., and Rosen, B.R., 1990. Conflicting isotopic and biotic evidence for tropical sea-surface temperatures during the Tertiary. *Paleogeogr., Palaeoclimatol., Palaeoecol.,* 77: 289–313.

Addicott, W.O., 1969. Tertiary climatic change in the marginal north-eastern Pacific Ocean. *Science,* 165: 583–586.

Albani, A.D., 1968. Recent Foraminiferida of the central coast of New South Wales. AMSA Handbook No. 1. *Aust. Mar. Sci. Assoc. Sydney.* 36 p.

Ballance, P.F., Gregory, M.R., and Gibson, G.W., 1981. Coconuts in Miocene turbidites in New Zealand: Possible evidence for tsunami origin of some turbidity currents. *Geology,* 9(12): 592–595.

Barker, P.F., Kennett, J.P. et al., 1990. Latest Cretaceous to Cenozoic climate and oceanographic developments in the Weddell Sea, Antarctica: An ocean drilling perspective. *Proceedings of the Ocean Drilling Program, Scientific Results* No. 113: 937–960.

Berry, E.W., 1926. *Cocos* and *Phymatocaryon* in the Pliocene of New Zealand. *Amer. J. Sci. Ser.,* 5, 12(69): 181–184.

Beu, A.G., 1966. Sea temperatures in New Zealand during the Cenozoic Era as indicated by molluscs. *Trans. Roy. Soc. N.Z. Geol.,* 4(9): 177–187.

Beu, A.G., 1974. Molluscan evidence of warm sea temperatures in New Zealand during Kapitean (Late Miocene) and Waipipian (Middle Pliocene) time. *N.Z. J. Geol. and Geophys.,* 17(2): 465–479.

Beu, A.G., 1975. Molluscan fossils and late Neogene paleotemperatures – Reply in correspondence. *N.Z. J. Geol. Geophys.,* 18(1): 198–202.

Beu, A.G., 1990. Molluscan generic diversity of New Zealand Neogene Stages: Extinction and biostratigraphic events. *Palaeogeogr., Palaeoclimatol., Palaeoecol.,* 77: 279–288.

Beu, A.G., Edwards, A.R., and Pillans, B.J., 1987. A review of New Zealand Pleistocene stratigraphy with emphasis on marine rocks. *In:* M. Itihara and T. Kamei (Eds.). *Proceedings of the First International Colloqium on Quaternary stratigraphy of Asia and Pacific Area, Osaka, 1986,* pp. 250–269.

Beu, A.G., Grant-Taylor, T.L., and Hornibrook, N. deB., 1977. Nukumaruan records of the subantarctic scallop *Chlamys delicatula* and crab *Jacquinotia edwardsii* in central Hawkes Bay. *N.Z. J. Geol. and Geophys.,* 20: 217–248.

Beu, A.G., and Maxwell, P.A., 1990. Cenozoic Mollusca of New Zealand. *N.Z. Geol. Surv. Paleont. Bull.,* 58: 518 p.

Boltovskoy, E. and Wright, R., 1976. *Recent Foraminifera.* Junk, the Hague. 515 p.

Burns, D.A. and Nelson, C.S., 1981. Oxygen isotope paleotemperatures across the Runangan-Whaingaroan (Eocene-Oligocene) boundary in a New Zealand shelf sequence. *N.Z. J. Geol. Geophys.,* 24(4): 529–538.

Cameron, A.A. and Waghorn, D.B., 1985. Bortonian foraminifera and calcareous nannofossils from Hampden Beach and McCulloch's Bridge. *N.Z. Geol. Surv. Record,* 9: 18–20.

Campbell, J.D., Fordyce, R.E., Grebneff, A., and Maxwell, P.A., 1991. Coconuts, Coconuts, Coconuts. *Geol. Soc. N.Z. Newsletter,* 92: 37–38.

Campbell, H.J. et al. (in press). The Cretaceous-Cenozoic geology and biostratigraphy of the Chatham Islands, New Zealand. *N.Z. Geol. Surv. Bulletin (n.s.),* 105.

Chaproniere, G.C.H., 1984a. The Neogene larger foraminiferal sequence in the Australian and New Zealand regions, and its relevance to the East Indies letter stage classification. *Palaeogeogr., Palaeoclimatatol., Palaeoecol.*, 46(1–3): 25–35.

Chaproniere, G.C.H., 1984b. Oligocene and Miocene larger Foraminiferida from Australia and New Zealand. *BMR Bulletin*, 188: 1–98.

Cole, W. Stores, 1967. Additional data on New Zealand *Asterocyclina* (Foraminifera). *Bull. Am. Paleont.*, 52(233): 5–18.

Collen, J.D., 1974. *Hyalinea* cf. *balthica* from Pliocene sediments, New Zealand. *N.Z. J. Geol. Geophys.*, 17(4): 907–912.

Crook, K.A.W. and Belbin, L., 1978. The Southwest Pacific Area during the last 90 million years. *J. Geol. Soc. Aust.*, 25: 23–40.

Devereux, I., 1967. Oxygen isotope paleotemperature measurements on New Zealand Tertiary fossils. *N.Z. J. Sci.*, 10(4): 988–1011.

Devereux, I., 1968. Oxygen isotope paleotemperatures from the Tertiary in New Zealand. *Tuatara*, 16(1): 41–45.

Devereux, I., Hendy, C.H., and Vella, P., 1970. Pliocene and Early Pleistocene sea temperature fluctuations, Mangaopari Stream, New Zealand. *Earth and Planetary Science Letters*, 8: 163–168.

Edwards, A.R., 1968a. The calcareous nannoplankton evidence for New Zealand Tertiary climate. *Tuatara*, 16(1): 26–31.

Edwards, A.R., 1968b. Marine climates in Oamaru district during Late Kaiatan to Early Whaingaroan time. *Tuatara*, 16(1): 75–79.

Edwards, A.R., 1987. An integrated biostratigraphy, magnetostratigraphy and oxygen isotope stratigraphy for the late Neogene of New Zealand. *N.Z. Geol. Surv. Record*, 23: 1–80.

Edwards, A.R., 1991. The Oamaru Diatomite. *N.Z. Geol. Surv. Paleont. Bull.*, 64 pp.

Edwards, A.R., Hornibrook, N. deB., Raine, J.I., Scott, G.H., Stevens, G.R., Strong, C.P., and Wilson, G.J., 1988. A New Zealand Cretaceous-Cenozoic Geological Time Scale. *N.Z. Geol. Surv. Record*, 35: 135–149.

Finlay, H.J., 1946. The microfaunas of the Oxford Chalk and Eyre River Beds. *Trans. Roy. Soc. N.Z.*, 76(2): 237–245.

Fordyce, R.E., 1977. The development of the Circum-Antarctic Current and the Evolution of the Mysticeti (Mammalia : Cetacea). *Palaeogeogr., Palaeoclimatol., Palaeoecol.*, 21(1977): 265–271.

Garner, D.M. and Ridgway, N.M., 1962. New Zealand Region, winter sea surface temperatures, 1956. *New Zealand Oceanographic Institute Chart, Miscell. Ser., No. 2.*

Haq, B.U., Hardenbol, J., and Vail, P.R., 1987. Chronology of fluctuating sea levels since the Triassic. *Science*, 235: 1156–1167.

Hayward, B.W., 1977. Lower Miocene corals from teh Waitakere Ranges, North Auckland, New Zealand. *J. Roy. Soc. N.Z.*, 7(1): 99–111.

Hayward, B.W., 1980. New records of warm-water foraminifera from North-eastern New Zealand. *Tane*, 26, 1980: 183–188.

Hayward, B.W., Moore, P.R., and Gibson, G.W., 1990. How warm was the late Oligocene of New Zealand? Coconuts, reef corals and larger foraminifera. *Geol. Soc. N.Z. Newsletter*, 90: 39–41.

Haywick, D.W., Lowe, D.A., Beu, A.G., Henderson, R.A., and Carter, R.M., 1991. Pliocene-Pleistocene (Nukumaruan) lithostratigraphy of the Tangoio block and origin of sedimentary cyclicity, central Hawke's Bay. *N.Z. J. Geol. Geophys.*, 34(2): 213–225.

Hornibrook, N. de B., 1968. Distribution of some warm water benthic foraminifera in the New Zealand Tertiary. *Tuatara,* 16(1): 11–15.

Hornibrook, N. de B., *Globorotalia truncatulinoides* and the Pliocene-Pleistocene boundary in northern Hawkes Bay, New Zealand. *In:* Y. Takayanagi, T. Saito (Eds.). *Progress in micropaleontology.* Micropaleont. Press, N.Y., pp. 83–102.

Hornibrook, N. de B., 1978. Tertiary climate. *In:* R.P. Suggate (Ed.). *"The Geology of New Zealand".* New Zealand Government Printer, Wellington, Vol. 2, pp. 436–443.

Hornibrook, N. de B., 1984. *Globorotalia* (planktic foraminifera) at the Miocene/Pliocene boundary in New Zealand. *Palaeogeogr., Palaeoclimatol., Palaeoecol.,* 46(1984): 107–117.

Hornibrook, N. de B. and Edwards, A.R., 1971. Integrated planktonic foraminiferal and calcareous nannoplankton datum levels in the New Zealand Cenozoic, *In:* A. Farinacci, (Ed.). *Planktonic Conference, Roma, 1970, Proceedings,* 1: 649–657.

Hornibrook, N. de B., Brazier, R.C., and Strong, C.P., 1989. Manual of New Zealand Permian to Pleistocene foraminiferal biostratigraphy. *N.Z. Geol. Surv. Paleont. Bull.,* 56: 1–175.

Hoskins, R.H., 1974. *In:* Officers of teh Geological Survey. Outline of the paleontology of the Greymouth District. *Report NZGS,* 67: 1–61.

Jenkins, D.G., 1968. Planktonic foraminifera as indicators of New Zealand Tertiary Climate. *Tuatara,* 16(1): 32–37.

Jenkins, D.G., 1973. Diversity changes in New Zealand Cenozoic Foraminifera. *J. Foraminif. Res.,* 3(2): 78–88.

Kennett, J.P., 1967. Recognition and correlation of the Kapitean Stage (Upper Miocene, New Zealand). *N.Z. J. Geol. Geophys.,* 10(4): 1051–1063.

Kennett, J.P., 1968. Paleo-oceanographic aspects of the foraminiferal zonation in the upper Miocene and lower Pliocene of New Zealand. Committee on Mediterranean Neogene Stratigraphy Proceedings IV Session Bologna 1967. *Giornale di Geologia,* Ser, 2, 35(33): 143–156.

Kennett, J.P., 1986. Miocene to early Pliocene oxygen and carbon isotope stratigraphy in the southwest Pacific, Deep Sea Drilling Project, Leg 90. *In:* J.P. Kennett, C.C. von der Borch, P.A. Baker, et al. *Initial Reports of the Deep Sea Drilling Project,* vol. 90, part 2, pp. 1383–1411.

Kennett, J.P., Margolis, N.S., Goddney, N.S., Dudley, D.E., and Kroopnick, P.M., 1979. Late Cenozoic oxygen and carbonate isotope history and volcanic ash stratigraphy, DSDP Site 284 South Pacific. *Am. J. Sci.,* 279: 57–69.

Kennett, J.P., von der Borch, C.C., et al., 1986a. *Init. Repts. DSDP,* 90(1): 653–744.

Kennett, J.P. and von der Borch, C.C., 1986b. Southwest Pacific Cenozoic paleoceanography. *Init. Repts. DSDP,* 90(2): 1493–1517.

Kennett, J.P. and Watkins, N.D., 1974. Late Miocene-Early Pliocene. Paleomagnetic stratigraphy, paleoclimatology and biostratigraphy in New Zealand. *Geol. Soc. Am. Bull.,* 85: 1385–1398.

Keyes, I.W., 1968. Cenozoic marine temperatures indicated by the scleractinian coral fauna of New Zealand. *Tuatara,* 16(1): 21–25.

Kingma, J.T., 1971. Geology of the Te Aute subdivision. *N.Z. Geol. Surv. Bull. n.s.,* 70: 173 pp.

Larsen, A.R., 1978a. A new species of *Amphistegina* from the Early Eocene of New Zealand. *N.Z. J. Geol. Geophys.,* 21(4): 501–504.

Larsen, A.R., 1978b. Phylogenetic and paleogeographical trends in the foraminiferal genus *Amphistegina*. *Revista Espanola de Micropaleontologia*, 9(2): 217–243.

Lee, D.E. and Campbell, J.D., 1987. Cenozoic records of the genus *Lingula* (Brachiopoda: Inarticulata) in New Zealand. *J. Roy. Soc. N.Z.*, 17(1): 17–30.

Leitch, E.C., Grant-Mackie, J.A., and Hornibrook, N. de B., 1969. Contributions to the geology of northernmost New Zealand: 1 – the Mid-Miocene Waikuku Limestone. *Trans. Roy. Soc. N.Z. Geol.*, 7: 21–32.

Maxwell, P.A., 1968. Two new records of larger foraminifera from the New Zealand Eocene. *N.Z. J. Geol. Geophys.*, 11(1): 236–264.

Maxwell, P.A., 1991. Late Oligocene Climate in New Zealand-Evidence from the Mollusca. *Geol. Soc. N.Z. Newsletter*, 92: 69–70.

Mildenhall, D.C., 1980. New Zealand Late Cretaceous and Cenozoic plant biogeography: a contribution. *Palaeogeogr., Palaeoclimatol., Palaeoecol.*, 31(1980): 197–233.

Mildenhall, D.C., 1989. Summary of the age and paleoecology of the Miocene Manuherikia Group, Central Otago, New Zealand. *J. Roy. Soc. N.Z.*, 19(1): 19–29.

Mildenhall, D.C. and Pocknall, D.T., 1984. Palaeobotanical evidence for changes in Miocene and Pliocene climates in New Zealand. *In:* J.C. Vogel (Ed.), "Late Cainozoic palaeoclimates of the Southern Hemisphere. *South African Society for Quaternary Research International Symposium, Swaziland 1983, A.A. Balkema, Rotterdam,* pp. 159–171.

Mildenhall, D.C. and Pocknall, D.T., 1989. Miocene-Pleistocene spores and pollen from Central Otago, South Island, New Zealand, *New Zealand Geol. Surv. Paleont. Bull.*, 59, 125 p.

Mildenhall, D.C. and Pocknall, D.T., 1990. *In:* M.J. Isaac, and J.K. Lindqvist (Eds.), Geology and lignite resources of the East Southland Group, New Zealand. *New Zealand Geol. Surv. Bull.*, 101, 202 p.

Nelson, C.S. and Burns, D.A., 1982. Effect of sampling interval on the resolution of oxygen isotopic paleotemperature trends – and example from the New Zealand Early Miocene. *N.Z. J. Geol. Geophys.*, 25(1): 77–81.

Nelson, C.S., Hendy, C.H., Cuthbertson, A.M., and Jarrett, G.R., 1986. Late Quaternary carbonate and isotope stratigraphy. Subantarctic Site 594, Southwest Pacific. *In:* J.P. Kennett, C.C. von der Borch, et al. *Initial Reports Deep Sea Drilling Project* (US Government Printing Office, Washington DC), pp. 1425–1436.

Pocknall, D.T., 1989. Late Eocene to Early Miocene vegetation and climatic history of New Zealand. *J. Roy. Soc. N.Z.*, 19(11): 1–18.

Pocknall, D.T., 1990. Palynological evidence for the Early to Middle Eocene vegetation and climatic history of New Zealand. *Rev. Palaeobot. Palynol.*, 65(1990): 57–69.

Riddolls, B.W., 1966. Note on the occurrence of *Asterocyclina speighti* (Chapman). *N.Z. J. Geol. Geophys.*, 9(4): 471–473.

Shackelton, N.J. and Kennett, J.P., 1975. Paleotemperature history of the Cenozoic and the initiation of Antarctic glaciation. *In Initial Reports of the Deep Sea Drilling Project,* 29. pp. 743–755.

Squires, D.F., 1962. A scleractinian coral fauna from Cape Rodney. *N.Z. J. Geol. Geophys.*, 5(3): 508–514.

Squires, D.F., 1968. Corals, coral reefs and paleotemperatures. *Tuatara*, 16(6): 16–20.

Srinivasan, M.S., 1965. *Studies in Late Eocene and Early Oligocene foraminifera of New Zealand*. Unpublished PhD thesis, Victoria University of Wellington Library. 421 p.

Srinivasan, M.S. and Vella, P., 1968. Late Eocene and Early Oligocene planktonic foraminifera from Port Elizabeth and Cape Foulwind, New Zealand. *Contrib. Cushman Foundation, Foraminiferal Research,* 19(4): 142–159.

Stevens, G.R. and Suggate, R.P., 1978. Atlas of paleogeographic maps, *In:* R.P. Suggate, G.R. Stevens, M.T. Te Punga (Eds.). *The Geology of New Zealand, vol. 2.* Government Printer, Wellington. pp. 727–745.

Suggate, R.P., 1990. Late Pliocene and Quaternary glaciations of New Zealand. *Quatern. Sci. Rev.,* 9: 175–197.

Todd, R., 1976. Some observations about *Amphistegina* (foraminifera). *In:* Y. Takayanagi, and T. Saito (Eds.). *Progress in Micropaleontology.* Special publication American Museum Natural History, pp. 382–394.

Vella, P., 1973. Ocean paleotemperatures and oscillations of the subtropical convergence zone in the eastern side of New Zealand. *In:* R. Fraser (Ed.), *Oceanography of the South Pacific,* 19(2): 315–318.

Vella, P. and Kennett, J.P., 1975. Molluscan fossils and late Neogene paleotemperatures – comment in correspondence. *N.Z. J. Geol. Geophys.,* 18(1): 197–198.

Walcott, R.I., 1984. Reconstructions of the New Zealand region for the Neogene – *Palaeogeogr., Palaeoclimatol., Palaeoecol.,* 46: 217–231.

Walcott, R.I., 1987. Geodetic strain and deformational history of the North Island of New Zealand during the late Cenozoic. *Philosophical Trans. Roy. Soc. London,* A321: 163–181.

Wisely, B., 1959. Factors influencing the settling of the principal marine fouling organisms in Sydney Harbour. *Aust. J. Mar. and Freshwater Res.,* 10: 30–44.

Wood, B.L., 1956. The Geology of the Gore Subdivision. *N.Z. Geol. Surv. Bull. (n.s.),* 53: 1–128.

Wood, B.L., 1969. Geology of Tuatapere Subdivision, Western Southland. *N.Z. Geol. Surv. Bull. (n.s.),* 79: 1–161.

Wright, I.C. and Ashby, J.N., 1985. An age for the sudden disappearance of *Globoquadrina dehiscens* in Mangapoike River Valley, New Zealand. *In:* R.A. Cooper (Ed.), Hornibrook Symposium, 1985, extended abstracts. *N.Z. Geol. Surv. Record,* 9: 102–104.

Wright, I.C. and Vella, P., 1988: A new New Zealand late Miocene magnetic stratigraphy: glacioeustatic and biostratigraphic correlations. *Earth and Planetary Science Letters,* 87: 193–204.

Late Cenozoic Evolution of Global Climate

M. E. RAYMO

*Department of Earth, Atmospheric and Planetary Sciences,
Massachusetts Institute of Technology, Cambridge, MA 02139, U.S.A.*

Abstract

Global cooling and the growth of continental ice sheets in both hemispheres during the Cenozoic, may have been caused by the tectonic uplift of the Tibetan Plateau. This uplift influenced global climate by increasing the meridionality of atmospheric circulation in the northern hemisphere and intensifying the Southeast Asian monsoon. The stronger monsoon combined with the uplift and exhumation of the Himalayan front ranges led to increased rates of continental chemical erosion, which led to a drop in atmospheric CO_2 levels and global cooling. Evidence is presented for the paleo-elevation history of Tibet and for global chemical erosion rates in the Cenozoic. In particular, the implications of the $^{87}Sr/^{86}Sr$ record of marine carbonates for models of the global carbon cycle are evaluated.

Introduction

The cause of glaciations and global cooling in the Cenozoic has long been a topic of debate. Over the last few years, this debate has focused on the potential role tectonic movements may play in controlling global climate. Ruddiman and Raymo (1988) and Ruddiman and Kutzbach (1989) proposed that the uplift of high topography in the late Cenozoic, in particular the Tibetan and Colorado Plateaus, had a profound effect on atmospheric circulation and global climate. In addition, Raymo et al. (1988) and Raymo (1991) proposed that tectonically driven increases in chemical weathering rates over the same interval caused a downdraw of atmospheric CO_2, furthering global cooling. In this paper, evidence for Cenozoic climate trends is presented with a discussion of the uplift-erosion hypotheses. These hypotheses are shown to be consistent with available late Cenozoic geologic data.

Due to the continuous nature of sedimentation in the deep sea, the best records of Cenozoic climate trends come from the ocean. The oxygen isotopic composition of calcite foraminiferal tests, sensitive to both ice volume and ocean temperature, has been a particularly valuable tool for reconstructing past climate variations. In Fig. 1, a composite $\delta^{18}O$ record of the last 70 m.y. (from Miller et al., 1987) shows the pattern of global ice volume and deep ocean temperature variations since the Cretaceous. Beginning in the Eocene, approximately 55 Ma BP (before present), a long-term cooling trend is indicated by a steady increase in $\delta^{18}O$ values punctuated by times of

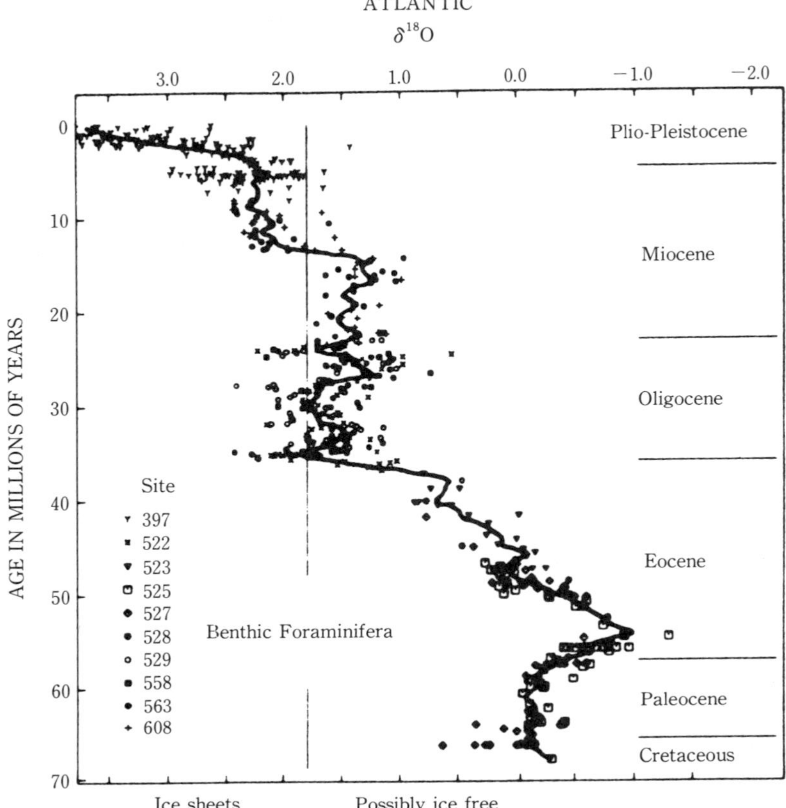

Fig. 1. Compilation of benthic $\delta^{18}O$ measurements from DSDP sites spanning the last 70 million years. The long term increase in $\delta^{18}O$ reflects cooling of the deep ocean and growth of ice sheets at high latitudes (after Miller et al., 1987).

relatively rapid change. The sharp increase in isotopic values in the early Oligocene (~36 Ma BP) may reflect the first major ice growth event of the Cenozoic (e.g. Miller et al., 1987; Barrett et al., 1987). Further increases in $\delta^{18}O$ in the middle Miocene and late Pliocene represent subsequent increases in ice volume in Antarctica and in the Northern Hemisphere, respectively, although the exact portion attributable to ice volume versus ocean temperature remains uncertain. Miller et al. (1987) infer deep water temperatures up to 12°C warmer than present in the early Eocene, while Shackleton and Kennett (1975) report data from the southern Pacific Ocean indicating deep and surface water temperatures up to 15°C warmer in the early Eocene than at present.

In addition to the long-term cooling inferred from marine records, the terrestrial fossil record of the Cenozoic shows enhanced differentiation of the mid-latitudes into areas of wetter and drier climates. At northern mid-latitudes, long-term climate trends inferred from fossil vegetation include: development of "Mediterranean" climates characterized by summer drought along the North American west coast and in the Mediterranean; drier winters in the American western plains and Asian interior; and wetter summers and winters in Southeast Asia and in Southeastern United States (see Ruddiman et al. (1989) for a summary of this evidence). At higher latitudes, subtropical flora have been identified from the Paleocene in Alaska. By the middle Oligocene, vegetation at this latitude reflected a temperate climate, with continued cooling observed up to the Pliocene, when essentially modern flora became established (Wolfe, 1971). Likewise, progressive cooling of the polar regions is suggested by faunal macrofossil evidence from Ellsmere Island (McKenna, 1980). At high southern latitudes, the presence of Oligocene fossil pollen assemblages in Antarctica and Australia (Kemp, 1978) suggest climates much warmer than at present.

Overall, oceanic and terrestrial evidence points to a marked, progressive cooling of global climate beginning in the early Eocene.

Mechanisms of Climate Change

The uplift of the Tibetan Plateau may have been a major driving force behind Cenozoic climate change. With a mean elevation of almost five kilometers and an area the size of France, this plateau is the most imposing topographic feature on the surface of the Earth. It formed as a result of the collision of the Indo-Australian plate with the Asian plate, a collision which began in earnest in the middle Eocene (40–45 m.y. BP; Mercier et al., 1987), and continues today. Figure 2 summarizes paleobotanical evidence for elevation with time (from Mercier et al., 1987). It is clear that a significant range of "uplift scenarios" is encompassed by this figure; from a rapid, almost exponential late Neogene increase in elevation (dashed curve) to a more gradual uplift history with a significant increase in plateau elevation by the middle Miocene (dotted curve).

Clearly, a more detailed knowledge of the time-elevation history of the Tibetan Plateau is needed to evaluate hypotheses which propose a link between climatic change and plateau elevation. One such hypothesis is described by Ruddiman and Raymo (1988), Ruddiman et al. (1989), and Ruddiman and Kutzbach (1989). They note that the modern Tibetan Plateau is at such a high altitude and is so broad that it not only drives a regionally intense monsoon circulation, but also perturbs atmospheric circulation on a hemispheric scale. They propose that the absence of this plateau prior to the Eocene would have resulted in significant climatological differences around the northern hemisphere.

To test this idea, Ruddiman and Kutzbach (1989) initiated a series of experiments using the NCAR general circulation model of the Earth's atmosphere in which they changed only the topographic boundary conditions. Their results (see also Kutzbach et al., 1989) showed that many of the global changes in precipitation and temperature inferred by geologists and paleobotanists to have occurred over the last 40 million years are consistent with changes in climate which occurred in the model as the elevation of the Tibetan Plateau (and other regions) was increased. Specific changes

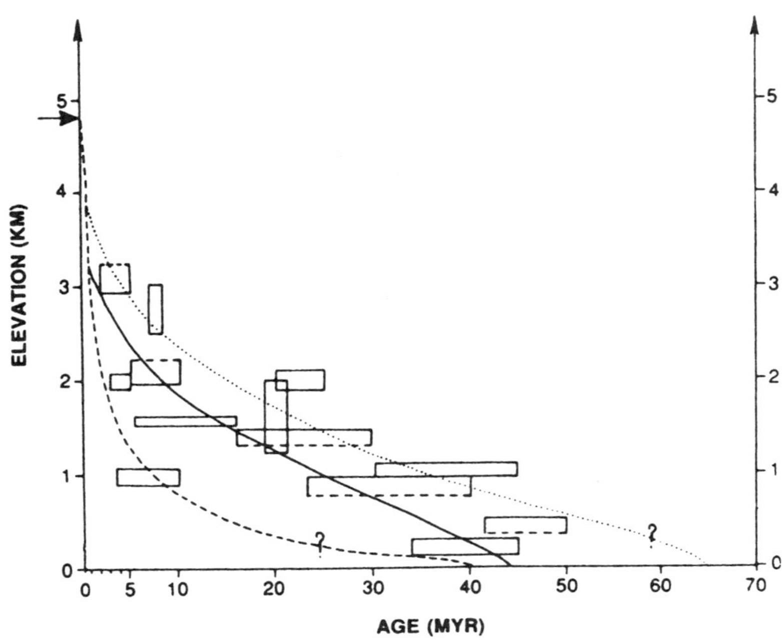

Fig. 2. Estimated paleoelevation of the Tibetan Plateau based on paleobotanical studies (after Mercier et al., 1987). Upper (dotted) curve is inferred uplift history in southern Tibet while solid curve is inferred uplift history for northern Tibet. The lower (dashed) curve reflects a view favoring very recent uplift. Boxes represent uncertainties in the age and elevation of the fossil flora.

(with uplift) predicted by the model include: a drying of the Asian interior north and west of the plateau due to the blockage, by subsiding air masses, of warm moist air from the Indian Ocean; an increase in precipitation in southeastern Asia and India due to the intensification of the summer monsoon; cooler summers and winters in Europe because of changes in prevailing wind patterns; a decrease in summer precipitation around the Mediterranean due to increased subsidence of dry air masses and stronger northeasterly winds. Geobotanical and sedimentological evidence from these regions agree with these predicted trends: flora in central Asia evolved from forest to steppe and desert-like vegetation; warm wet adapted vegetation has persisted in southeast Asia and India for well over 40 million years; subtropical vegetation gradually disappeared from Europe over the last 20 million years; and the spread of summer-dry vegetation and increased dustiness is observed in the Mediterranean region since the Miocene.

Overall, both the GCM uplift experiments and the paleobotanical record reflect a trend toward increased seasonality and regional differentiation of climate: from the more equable, moist temperate climates of the early and mid Cenozoic to the colder/ warmer and drier/wetter regional patterns observed in the northern hemisphere today. However, as discussed by Ruddiman and Kutzbach (1989), these GCM experiments do not predict the pronounced drop in high latitudes temperature observed in the geologic record ($\sim 10°C$), and because of this, they do not predict the growth of large terrestrial ice sheets in the northern and southern hemispheres. This indicates that other factors, such as atmospheric CO_2 variations, may be required to explain global Cenozoic cooling.

The Cenozoic "Icehouse Effect"

The concentration of radiatively important trace gases in the atmosphere (in particular CO_2) can have an important influence on Earth's climate. A direct link between atmospheric temperature and CO_2 concentration is observed throughout the 140,000-year Vostok ice core record of the last glacial-interglacial cycle (Barnola, 1987), and many investigators have proposed a link between the evolution of global climate over the Cenozoic and changes in the composition of the Earth's atmosphere (e.g. Berner et al., 1983; Barron and Washington, 1985). On time scales longer than a million years, atmospheric CO_2 levels are controlled primarily by the balance between rates of mantle degassing and the rate of output via chemical weathering of rocks at the Earth's surface. Two types of climate hypotheses have been proposed: ones in which variations in the atmospheric CO_2 levels are driven primarily by changes in volcanic outgassing, the main input term (e.g. Walker et al., 1981; Berner et al., 1983); and ones in which CO_2 variations are driven by changes in chemical erosion rates, the main output term (Chamberlin, 1899; Raymo et al., 1988; Raymo, 1991). In the first class of models, times of higher CO_2 levels (due to greater outgassing rates) are associated with higher rates of chemical weathering (which prevent a runaway greenhouse); in constrast, the erosion hypotheses predict an association between colder global temperatures/lowered CO_2 levels and increased chemical weathering rates.

Berner et al. (1983) (or BLAG for Berner, Lasaga, and Garrels) provided a major advance in paleoclimatology by developing a numerical model which quantified CO_2 inputs (via metamorphism and decarbonation reactions) and outputs (via chemical weathering) to the atmosphere over the last 100 million years. The flux of CO_2 into the atmosphere was linearly tied to the rate of sea-floor production. The output flux of CO_2 from the atmosphere was a function of the areal extent of continents available for chemical weathering, with atmospheric temperatures acting as a strong negative feedback (with higher temperatures causing more rapid removal of CO_2). As formulated, their model produced higher CO_2 levels in the Cretaceous because of (a) more rapid sea-floor spreading and volcanic outgassing of CO_2 at that time, and (b) a decrease in the land area available to chemical erosion due to globally higher sea levels. The higher sea levels are in large part related to faster sea floor spreading rates and increased ridge volume.

The BLAG model and subsequent modifications (e.g. Lasaga et al., 1985; Volk, 1989; Berner, 1990) correctly predict higher global temperatures in the Cretaceous. However, the timing of the major CO_2 and temperature changes suggested for the Cenozoic by this model does not match the timing of major climate cooling inferred from the geologic record. In the BLAG models, the largest decline in atmospheric

CO_2 and global temperatures occurs between 100 and 50 Ma, driven primarily by decreases in global sea-floor-spreading rates which control mantle outgassing and, indirectly, sea level. However, geologic evidence summarized by Crowley and North (1991) suggests that the greatest climatic cooling occurred primarily between 50 Ma and the present.

A potential explanation for this mismatch was suggested by Raymo et al. (1988) and Raymo (1991). We proposed that the rate of removal of CO_2 from the atmosphere by chemical erosion is a strong function of continental relief rather than area, and that atmospheric temperature is a very weak control on chemical erosion rates. Much of the Earth's surface is characterized by low-lying, deeply weathered shield areas which contribute very little to the global river flux of solutes. Even though chemical weathering is pervasive in these regions, absolute rates are very low. In contrast, mountainous regions are dominated by tectonically and isostatically induced uplift and exhumation, which result in tremendous amounts of mechanical erosion which, in turn, increases the surface area of fresh minerals available for chemical attack. Three factors further enhance chemical breakdown of detrital grains in mountainous regions: the abundance of easily weathered sedimentary silicates derived from uplifted passive margin sequences (weathering of carbonates does not ultimately alter atmospheric CO_2), the orographic concentration of rainfall on mountain slopes, and steep slopes which flush away chemical erosion products and maintain the constant exposure of new minerals.

These effects are most pronounced in the Himalayas, where the collision of India with Asia has resulted not only in the uplift, deformation, and erosion of an antecedent passive margin but also in the formation of the Tibetan Plateau where incident solar heating in summer drives the strong atmospheric convection which results in the Asian monsoon (Hahn and Manabe, 1975). Data from the eight largest rivers draining the Himalayan-Tibetan region show that almost 25% of the total dissolved load reaching the ocean today is from a watershed area that represents only 5% of the Earth's surface (Pinet and Souriau, 1992). For reasons which appear to be intimately related to the plateau-building process and the presence of an intense monsoonal circulation, a significant fraction of the Earth's chemical weathering is occurring in this relatively small region in Asia.

Based on these and similar observations, we proposed that late Cenozoic uplift of the Himalayan region and the Tibetan Plateau has resulted in both regionally and, thus, globally higher chemical erosion rates, resulting in a drawdown of atmospheric CO_2 and global cooling. The timing of this tectonically driven CO_2 decrease was post-Eocene, when the collision of India with the Asian subcontinent began. Such a model is in agreement with geologic evidence for rapid post-Eocene global cooling and is consistent with the conclusions of many climate modelers: namely, the need for higher atmospheric CO_2 levels to explain Cretaceous to Eocene warmth at high latitudes (Barron and Washington, 1985; Crowley and North, 1991; Ruddiman and Kutzbach, 1989). One important difference between the two types of hypotheses described above is the prediction of when chemical weathering rates are greatest; in

BLAG-type models weathering rates increase as global temperatures rise, while in our model weathering rate increases are associated with falling global temperatures. To test these predictions, we can use the $^{87}Sr/^{86}Sr$ isotopic record preserved in marine carbonates as a proxy for global chemical erosion rates.

The seawater $^{87}Sr/^{86}Sr$ ratio recorded by marine carbonates (Fig. 3) reflects primarily a balance between input of radiogenic, high $^{87}Sr/^{86}Sr$ material weathered from continents (average runoff value = 0.7119; Palmer and Edmond, 1989) and non-radiogenic, low $^{87}Sr/^{86}Sr$ material introduced by hydrothermal activity (average value = 0.7035). The isotopic composition of the river strontium flux is not invariant but depends on the proportion of granitic rocks relative to the igneous and carbonate rocks being weathered on land (e.g., Palmer and Elderfield, 1985); however, substantial changes in the river $^{87}Sr/^{86}Sr$ ratio through time would require that the types of rocks exposed at Earth's surface vary significantly. A third strontium input to the ocean, from redissolution of marine carbonates (average value = 0.7084; Palmer and

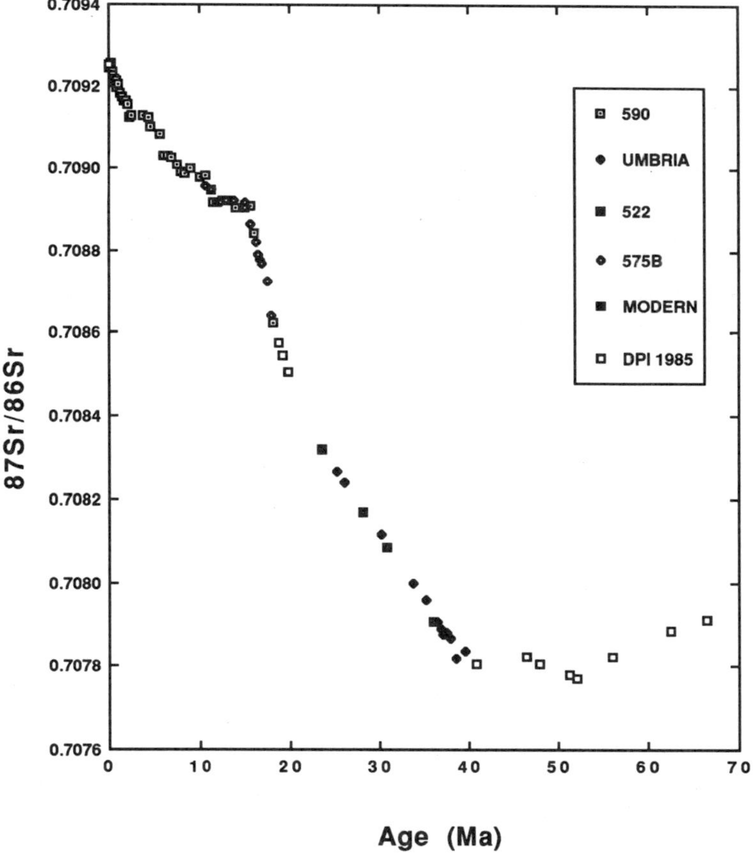

Fig. 3. Strontium isotopic composition of seawater for the last 70 million years based on analysis of marine carbonates (after Richter et al., 1992).

Edmond, 1989), serves to buffer the isotopic value of the oceanic strontium reservoir towards the integrated mean value of the preceeding 10 to 15 m.y.

The simplest interpretation of the dramatic post-Eocene rise in $^{87}Sr/^{86}Sr$ ratios shown in Fig. 3 is either an increase in the rate of continental chemical erosion or a decrease in seafloor hydrothermal activity (e.g. Brass, 1976; Burke et al., 1982). The widely held assumption that hydrothermal activity is a function of seafloor spreading rates, which are estimated using marine magnetic anomalies, can be invoked to estimate the hydrothermal flux of non-radiogenic strontium through time. Sea floor spreading rates appear to have changed little over the last 30–40 m.y. and thus cannot account for the observed $^{87}Sr/^{86}Sr$ changes. It is widely agreed that the pronounced late Cenozoic increase in the oceanic $^{87}Sr/^{86}Sr$ ratio can only be explained by a dramatic increase in the delivery of radiogenic strontium from land driven by an increase in global chemical weathering rates (Raymo et al., 1988; François and Walker, 1991; Richter et al., 1992).

Conclusions

The Ruddiman-Kutzbach "atmospheric perturbation" and Chamberlin-Raymo "chemical weathering" hypotheses work hand in hand. We have proposed that over the last 40 m.y. uplift of the Tibetan Plateau has resulted in progressively stronger deflections of the atmospheric jet stream, an intensification of the Asian monsoon, and increased rainfall on the front slopes of the Himalayas, leading to greater rates of chemical weathering and, ultimately, lower atmospheric CO_2 levels.

Although this model is conceptually simple and is supported by results from GCM modeling and data from geochemical tracers, major uncertainties and questions remain. To evaluate the link between uplift and climate variations, we need a better understanding of the elevation history of the Tibetan Plateau. In addition, can the mechanistic link between tectonism, relief, and chemical weathering rates be quantified? What factors in mountainous regions are most critical to the high rates of chemical erosion observed today: mechanical breakup which increases surface areas, orographic concentration of precipitation on slopes, runoff, rock composition, etc.? To improve estimates of past chemical weathering rates, the development of new proxies such as Ge/Si ratios (Shemesh et al., 1988) is needed. It will also be important to find means of linking changes in weathering-sensitive ocean chemical reservoirs to specific regions, such as the Himalayas. Ultimately, direct determination of atmospheric CO_2 levels over the last 100 million years, possibly through the measurement of $\delta^{13}C$ of marine organic matter (e.g. Rau et al., 1991; Jasper and Hays, 1990), will be critical to evaluating the uplift-erosion model proposed by Raymo et al. (1988) relative to the earlier carbon cycle-climate models of Walker et al. (1981), Berner et al. (1983), and others.

Acknowledgements
I thank R. Tsuchi for generously supporting my attendence at the V-CPNS-

IGCP-246 Congress and N. Shackleton for encouraging me to talk about this topic. It was a pleasure and honor to take part in this meeting. Much of this paper reviews work accomplished with W.F. Ruddiman at LDGO and supported by NSF grant OCE90-12279.

References

Barnola, J.M., Raynaud, D., Korotevich, Y.S., and Lorius, C., 1987. Vostok ice core provides 160,000 year record of Atmospheric CO_2. *Nature,* 329: 408–414.

Barron, E.J. and Washington, W.M., 1985. Warm Cretaceous Climates: high atmospheric CO_2 as a plausible mechanism. *In:* E.T. Sundquist and W.S. Broecker (Eds.), The Carbon Cycle and Atmospheric CO_2: Natural Variations Archean to Present. *Geophys. Mono.* 32, AGU, Washington, D.C., pp. 546–553.

Barrett, P.J., Elston, D.P., Harwood, D.M., McKelvey, B.C., and Webb, P.-N., 1987. Mid-Cenozoic record of glaciation and sea-level change on the margin of the Victorian Land Basin, Antarctica. *Geology,* 15: 634–637.

Berner, R.A., 1990. Atmospheric carbon dioxide levels over Phanerozoic time. *Science,* 249: 1382–1386.

Berner, R.A., Lasaga, A.C., and Garrels, R.M., 1983. The carbonate-silicate geochemical cycle and its effect on atmospheric carbon dioxide over the past 100 million years. *American Journal of Science,* 283: 641–683.

Brass, G.W., 1976. The variation of the marine $^{87}Sr/^{86}Sr$ ratio during Phanerozoic time: interpretation using a flux model. *Geochim. et Cosmochim. Acta,* 40: 721–730.

Burke, W.H., Denison, R.E., Hetherington, E.A., Koepnick, R.B., Nelson, H.F., and Otto, J.B., 1982. Variation of seawater $^{87}Sr/^{86}Sr$ throughout Phanerozoic time. *Geology,* 10: 516–519.

Chamberlin, T.C., 1899. An attempt to frame a working hypothesis of the cause of glacial periods on an atmospheric basis. *Journal of Geology,* 7: 545–584, 667–685, 751–787.

Crowley, T.J. and North, G.N., 1991. Paleoclimatology. Oxford University Press, New York, 339 pp.

François, L.M. and Walker, J.C.G., 1991. Modeling the Phanerozoic carbon cycle and climate: constraints from the $^{87}Sr/^{86}Sr$ isotopic ratio of seawater. *Am. J. Sci.,* 292: 81-135.

Hahn, D.G. and Manabe, S., 1975. The role of mountains in the south Asian monsoon circulation, *J. Atmos. Sci.,* 32: 1515–1541.

Jasper, J.P. and Hayes, J.M., 1990. A carbon isotope record of CO_2 levels during the late Quaternary. *Nature,* 347: 462–464.

Kemp, E.M., 1978. Tertiary climatic evolution and vegetation history in the southeast Indian Ocean. *Palaeogeog., Palaeoclimatol., Palaeoecol.,* 24: 169–208.

Kutzbach, J.E., Guetter, P.J., Ruddiman, W.F., and Prell, W.L., 1989. Sensitivity of climate to late Cenozoic uplift in southern Asia and the American West: numerical experiments. *JGR,* 94: 18393–18407.

Lasaga, A.C., Berner, R.A. and Garrels, R.M., 1985. An improved geochemical model of atmospheric CO_2 fluctuations over the past 100 million years. *In:* E.T. Sundquist and W.S. Broecker (Eds.), The Carbon Cycle and Atmospheric CO_2: Natural Variations Archean to Present. Geophys. Mono. 32, AGU, Washington, D.C., pp. 397–410.

McKenna, M.C., 1980. Eocene paleolatitude, climate, and mammals of Ellesmere Island. *Palaeo. Palaeo., Palaeo.,* 30: 349–362.

Mercier, J.-L., Armijo, R., Tapponnier, P., Carey-Gailhardis, E., and Lin, H.T., 1987. Change from late Tertiary compression to Quaternary extension in southern Tibet during the India-Asia collision. *Tectonics,* 6: 275–304.

Miller, K.G., Fairbanks, R.G., and Mountain, G.S., 1987. Tertiary oxygen isotope synthesis, sea level history, and continental margin erosion. *Paleoceanography,* 2: 1–19.

Palmer, M.R. and Edmond, J.M., 1989. The strontium isotope budget of the modern ocean. *EPSL,* 92: 11–26.

Palmer, M.R. and Elderfield, H., 1985. Sr isotope composition of sea water over the past 75 Myr. *Nature,* 314: 526–528.

Pinet, P. and Souriau, M., 1992. Continental erosion and large scale relief, in press.

Rau, G.H., Froelich, P.N., Takahashi, T., and DesMarais, D.J., 1991. Does sedimentary organic $\delta^{13}C$ record variations in Quaternary ocean $[CO_2(aq)]$? *Paleoceanography,* 6: 335–347.

Raymo, M.E., 1991. Geochemical evidence supporting T.C. Chamberlin's theory of glaciation. *Geology,* 19: 344–347.

Raymo, M.E., Ruddiman, W.F., and Froelich, P.N., 1988. Influence of late Cenozoic mountain building on ocean geochemical cycles. *Geology,* 16: 649–653.

Richter, F.M., Rowley, D.B., and DePaolo, D.J., 1992. Sr Isotope Evolution of sea water: The role of tectonics. *Earth and Planetary Science Letters*, 109: 11-23.

Ruddiman, W.F. and Raymo, M.E., 1988. Northern hemisphere climate regimes during the past 3 Ma: possible tectonic connections. *In:* N.J. Shackleton, R.G. West, and D.Q. Bowen (Ed.), The past three million years: evolution of climatic variability in the North Atlantic region, University Press, Cambridge, pp. 227–234.

Ruddiman, W.F. and Kutzbach, J.E., 1989. Forcing of late Cenozoic northern hemisphere climate by plateau uplift in southeast Asia and the American Southwest. *Jour. Geophys. Res.,* 94, no. D15: 18409–18427.

Ruddiman, W.F., Prell, W.L., and Raymo, M.E., 1989. History of late Cenozoic uplift on Southeast Asia and the American Southwest: rationale for general circulation modeling experiments. *Jour. Geophys. Res.,* 94: 18379–18391.

Shackleton, N.J. and Kennett, J.P., 1975. Paleotemperature history of the Cenozoic and the initiation of Antarctic glaciation: oxygen and carbon isotope analysis in DSDP sites 277, 279, and 281. *In:* J.P. Kennett et al. (Eds.), *Init. Rep. Deep Sea Drilling Proj., 29.,* U.S. Gov. Print. Office, Washington, D.C., pp. 801–807.

Shemesh, A., Mortlock, R., Smith, and Froelich, P.N., 1988. Determination of Ge/Si in marine siliceous microfossils; separation, cleaning, and dissolution of diatoms and radiolaria. *Mar. Chem.,* 25(4): 305–323.

Volk, T., 1989. Rise of angiosperms as a factor in long-term climatic cooling. *Geology,* 17: 107–110.

Walker, J.C.G., Hays, P.B., and Kasting, J.F., 1981. A negative feedback mechanism for the long-term stabilization of Earth's surface temperature. *J. Geophys. Res.,* 86: 9976–9782.

Wolfe, J.A., 1971. Tertiary climate fluctuations. *Palaeo., Palaeo., Palaeo.,* 9: 25.

Isotopic Evidence for Depth Stratification and Paleoecology of Miocene Planktonic Foraminifera: Western Equatorial Pacific DSDP Site 289

Joseph T. Gasperi and James P. Kennett

Department of Geological Sciences and
Marine Science Institute, University of California, Santa Barbara, CA 93106, U.S.A.

Abstract

Modern planktonic foraminifera are depth stratified within the water column, their calcite tests often exhibiting the oxygen and carbon isotopic compositional differences related to the depth differences. The identification of depth stratification in fossil assemblages of planktonic foraminifera is required for studies of the evolution of structure in the upper parts of the ocean. Relative depth rankings have been determined for 13 Miocene ($\sim$18.7 to 5.8 Ma) planktonic foraminiferal species from the western equatorial Pacific (DSDP Site 289) based upon more than 600 oxygen and carbon isotopic analyses of their shells.

Depth rankings based upon the carbon isotopic data do not always agree with those inferred from the oxygen isotopic data. The distribution of the oxygen isotopic data suggests that, as with modern species, there was no significant disequilibrium fractionation of the oxygen isotopes during calcification of any species. We suggest instead that five inferred shallow-dwelling, spinose species with high $\delta^{13}C$ values (*Globigerinoides immaturus, Globigerinoides sacculifer, Globigerinoides obliquus, Dentoglobigerina altispira* and *Globoquadrina baroemoenensis*) were hosts to large concentrations of algal symbionts that, through photosynthesis, led to carbon isotopic fractionation and relatively high $\delta^{13}C$ values of the host calcite. Depth rankings based upon $\delta^{18}O$ and $\delta^{13}C$ values for all measured species agree if the high carbon isotopic values exhibited by these five species resulted from metabolically related isotopic fractionation.

From $\sim$18.7 to 7.5 Ma, *Globigerinoides obliquus, Globigerinoides sacculifer, Globorotalia (Jenkinsella) siakensis, Globorotalia (Jenkinsella) mayeri, Dentoglobigerina altispira, Globoquadrina baroemoenensis* and *Globigerinoides immaturus* were shallow-dwelling forms that are inferred either to have had moderately weak depth stratification from shallower to deeper as listed, or to record slight variations in seasonal (temperature) distributions. Deep-dwelling forms include *Globoquadrina dehiscens, Globorotalia (Menardella) limbata, Globorotalia (Menardella) menardii, Globoquadrina venezuelana* and *Globorotalia (Menardella) praemenardii*. No inferred intermediate-dwelling forms are recognized. The latest Miocene interval, from $\sim$7.5 to 5.8 Ma, marks a change in depth distribution of the *Globorotalia (Menardella)* group and the appearance for the first time of intermediate-dwelling forms. During this interval, shallow-dwelling forms include *Globigerinoides obliquus, Globigerinoides sacculifer, Dentoglobigerina altispira, Globoquadrina baroemoenensis, Globorotalia (Menardella) limbata* and *Globigerinoides immaturus.* Intermediate-dwelling forms were *Globorotalia (Menardella) menardii* and *Globorotalia (Globorotalia) merotumida,* while *Globoquadrina venezuelana* was a deep-dwelling form.

A time-series of the isotopic data for individual species strongly supports the idea that planktonic foraminifera have a preferred depth habitat and that their depth relationships remained largely unchanged during the Miocene. However, the relative depth rankings of *Globorotalia (Menardella) menardii* and *Globorotalia (Menardella) limbata* changed from deep to intermediate and from deep to shallow, respectively, during the late Miocene between ~9.9 and 7.5 Ma.

Introduction

Modern planktonic foraminiferal species are depth-stratified in the water column and/or exhibit a seasonal distribution depending on the hydrographic regime. These observations are based upon studies of plankton tows (Berger, 1969; Bé et al., 1973), sediment traps (Kahn and Williams, 1981; Fairbanks et al., 1982; Curry, Thunell, and Honjo, 1983; Thunell, Curry, and Honjo, 1983a,b; Thunell and Reynolds, 1984; Reiss and Hottinger, 1984; Deuser, 1987; Sautter and Thunell, 1991), analysis of core-top samples (Williams et al., 1977; Curry and Matthews, 1981a,b), and laboratory cultures (Spero and DeNiro, 1987; Spero and Williams, 1988; Hemleben et al., 1989; Bijma et al., 1990). Better understanding of the oceanographic and biological controls on planktonic foraminiferal distributions, morphology, depth habitat, and nutrient, temperature and salinity requirements, and how these variables affect the oxygen and carbon isotope ratios of the calcite test are crucial for paleoceanographic and paleo-climatic interpretations. This study presents an analysis of depth stratification among thirteen species of Miocene planktonic foraminifera from ~18.7 to 5.8 Ma at DSDP Site 289 (Fig. 1). Site 289 was located at or near the equator during the Miocene (Sclater et al., 1985) and preserves an isotopic record of paleoceanographic change of western equatorial Pacific surface waters.

Previous studies of depth stratification among fossil planktonic foraminifera have used $\delta^{18}O$ values to establish depth rankings (Douglas and Savin, 1978; Keller, 1985; Savin et al., 1985; Barrera et al., 1985). The $\delta^{18}O$ value of the foraminiferal test cal-cified in isotopic equilibrium is indicative of the temperature and isotopic composi-tion of the water in which it lived. Laboratory culture experiments with several species of living planktonic foraminifera have shown that the species secreted shell calcite at or close to isotopic equilibrium (Erez and Luz, 1983; Bouvier-Soumagnak et al., 1986; Spero, in press). Comparisons of $\delta^{18}O$ values from many planktonic species have enabled the asssignment of relative depth rankings of fossil planktonic foraminifera. Because the fractionation of oxygen isotopes between sea water and calcium carbonate is temperature-dependent and vertical thermal gradients exist in

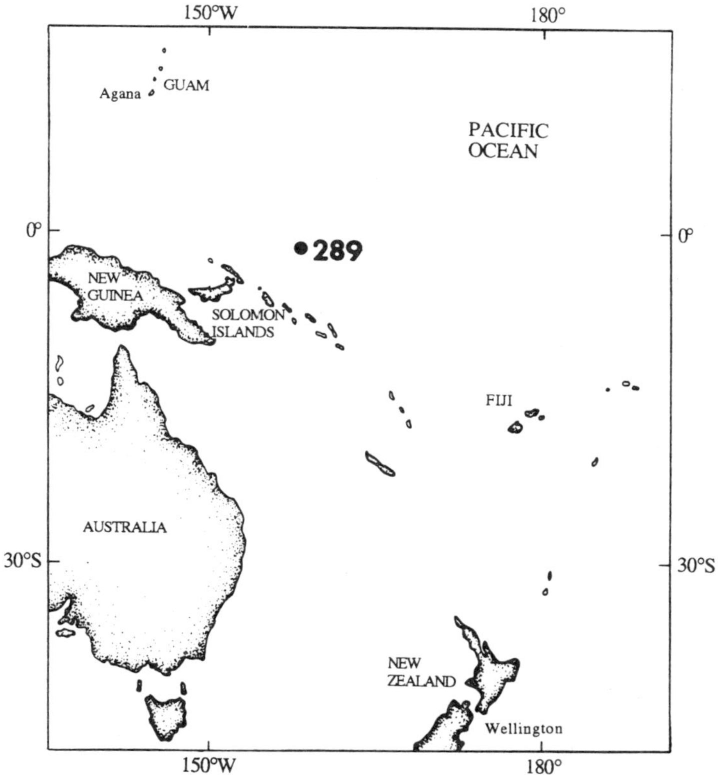

Fig. 1. Location of DSDP Site 289 in the western equatorial Pacific.

the upper water column where planktonic foraminifera live, comparison of $\delta^{18}O$ values between species provides a valuable tool for studying the depth distribution of species (Fig. 2a). Knowledge of temperature-dependent fractionation has made possible studies of vertical and latitudinal thermal gradients of the oceans and how they have changed through time in response to climatic variability (Fairbanks et al., 1982; Loutit et al., 1983).

Carbon isotopic data are also valuable in helping to determine relative depth rankings of planktonic foraminiferal species because of vertical gradients in the total dissolved CO_2 of the oceans (Kroopnick, 1985; Berger and Vincent, 1986). The ΣCO_2 gradient of sea water is a function of the biological productivity of surface waters. Regions of high surface water productivity display well developed vertical gradients of ΣCO_2. Surface waters in such areas have relatively low levels of ΣCO_2 because of utilization of much of the available carbon. Deeper waters contain increasingly higher concentrations of ΣCO_2 due to respiration and bacterial decomposition. The $^{13}C/^{12}C$ ratio is highest in surface waters, where ^{12}C is preferentially used in cell growth, and decreases with increasing depth as ^{12}C is returned to the water due to the respiration and decomposition of the organic matter (Fig. 2b). Therefore, carbon

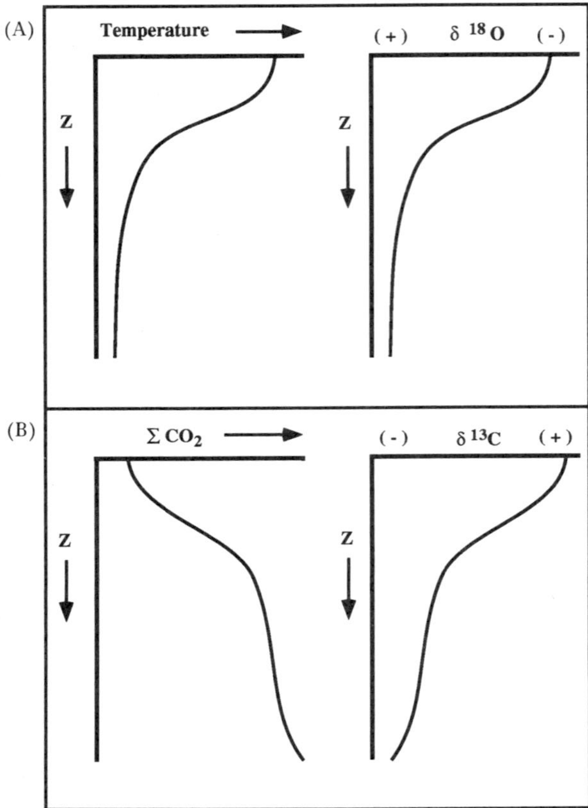

Fig. 2. A) Schematic illustration of equilibrium oxygen isotopic fractionation temperature dependence and relationships between $\delta^{18}O$ values measured in planktonic foraminiferal tests and vertical thermal gradients in surface waters.

    B) Schematic illustration of relations between equilibrium carbon isotopic values measured in planktonic foraminiferal tests and total dissolved CO_2 in surface waters.

isotopic gradients within the upper water column are recorded by planktonic foraminifera in isotopic equilibrium with sea water and can be used to establish depth rankings among these forms.

In areas or during seasons of low productivity and a well-developed thermocline, carbon isotopic gradients are smaller (Fairbanks et al., 1982; Curry et al., 1983) and hence planktonic foraminifera living at different depths exhibit only small $\delta^{13}C$ differences (Berger et al., 1978). Thus, $\delta^{13}C$ differences among planktonic foraminiferal species in these areas are of little value in establishing inter-specific depth rankings. However, the vertical thermal gradients are strong and planktonic foraminifera living at different depths exhibit large $\delta^{18}O$ differences (Berger et al., 1978). These differences make oxygen isotopic values a powerful tool for determining interspecific depth rankings (Fig. 3a). Such depth relationships, for the modern eastern equatorial Pacific (1°S; 170°E) in the region of Site 289, are illustrated by Berger et al. (1978).

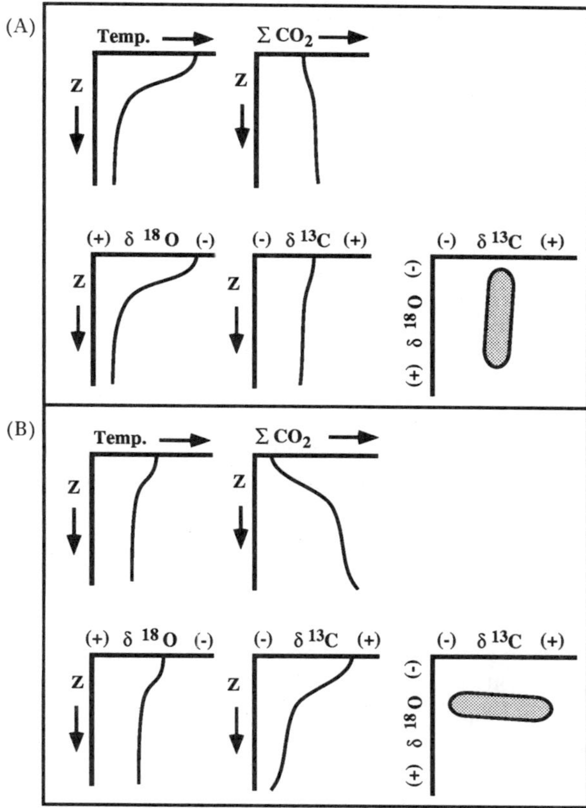

Fig. 3. A) Schematic illustration showing the thermal gradient and ΣCO_2 levels in surface waters of a region with well-developed vertical thermal gradients and low surface water biological productivity, and the predicted $\delta^{18}O$ and $\delta^{13}C$ values of planktonic foraminiferal species living in these waters. In this region of strong oxygen isotopic gradients and weak carbon isotopic gradients, $\delta^{18}O$ values are more useful than $\delta^{13}C$ values for assigning relative depth rankings of planktonic foraminiferal species.

  B) Schematic illustration showing the thermal gradient and ΣCO_2 levels in surface waters of an upwelling region with high surface water biological productivity, and the predicted $\delta^{18}O$ and $\delta^{13}C$ values of planktonic foraminiferal species living in these waters. In this region of strong carbon isotopic gradients and weak oxygen isotopic gradients, $\delta^{13}C$ values are more useful than $\delta^{18}O$ values for assigning relative depth rankings of planktonic foraminiferal species.

In contrast, regions of high productivity, where upwelling reduces the vertical thermal gradients in the upper water column, may have much stronger carbon isotopic gradients than temperature gradients. Under these conditions, carbon isotopic values are more useful for determining interspecific depth rankings (Fig. 3b). In all cases, oxygen and carbon isotopic rankings should agree unless there is isotopic fractionation of either oxygen or carbon, or both.

The purposes of this study are to: 1) establish relative depth rankings of species based upon $\delta^{18}O$ and $\delta^{13}C$ isotopic values; 2) determine how consistent the relative depth rankings were throughout the Miocene; and 3) determine if any of the taxa exhibit significant changes in depth ranking during the Miocene.

Location

Deep Sea Drilling Project Site 289 (00°29.92'S, 158°38.69'E) is located close to the equator on the Ontong-Java Plateau in the western equatorial Pacific (Fig. 1). This site lies near the crest of the plateau at a present water depth of 2206 m (Andrews, Packham et al., 1975). During the early Miocene (~22 Ma), Site 289 had an estimated paleodepth of 2398 m (Sclater et al., 1985), and was located well above the CCD throughout the Miocene (Berger and Winterer, 1973). Planktonic foraminifera are well preserved with minimal dissolution (Woodruff and Douglas, 1981). The Miocene sequence is continuous except for a brief hiatus within core 58 between sections 2 and 3 (Keller and Barron, 1983), but otherwise is believed to be complete. Rotary coring of the site produced ~80% core recovery during the Miocene; total core recovery was 56% (Andrews et al., 1975). The Miocene age sediment consists of nanno-foram ooze to interbedded nanno-foram ooze and nanno-foram chalk (Andrews et al., 1975). Investigations of interstitial water chemistry of the Site 289 sediments (Elderfield et al., 1982) suggest that there should be substantial calcite recrystallization deeper than 500 mbsf (Core 54). However, the foraminiferal specimens picked for isotopic analysis are well preserved and show no evidence of having undergone diagenetic alteration. Samples from between 504 mbsf (Core 54) and 570m mbsf (Core 61) are from an interval transitional between carbonate ooze and chalk. Scanning electron microscope examination of several foraminiferal specimens from this interval reveal no significant evidence for recrystallization on the surfaces of the well preserved specimens picked for isotopic analysis. Furthermore, in this interval, distinct oxygen isotopic differences are retained between shallow and deep dwelling individuals, differences which would be unlikely if the foraminiferal tests had undergone recrystallization.

Methods

Samples from the Deep Sea Drilling Program were washed with deionized water over a 74μm sieve and the fine fraction was saved. Planktonic foraminifera were picked from the >250μm fraction if they were well preserved and present in sufficient numbers for isotopic analysis. The taxonomy of the planktonic foraminifera used in this study follows Kennett and Srinivasan (1983). Twenty-three species of planktonic foraminifera were picked from 51 samples at DSDP Site 289 for isotopic analysis (Table 1). Picked samples were washed in methanol, sonified, and dried prior to roasting under vacuum for 1 hour at 375°C. For each species, 10 to 40 individuals (depending on mass) were run for isotopic analysis.

Isotopic analyses were conducted using a Finnigan Mat 251 mass spectrometer equipped with an automated carousel device located at the University of California, Santa Barbara. Values are given in delta notation as per mil deviations of the $^{18}O/^{16}O$

Table 1. Planktonic foraminiferal species isotopically analyzed ($\delta^{18}O$ and $\delta^{13}C$) in this investigation.

Globigerina (Zeaglobigerina) woodi Jenkins
Globigerinoides obliquus Bolli
Globigerinoides triloba (Reuss)
Globigerinoides immaturus Leroy
Globigerinoides sacculifer (Brady)
Globigerinoides parawoodi Keller
Globigerinoides subquadratus Brönnimann
Orbulina universa d'Orbigny
Globorotalia (Fohsella) peripheroronda Blow and Banner
Globorotalia (Fohsella) fohsi lobata Bermúdez
Globorotalia (Menardella) praemenardii Cushman and Stainforth
Globorotalia (Menardella) menardii (Parker, Jones and Brady)
Globorotalia (Menardella) limbata (Fornasini)
Globorotalia (Globorotalia) merotumida Blow and Banner
Globorotalia (Jenkinsella) siakensis Leroy
Globorotalia (Jenkinsella) mayeri Cushman and Ellisor
Globoquadrina venezeulana (Hedberg)
Globoquadrina dehiscens (Chapman, Parr and Collins)
Globoquadrina baroemoenensis (Leroy)
Dentoglobigerina altispira altispira (Cushman and Jarvis)
Dentoglobigerina altispira globosa (Bolli)
Sphaeroidinellopsis kochi Caudri
Globigerinita glutinata (Egger)

---

or $^{13}C/^{12}C$ ratio of the sample from that of the PDB standard. Isotopic analyses were related to PDB through repeated analyses of NBS-20 with values following Craig (1957) of: $\delta^{18}O = -4.14\%o$ and $\delta^{13}C = -1.06\%o$. Measurement of oxygen and carbon isotopic ratios have an analytical precision of $\pm0.1\%o$.

Stratigraphy and Chronology

The age model developed for DSDP Site 289 is based on paleomagnetically dated planktonic foraminiferal and calcareous nannofossil datum levels as compiled by Johnson and Nigrini (1985). The chronology follows the time scale of Berggren et al. (1985) whereby paleomagnetic anomaly 5 is correlated with paleomagnetic Chron 11 (Table 2). The foraminiferal datums are compiled from Saito (1975) and Srinivasan and Kennett (1981a,b); ages, from Keller (1980), Berggren et al. (1985), and Barron et al. (1985); and magnetostratigraphic calibrations, from Saito et al. (1975), Keigwin (1982), Ryan et al. (1974), Miller et al. (1985), Poore et al. (1983), Berggren et al. (1983), and Barron et al. (1985). The calcareous nannofossil datums are compiled from Bukry (1975) and Shafik (1975), ages from Berggren et al. (1985) and Barron et al. (1985), and magnetostratigraphic calibrations from Haq et al. (1980) and Miller et al. (1985).

Table 2. Paleomagnetically calibrated planktonic foraminiferal and calcareous nannofossil datum levels, compiled by Johnson and Nigrini (1985), used to develop the age model for DSDP Site 289. Assigned ages follow the time scale of Berggren et al. (1985) using the correlation of paleomagnetic anomaly 5 to paleomagnetic Chron 11. B = bottom of biostratigraphic range; T = top of biostratigraphic range.

Microfossil	Datum	Sample	Depth, m	Age, Ma
foram	B G. tumida	18–4/19–2	166.5–173.0	5.2
foram	B P. primalis	22–1/22–2	201.0–202.0	5.8
nanno	B A. primus	25–1/25–2	229.0–230.5	6.5
nanno	B D. quinqueramus	27–3/27–6	250.5–255.0	7.3
foram	B N. acostaensis	33–4/34–2	309.0–315.5	8.6
nanno	T D. hamatus	34–1/34–3	314.0–317.0	8.9
foram	T G. siakensis	36–2/36–4	334.5–337.5	9.9
nanno	B D. hamatus	36–CC/34–1	342.0–343.0	10.0
nanno	B C. coalitus	37–6,35–120	350.5–351.5	10.8
foram	T G. fohsi robusta, G. fohsi lobata	40–4/40–5	375.5–377.0	11.5
foram	B G. fohsi robusta	43–6/44–1	407.0–409.0	12.6
foram	B G. fohsi lobata	45–2/45–4	420.0–423.0	13.1
foram	B G. fohsi praefohsi	47–6/48–1	445.0–447.0	13.9
foram	B O. suturalis	51–6/52–2	483.0–487.0	15.2
foram	T C. dissimilis	58–2,60–92	543.5–544.0	17.6
foram	B G. insueta	59–1/59–2	551.5–553.0	18.1
foram	T G. kugleri	65–3/65–CC	611.6–617.5	21.7

The 51 samples from Site 289 used in this study range between ~18.7 and 5.8 Ma. Samples were taken every 5 to 10 m between 571 and 201m below the sea floor, representing time intervals between 1×10^5 and 5×10^5 years.

Approach to Isotopic Ranking

Twenty-three of the most abundant Miocene planktonic foraminiferal species in Site 289 were picked for isotopic analysis (Table 1). However, the stratigraphic ranges of nine of these species are too restricted and hence are not considered in this report. Also, *Globigerinita glutinata* is not discussed because its isotopic values are extremely variable. The remaining thirteen species are abundant and exhibit sufficiently long stratigraphic ranges for comparative studies (Table 3). The original isotopic data is available in tabular from upon request from the authors.

None of the species examined occurred in all samples. Also, because of biotic evolution and extinction, most of the species do not range throughout the entire Miocene interval between ~18.7 and 5.8 Ma. The relative abundance of individual species also varies throughout the sequence, and in some intervals some species occur in insufficient numbers for isotopic analysis.

An isotopic depth ranking for the 13 planktonic species was established based upon a number of assumptions: 1) each species calcified its test in isotopic equilibrium; 2)

Table 3. List of 13 planktonic foraminiferal species used for isotopic comparative studies in this investigation.

Globigerinoides obliquus Bolli
Globigerinoides immaturus Leroy
Globigerinoides sacculifer (Brady)
Globorotalia (Menardella) praemenardii Cushman and Stainforth
Globorotalia (Menardella) menardii (Parker, Jones and Brady)
Globorotalia (Menardella) limbata (Fornasini)
Globorotalia (Globorotalia) merotumida Blow and Banner
Globorotalia (Jenkinsella) siakensis Leroy
Globorotalia (Jenkinsella) mayeri Cushman and Ellisor
Globoquadrina venezuelana (Hedberg)
Globoquadrina dehiscens (Chapman, Parr and Collins)
Globoquadrina baroemoenensis (Leroy)
Dentoglobigerina altispira altispira (Cushman and Jarvis)

---

water temperatures are highest at the surface and decrease with depth; 3) oxygen isotopic fractionation is temperature-dependent, so that shallower-dwelling species have lower $\delta^{18}O$ values relative to deeper-dwelling species; and 4) carbon isotopic values in planktonic foraminifera change vertically in the upper part of the water column, such that $\delta^{13}C$ values are higher in surface waters due to higher biological productivity and decrease with depth as respiration and decay of organic matter recycles ^{12}C-rich CO_2 back into the water.

Thus, species were ranked using 1) their relative oxygen isotopic values from lowest (shallow) to highest (deep) and 2) their relative carbon isotopic values from highest (shallow) to lowest (deep). However, much variability exists in both the oxygen and carbon isotopic rankings of species between individual samples.

This isotopic variability complicates the development of a relative depth ranking for each species. Nevertheless, it is possible to produce a general depth ranking by comparing interspecies isotopic values in all samples where any two given species co-occur. If, for example, *Globigerinoides immaturus* and *Dentoglobigerina altispira* co-occur in 12 samples, in how many samples did *Gs. immaturus* exhibit higher $\delta^{18}O$ values, lower $\delta^{18}O$ values, or values that were essentially the same (i.e. $\Delta\delta^{18}O$ <0.05‰)? Comparisons with $\Delta\delta$ values <0.2‰ fall within the range of analytical precision of the measurements (±0.1 ‰) and are considered to be equal. Values >0.2 ‰ but <0.4 ‰ lie within the analytical error of the measurements plus one sigma and indicate slight differences between species. Those comparisons, which differ by >0.4 ‰, represent a significant isotopic difference and indicate that calcification occurred under different environmental conditions.

Using these procedures, we compared each of the thirteen species to the others for each sample in which they co-occurred. The $\Delta\delta^{18}O$ comparisons are shown in Tables 4 to 8, and the $\Delta\delta^{13}O$ comparisons are shown in Tables 9 and 10. Furthermore, the relative depth rankings of the species were examined throughout the Miocene (~18.7

to 5.8 Ma) for evidence of any change resulting from biological evolution or paleoenvironmental change.

Results

Oxygen Isotopes

Comparison of the $\delta^{18}O$ values of the thirteen planktonic foraminiferal species indicates the existence of two distinct groups with no overlap. One group with relatively low $\delta^{18}O$ values consists of seven species (*Globigerinoides immaturus, Gs. saccculifer, Gs. obliquus, Dentoglobigerina altispira, Globoquadrina baroemoenensis, Globorotalia (Jenkinsella) mayeri* and *Gr. (J.) siakensis*). The other group, with relatively high $\delta^{18}O$ values, includes *Globoquadrina venezuelana* and *Gq. dehiscens*. Members of these two groups of planktonic species remain isotopically distinct throughout the Miocene. *Gq. venezuelana* has no $\delta^{18}O$ values lower than the group of seven species; *Gq. dehiscens* exhibits only one value lower than *Dg. altispira* and one equal to *Gq. baroemoenensis*.

Taxa with Low $\delta^{18}O$ Values

Each of the seven species with low $\delta^{18}O$ values was compared to the other six, and the data are listed in Tables 4A–G. These comparisons suggest that all seven species lived within similar water masses. However, small differences exist in the long-term average isotopic values between these forms that suggest slight vertical water mass segregation or calcification during different seasons. *Gs. obliquus* appears to have the lowest $\delta^{18}O$ values, followed by *Gs. sacculifer* with slightly higher values. A group of three species, *Gr. (J.) siakensis, Gr. (J.) mayeri,* and *Dg. altispira*, generally have higher values than *Gs. sacculifer, Gq. baroemoenensis*, and *Gs. immaturus* exhibit the highest values.

Taxa with High $\delta^{18}O$ Values

Isotopic comparisons between the two species which exhibit high $\delta^{18}O$ values, *Gq. venezuelana* and *Gq. dehiscens,* show that they co-occur in 9 samples and that *Gq. venezuelana* has higher $\delta^{18}O$ values in six samples and lower values in three (Table 5). These two species appear to be isotopically similar. However, *Gq. venezuelana* exhibits a much higher amplitude of $\delta^{18}O$ change over time, ranging from -1.2 to 0.1 ‰, compared to *Gq. dehiscens* which varies between -0.9 and -0.3 ‰ (Fig. 4). Thus, these species reacted differently to temperature changes in the water column, although this may partly be an artifact related to larger sample size (44 samples) of *Gq. venezuelana*, compared with *Gq. dehiscens* (10 samples).

Gr. (Menardella) Group

Members of the *Globorotalia (Menardella)* lineage (*Gr. praemenardii, Gr. menardii* and *Gr. limbata)* exhibit an oxygen isotopic record that is distinctly different from the previous nine species, and as a consequence these are treated separately. The time-

Table 4. Inter-species $\delta^{18}O$ comparisons of planktonic foraminiferal species with low $\delta^{18}O$ values (*Gs. immaturus, Gs. sacculifer, Gs. obliquus, Gq. baroemoenensis, Dg. altispira, Gq. baroemoenensis, Gr. mayeri* and *Gr. siakensis*). $\Delta\delta^{18}O$ values <0.2 ‰ are not significant, $\Delta\delta^{18}O$ values >0.2 ‰, but <0.4 ‰ represent slight differences, and $\Delta\delta^{18}O$ values >0.4 ‰ are significantly different. The shaded area shows that there is little oxygen isotopic similarity between planktonic foraminiferal species with low $\delta^{18}O$ values and those with high $\delta^{18}O$ values (*Gq. dehiscens, Gq. venezuelana, Gr. praemenardii, Gr. menardii, Gr. limbata,* and *Gr. merotumida*).

A

	Gs. immaturus	*Gs. sacculifer*	*Gs. obliquus*	*Dg. altispira*	*Gq. baroemoenensis*	*Gr. mayeri*	*Gr. siakensis*	*Gq. dehiscens*	*Gq. venezuelana*	*Gr. praemenardii*	*Gr. menardii*	*Gr. limbata*	*Gr. merotumida*
Co-occur	0	10	12	27	31	19	13	5	30	4	10	7	6
Gs. immaturus higher	0	6	10	17	15	14	11	0	0	0	0	0	0
Δδ18O >0.2‰	0	0	6	12	7	13	6	0	0	0	0	0	0
Δδ18O >0.4‰	0	0	1	2	2	4	3	0	0	0	0	0	0
Gs. immaturus lower	0	1	1	7	12	1	1	5	30	4	9	4	6
Δδ18O >0.2‰	0	0	0	3	3	0	1	3	27	4	8	3	2
Δδ18O >0.4‰	0	0	0	0	0	0	0	1	21	4	7	2	2
Equal	0	3	1	3	4	4	1	0	0	0	1	3	0

B

	Gs. immaturus	*Gs. sacculifer*	*Gs. obliquus*	*Dg. altispira*	*Gq. baroemoenensis*	*Gr. mayeri*	*Gr. siakensis*	*Gq. dehiscens*	*Gq. venezuelana*	*Gr. praemenardii*	*Gr. menardii*	*Gr. limbata*	*Gr. merotumida*
Co-occur	10	0	9	11	11	1	1	1	10	0	7	7	6
Gs. sacculifer higher	1	0	5	2	1	1	1	0	0	0	0	0	0
Δδ18O >0.2‰	0	0	2	2	0	0	0	0	0	0	0	0	0
Δδ18O >0.4‰	0	0	0	0	0	0	0	0	0	0	0	0	0
Gs. sacculifer lower	6	0	3	6	9	0	0	1	10	0	7	6	6
Δδ18O >0.2‰	0	0	0	5	3	0	0	1	9	0	6	4	6
Δδ18O >0.4‰	0	0	0	0	1	0	0	0	8	0	4	1	1
Equal	3	0	1	3	1	0	0	0	0	0	0	1	0

C

	Gs. immaturus	*Gs. sacculifer*	*Gs. obliquus*	*Dg. altispira*	*Gq. baroemoenensis*	*Gr. mayeri*	*Gr. siakensis*	*Gq. dehiscens*	*Gq. venezuelana*	*Gr. praemenardii*	*Gr. menardii*	*Gr. limbata*	*Gr. merotumida*
Co-occur	12	9	0	14	13	3	3	1	14	1	11	7	6
Gs. obliquus higher	1	3	0	1	0	0	1	0	0	0	0	0	0
Δδ18O >0.2‰	0	0	0	0	0	0	0	0	0	0	0	0	0
Δδ18O >0.4‰	0	0	0	0	0	0	0	0	0	0	0	0	0
Gs. obliquus lower	10	5	0	11	12	3	2	1	14	1	11	7	6
Δδ18O >0.2‰	6	2	0	7	7	1	2	1	14	1	11	4	6
Δδ18O >0.4‰	1	0	0	0	2	1	0	1	14	1	8	2	3
Equal	1	1	0	2	1	0	0	0	0	0	0	0	0

D

	Gs. immaturus	*Gs. sacculifer*	*Gs. obliquus*	*Dg. altispira*	*Gq. baroemoenensis*	*Gr. mayeri*	*Gr. siakensis*	*Gq. dehiscens*	*Gq. venezuelana*	*Gr. praemenardii*	*Gr. menardii*	*Gr. limbata*	*Gr. merotumida*
Co-occur	31	11	13	38	0	32	17	11	43	5	13	8	6
Gq. baroemoenensis higher	12	9	12	22	0	20	11	1	0	0	0	1	0
Δδ18O >0.2‰	3	3	7	10	0	7	2	0	0	0	0	0	0
Δδ18O >0.4‰	0	1	2	0	0	1	1	0	0	0	0	0	0
Gq. baroemoenensis lower	15	1	0	15	0	8	3	10	43	5	13	6	5
Δδ18O >0.2‰	7	0	0	1	0	4	0	10	42	5	11	4	3
Δδ18O >0.4‰	2	0	0	0	0	2	0	3	31	4	8	1	0
Equal	4	1	1	1	0	4	3	1	0	0	0	1	1

E

	Gs. immaturus	*Gs. sacculifer*	*Gs. obliquus*	*Dg. altispira*	*Gq. baroemoenensis*	*Gr. mayeri*	*Gr. siakensis*	*Gq. dehiscens*	*Gq. venezuelana*	*Gr. praemenardii*	*Gr. menardii*	*Gr. limbata*	*Gr. merotumida*
Co-occur	27	11	14	0	38	22	15	8	37	5	14	8	6
Dg. altispira higher	7	6	11	0	15	11	8	1	0	0	0	2	0
Δδ18O >0.2‰	3	5	7	0	1	2	2	0	0	0	0	0	0
Δδ18O >0.4‰	0	0	0	0	0	1	1	0	0	0	0	0	0
Dg. altispira lower	17	2	1	0	22	9	4	7	37	5	14	5	5
Δδ18O >0.2‰	12	2	0	0	10	2	3	7	37	5	12	3	3
Δδ18O >0.4‰	2	0	0	0	0	1	1	6	33	4	8	2	2
Equal	3	3	2	0	1	4	3	0	0	0	0	1	1

F

	Gs. immaturus	*Gs. sacculifer*	*Gs. obliquus*	*Dg. altispira*	*Gq. baroemoenensis*	*Gr. mayeri*	*Gr. siakensis*	*Gq. dehiscens*	*Gq. venezuelana*	*Gr. praemenardii*	*Gr. menardii*	*Gr. limbata*	*Gr. merotumida*
Co-occur	19	1	3	22	32	0	18	9	29	5	3	0	0
Gr. mayeri higher	1	0	3	9	8	0	7	0	0	0	0	0	0
Δδ18O >0.2‰	0	0	1	2	4	0	1	0	0	0	0	0	0
Δδ18O >0.4‰	0	0	0	1	2	0	0	0	0	0	0	0	0
Gr. mayeri lower	14	1	0	11	20	0	8	9	29	5	3	0	0
Δδ18O >0.2‰	13	0	0	2	7	0	4	7	29	5	3	0	0
Δδ18O >0.4‰	4	0	0	1	1	0	0	4	28	5	3	0	0
Equal	4	0	0	2	4	0	3	0	0	0	0	0	0

G

	Gs. immaturus	*Gs. sacculifer*	*Gs. obliquus*	*Dg. altispira*	*Gq. baroemoenensis*	*Gr. mayeri*	*Gr. siakensis*	*Gq. dehiscens*	*Gq. venezuelana*	*Gr. praemenardii*	*Gr. menardii*	*Gr. limbata*	*Gr. merotumida*
Co-occur	13	1	3	15	17	18	0	4	18	4	3	0	0
Gr. siakensis higher	1	0	2	4	3	8	0	0	0	0	0	0	0
Δδ18O >0.2‰	1	0	2	3	0	4	0	0	0	0	0	0	0
Δδ18O >0.4‰	0	0	1	1	0	0	0	0	0	0	0	0	0
Gr. siakensis lower	11	1	1	8	11	7	0	4	18	4	3	0	0
Δδ18O >0.2‰	6	0	0	2	2	1	0	4	17	4	3	0	0
Δδ18O >0.4‰	3	0	0	1	1	0	0	3	17	4	3	0	0
Equal	1	0	0	3	3	3	0	0	0	0	0	0	0

Table 5. Inter-species $\delta^{18}O$ comparisons of planktonic foraminiferal species with high $\delta^{18}O$ values (*Gq. dehiscens* and *Gq. venezuelana*). $\Delta\delta^{18}O$ values <0.2 ‰ are not significant, $\Delta\delta^{18}O$ values >0.2 ‰, but <0.4 ‰ represent slight differences, and $\Delta\delta^{18}O$ values >0.4 ‰ are significantly different. The shaded area shows that there is little oxygen isotopic similarity between planktonic foraminiferal species with high $\delta^{18}O$ values and those with low $\delta^{18}O$ values (*Gs. immaturus, Gs. sacculifer, Gs. obliquus, Dg. altispira, Gq. baroemoenensis, Gr. mayeri,* and *Gr. siakensis*).

A

	Gs. immaturus	*Gs. sacculifer*	*Gs. obliquus*	*Dg. altispira*	*Gq. baroemoenensis*	*Gr. mayeri*	*Gr. siakensis*	*Gq. dehiscens*	*Gq. venezuelana*	*Gr. praemenardii*	*Gr. menardii*	*Gr. limbata*	*Gr. merotumida*
Co-occur	5	1	1	8	11	9	4	0	9	2	1	0	0
Gq. dehiscens higher	5	1	1	7	10	9	4	0	3	0	0	0	0
$\Delta\delta^{18}O$ >0.2 ‰	3	1	1	7	10	7	4	0	2	0	0	0	0
$\Delta\delta^{18}O$ >0.4 ‰	1	0	1	6	3	4	3	0	0	0	0	0	0
Gq. dehiscens lower	0	0	0	1	0	0	0	0	6	2	1	0	0
$\Delta\delta^{18}O$ >0.2 ‰	0	0	0	0	0	0	0	0	4	2	0	0	0
$\Delta\delta^{18}O$ >0.4 ‰	0	0	0	0	0	0	0	0	2	1	0	0	0
Equal	0	0	0	0	1	0	0	0	0	0	0	0	0

B

	Gs. immaturus	*Gs. sacculifer*	*Gs. obliquus*	*Dg. altispira*	*Gq. baroemoenensis*	*Gr. mayeri*	*Gr. siakensis*	*Gq. dehiscens*	*Gq. venezuelana*	*Gr. praemenardii*	*Gr. menardii*	*Gr. limbata*	*Gr. merotumida*
Co-occur	30	10	14	37	43	29	18	9	0	5	12	7	6
Gq. venezuelana higher	30	10	14	37	43	29	18	6	0	2	6	6	5
$\Delta\delta^{18}O$ >0.2 ‰	27	9	14	37	42	29	17	4	0	0	4	5	5
$\Delta\delta^{18}O$ >0.4 ‰	21	8	14	33	31	28	17	2	0	0	1	5	1
Gq. venezuelana lower	0	0	0	0	0	0	0	3	0	3	5	1	1
$\Delta\delta^{18}O$ >0.2 ‰	0	0	0	0	0	0	0	2	0	2	1	0	0
$\Delta\delta^{18}O$ >0.4 ‰	0	0	0	0	0	0	0	0	0	1	0	0	0
Equal	0	0	0	0	0	0	0	0	0	0	1	0	0

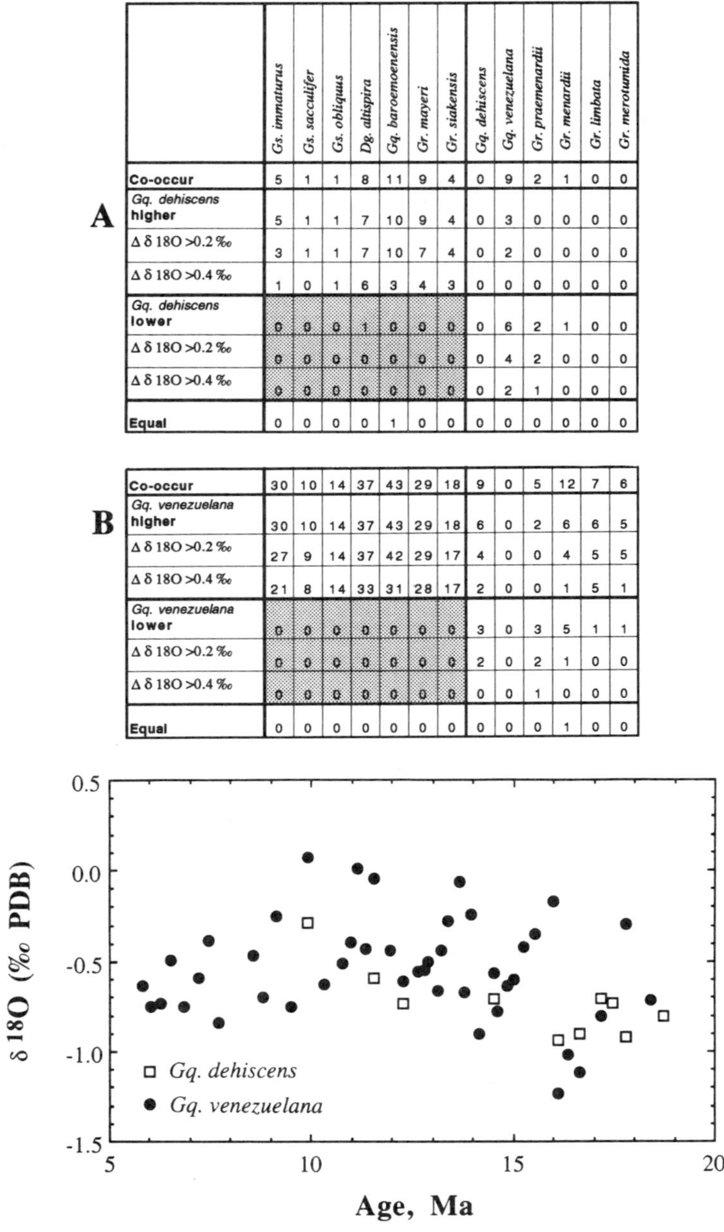

Fig. 4. Plot of $\delta^{18}O$ values for *Gq. venezuelana* and *Gq. dehiscens* v. Age (Ma) for DSDP Site 289.

Table 6. A) Inter-species $\delta^{18}O$ comparisons with *Gr. praemenardii,* a planktonic foraminiferal species with high $\delta^{18}O$ values. $\Delta\delta^{18}O$ values <0.2 ‰ are not significant, $\Delta\delta^{18}O$ values >0.2 ‰, but <0.4 ‰ represent slight differences, and $\Delta\delta^{18}O$ values >0.4 ‰ are significantly different. The shaded area shows that there is little oxygen isotopic similarity between *Gr. praemenardii* and planktonic foraminifera with low $\delta^{18}O$ values (*Gs. immaturus, Gs. sacculifer, Gs. obliquus, Dg. altispira, Gq. baroemoenensis, Gr. mayeri,* and *Gr. siakensis*).

B) Inter-species $\delta^{18}O$ comparisons with *Gr. menardii,* a planktonic foraminiferal species with high to intermediate $\delta^{18}O$ values. $\Delta\delta^{18}O$ values <0.2 ‰ are not significant, $\Delta\delta^{18}O$ values >0.2 ‰, but <0.4 ‰ represent slight differences, and $\Delta\delta^{18}O$ values >0.4 ‰ are significantly different. The shaded area shows that there is little oxygen isotopic similarity between *Gr. menardii* and planktonic foraminifera with low $\delta^{18}O$ values (*Gs. immaturus, Gs. sacculifer, Gs. obliquus, Dg. altispira, Gq. baroemoenensis, Gr. mayeri,* and *Gr. siakensis*).

C) Inter-species $\delta^{18}O$ comparisons with *Gr. limbata,* a planktonic foraminiferal species with high to low $\delta^{18}O$ values. $\Delta\delta^{18}O$ values <0.2 ‰ are not significant, $\Delta\delta^{18}O$ values >0.2 ‰, but <0.4 ‰ represent slight differences, and $\Delta\delta^{18}O$ values >0.4 ‰ are significantly different.

A)

	Gs. immaturus	*Gs. sacculifer*	*Gs. obliquus*	*Dg. altispira*	*Gq. baroemoenensis*	*Gr. mayeri*	*Gr. siakensis*	*Gq. dehiscens*	*Gq. venezuelana*	*Gr. praemenardii*	*Gr. menardii*	*Gr. limbata*	*Gr. merotumida*
Co-occur	4	0	1	5	5	5	4	2	5	0	1	0	0
Gr. praemenardii higher	4	0	1	5	5	5	4	2	3	0	1	0	0
$\Delta\delta$ 18O >0.2‰	4	0	1	5	5	5	4	2	2	0	0	0	0
$\Delta\delta$ 18O >0.4‰	4	0	1	4	4	5	4	1	1	0	0	0	0
Gr. praemenardii lower	0	0	0	0	0	0	0	0	2	0	0	0	0
$\Delta\delta$ 18O >0.2‰	0	0	0	0	0	0	0	0	0	0	0	0	0
$\Delta\delta$ 18O >0.4‰	0	0	0	0	0	0	0	0	0	0	0	0	0
Equal	0	0	0	0	0	0	0	0	0	0	0	0	0

B)

	Gs. immaturus	*Gs. sacculifer*	*Gs. obliquus*	*Dg. altispira*	*Gq. baroemoenensis*	*Gr. mayeri*	*Gr. siakensis*	*Gq. dehiscens*	*Gq. venezuelana*	*Gr. praemenardii*	*Gr. menardii*	*Gr. limbata*	*Gr. merotumida*
Co-occur	10	7	11	14	13	3	3	1	12	1	0	7	5
Gr. menardii higher	9	7	11	14	13	3	3	1	5	0	0	4	2
$\Delta\delta$ 18O >0.2‰	8	6	11	12	11	3	3	0	1	0	0	2	1
$\Delta\delta$ 18O >0.4‰	7	4	8	8	8	3	3	0	0	0	0	1	0
Gr. menardii lower	0	0	0	0	0	0	0	0	6	1	0	1	1
$\Delta\delta$ 18O >0.2‰	0	0	0	0	0	0	0	0	4	0	0	0	0
$\Delta\delta$ 18O >0.4‰	0	0	0	0	0	0	0	0	1	0	0	0	0
Equal	1	0	0	0	0	0	0	0	1	0	0	2	2

C)

	Gs. immaturus	*Gs. sacculifer*	*Gs. obliquus*	*Dg. altispira*	*Gq. baroemoenensis*	*Gr. mayeri*	*Gr. siakensis*	*Gq. dehiscens*	*Gq. venezuelana*	*Gr. praemenardii*	*Gr. menardii*	*Gr. limbata*	*Gr. merotumida*
Co-occur	7	7	7	8	8	0	0	0	7	0	7	0	6
Gr. limbata higher	4	6	7	5	6	0	0	0	1	0	1	0	0
$\Delta\delta$ 18O >0.2‰	3	4	4	3	4	0	0	0	0	0	0	0	0
$\Delta\delta$ 18O >0.4‰	2	1	2	2	1	0	0	0	0	0	0	0	0
Gr. limbata lower	0	0	0	2	1	0	0	0	6	0	4	0	5
$\Delta\delta$ 18O >0.2‰	0	0	0	0	0	0	0	0	5	0	2	0	1
$\Delta\delta$ 18O >0.4‰	0	0	0	0	0	0	0	0	5	0	1	0	0
Equal	3	1	0	1	1	0	0	0	0	0	2	0	1

series of oxygen isotopic rankings within the *Gr. (Menardella)* lineage indicate that before ~7.5 Ma *Gr. (M.) praemenardii, Gr. (M.) menardii*, and *Gr. (M.) limbata* exhibit high oxygen isotopic values, similar to *Gq. venezuelana.* The oxygen isotopic differences are relatively large (>0.4 ‰) compared with the group of species with low $\delta^{18}O$ prior to the extinction of the *Gr. (Jenkinsella)* group at ~9.9 Ma. However, from ~9.9 to 7.5 Ma, the differences are much smaller (Tables 6A–C and 8A–F).

It is clear that by ~7.5 Ma, *Gr. limbata* had changed from a form with high $\delta^{18}O$ values to one with low $\delta^{18}O$ values. Also at the same time, *Gr. menardii* changed from having high $\delta^{18}O$ values to a form intermediate between species with high and low $\delta^{18}O$ values. This led to the first forms with intermediate oxygen isotopic values during the Miocene in this region.

Globorotalia (Globorotalia) merotumida

Most of the isotope data for *Globorotalia (Globorotalia) merotumida* are from the interval following ~7.5 Ma. These data record intermediate $\delta^{18}O$ values similar to *Gr. menardii* (Table 7).

The oxygen isotopic depth rankings are summarized in Table 11. This table shows that from ~18.7 to 7.5 Ma the planktonic foraminiferal species at this site can be defined as having either low $\delta^{18}O$ values (shallow-dwelling) or high $\delta^{18}O$ values (deep-dwelling). After ~7.5 Ma, species with intermediate $\delta^{18}O$ values can also be recognized. Also by ~7.5 Ma, the $\delta^{18}O$ values of *Gr. limbata* changed from high to low, and the $\delta^{18}O$ values of *Gr. menardii* had changed from high to intermediate.

Table 7. Inter-species $\delta^{18}O$ comparisons with *Gr. merotumida,* a planktonic foraminiferal species with intermediate $\delta^{18}O$ values. $\Delta\delta^{18}O$ values <0.2 ‰ are not significant, $\Delta\delta^{18}O$ values >0.2 ‰, but <0.4 ‰ represent slight differences, and $\Delta\delta^{18}O$ values >0.4 ‰ are significantly different. The shaded area shows that there is little oxygen isotopic similarity between *Gr. merotumida* and planktonic foraminifera with low $\delta^{18}O$ values (*Gs. immaturus, Gs. sacculifer, Gs. obliquus, Dg. altispira, Gq. baroemoenensis, Gr. mayeri,* and *Gr. siakensis*) and high $\delta^{18}O$ values (*Gq. venezuelana*).

	Gs. immaturus	*Gs. sacculifer*	*Gs. obliquus*	*Dg. altispira*	*Gq. baroemoenensis*	*Gr. mayeri*	*Gr. siakensis*	*Gq. dehiscens*	*Gq. venezuelana*	*Gr. praemenardii*	*Gr. menardii*	*Gr. limbata*	*Gr. merotumida*
Co-occur	6	6	6	6	6	0	0	0	6	0	5	6	0
Gr. merotumida **higher**	6	6	6	5	5	0	0	0	1	0	1	5	0
$\Delta\delta$ 18O >0.2 ‰	2	6	6	3	3	0	0	0	0	0	0	1	0
$\Delta\delta$ 18O >0.4 ‰	2	1	3	2	0	0	0	0	0	0	0	0	0
Gr. merotumida **lower**	0	0	0	0	0	0	0	0	5	0	2	0	0
$\Delta\delta$ 18O >0.2 ‰	0	0	0	0	0	0	0	0	5	0	1	0	0
$\Delta\delta$ 18O >0.4 ‰	0	0	0	0	0	0	0	0	1	0	0	0	0
Equal	0	0	0	1	1	0	0	0	0	0	2	1	0

Table 8. Inter-species $\delta^{18}O$ comparisons of planktonic foraminiferal species with high $\delta^{18}O$ values for three time intervals: A. and B. are ~18.7 to 9.9 Ma; C. and D. are ~9.9 to 7.5 Ma; E. and F. are ~7.5 to 5.8 Ma. In each case $\Delta^{18}O$ values >0.2 ‰ are not significant, $\Delta\delta^{18}O$ values >0.2 ‰, but >0.4 ‰ represent slight differences, and $\Delta\delta^{18}O$ values >0.4 ‰ are significantly different. The shaded areas show that there is little oxygen isotopic similarity between indicated species with high $\delta^{18}O$ values and those species with low $\delta^{18}O$ values in each case (A–F).

A. From ~18.7 to 9.9 Ma

	Gs. immaturus	Gs. sacculifer	Gs. obliquus	Dg. altispira	Gq. baroemoenensis	Gr. mayeri	Gr. siakensis	Gq. dehiscens	Gq. venezuelana	Gr. praemenardii	Gr. menardii	Gr. limbata	Gr. merotumida
Co-occur	2	0	3	3	2	3	3	1	3	1	0	0	0
Gr. menardii heavier	2	0	3	3	2	3	3	1	2	0	0	0	0
$\Delta \delta^{18}O$ >0.2 per mil	2	0	3	3	2	3	3	0	1	0	0	0	0
$\Delta \delta^{18}O$ >0.4 per mil	2	0	3	3	2	3	3	0	0	0	0	0	0
Gr. menardii lighter	0	0	0	0	0	0	0	0	1	1	0	0	0
$\Delta \delta^{18}O$ >0.2 per mil	0	0	0	0	0	0	0	0	0	0	0	0	0
$\Delta \delta^{18}O$ >0.4 per mil	0	0	0	0	0	0	0	0	0	0	0	0	0
Equal	0	0	0	0	0	0	0	0	0	0	0	0	0

B. From ~18.7 to 9.9 Ma

	Gs. immaturus	Gs. sacculifer	Gs. obliquus	Dg. altispira	Gq. baroemoenensis	Gr. mayeri	Gr. siakensis	Gq. dehiscens	Gq. venezuelana	Gr. praemenardii	Gr. menardii	Gr. limbata	Gr. merotumida
Co-occur	4	0	1	4	5	5	4	2	5	0	1	0	0
Gr. praemenardii heavier	4	0	1	4	5	5	4	2	3	0	1	0	0
$\Delta \delta^{18}O$ >0.2 per mil	4	0	1	4	5	5	4	2	2	0	0	0	0
$\Delta \delta^{18}O$ >0.4 per mil	4	0	1	4	5	5	4	1	1	0	0	0	0
Gr. praemenardii lighter	0	0	0	0	0	0	0	0	2	0	0	0	0
$\Delta \delta^{18}O$ >0.2 per mil	0	0	0	0	0	0	0	0	0	0	0	0	0
$\Delta \delta^{18}O$ >0.4 per mil	0	0	0	0	0	0	0	0	0	0	0	0	0
Equal	0	0	0	0	0	0	0	0	0	0	0	0	0

C. From ~9.9 to 7.5 Ma

	Gs. immaturus	Gs. sacculifer	Gs. obliquus	Dg. altispira	Gq. baroemoenensis	Gr. mayeri	Gr. siakensis	Gq. dehiscens	Gq. venezuelana	Gr. praemenardii	Gr. menardii	Gr. limbata	Gr. merotumida
Co-occur	4	3	4	7	7	0	0	0	5	0	0	3	1
Gr. menardii heavier	4	3	4	7	7	0	0	0	3	0	0	1	0
$\Delta \delta^{18}O$ >0.2 per mil	3	3	4	7	7	0	0	0	0	0	0	0	0
$\Delta \delta^{18}O$ >0.4 per mil	3	2	4	4	5	0	0	0	0	0	0	0	0
Gr. menardii lighter	0	0	0	0	0	0	0	0	2	0	0	1	0
$\Delta \delta^{18}O$ >0.2 per mil	0	0	0	0	0	0	0	0	1	0	0	0	0
$\Delta \delta^{18}O$ >0.4 per mil	0	0	0	0	0	0	0	0	0	0	0	0	0
Equal	0	0	0	0	0	0	0	0	0	0	0	1	1

D. From ~9.9 to 7.5 Ma

	Gs. immaturus	Gs. sacculifer	Gs. obliquus	Dg. altispira	Gq. baroemoenensis	Gr. mayeri	Gr. siakensis	Gq. dehiscens	Gq. venezuelana	Gr. praemenardii	Gr. menardii	Gr. limbata	Gr. merotumida
Co-occur	2	2	2	3	3	0	0	0	2	0	3	0	1
Gr. limbata heavier	2	2	2	3	3	0	0	0	1	0	1	0	0
$\Delta \delta^{18}O$ >0.2 per mil	2	2	2	2	3	0	0	0	0	0	0	0	0
$\Delta \delta^{18}O$ >0.4 per mil	2	1	2	2	1	0	0	0	0	0	0	0	0
Gr. limbata lighter	0	0	0	0	0	0	0	0	1	0	1	0	0
$\Delta \delta^{18}O$ >0.2 per mil	0	0	0	0	0	0	0	0	0	0	0	0	0
$\Delta \delta^{18}O$ >0.4 per mil	0	0	0	0	0	0	0	0	0	0	0	0	0
Equal	0	0	0	0	0	0	0	0	0	0	1	0	1

E. From ~7.5 to 5.8 Ma

	Gs. immaturus	Gs. sacculifer	Gs. obliquus	Dg. altispira	Gq. baroemoenensis	Gr. mayeri	Gr. siakensis	Gq. dehiscens	Gq. venezuelana	Gr. praemenardii	Gr. menardii	Gr. limbata	Gr. merotumida
Co-occur	4	4	4	4	4	0	0	0	4	0	0	4	4
Gr. menardii heavier	3	4	4	4	4	0	0	0	0	0	0	3	2
$\Delta \delta^{18}O$ >0.2 per mil	3	3	4	2	2	0	0	0	0	0	0	2	1
$\Delta \delta^{18}O$ >0.4 per mil	2	2	1	1	1	0	0	0	0	0	0	1	0
Gr. menardii lighter	0	0	0	0	0	0	0	0	3	0	0	0	1
$\Delta \delta^{18}O$ >0.2 per mil	0	0	0	0	0	0	0	0	3	0	0	0	0
$\Delta \delta^{18}O$ >0.4 per mil	0	0	0	0	0	0	0	0	1	0	0	0	0
Equal	0	0	0	0	0	0	0	0	1	0	0	1	1

F. From ~7.5 to 5.8 Ma

	Gs. immaturus	Gs. sacculifer	Gs. obliquus	Dg. altispira	Gq. baroemoenensis	Gr. mayeri	Gr. siakensis	Gq. dehiscens	Gq. venezuelana	Gr. praemenardii	Gr. menardii	Gr. limbata	Gr. merotumida
Co-occur	5	5	5	5	5	0	0	0	5	0	4	0	5
Gr. limbata heavier	2	4	5	2	3	0	0	0	0	0	0	0	0
$\Delta \delta^{18}O$ >0.2 per mil	1	2	2	1	1	0	0	0	0	0	0	0	0
$\Delta \delta^{18}O$ >0.4 per mil	0	0	0	0	0	0	0	0	0	0	0	0	0
Gr. limbata lighter	0	0	0	2	1	0	0	0	5	0	3	0	5
$\Delta \delta^{18}O$ >0.2 per mil	0	0	0	0	0	0	0	0	5	0	2	0	1
$\Delta \delta^{18}O$ >0.4 per mil	0	0	0	0	0	0	0	0	5	0	1	0	0
Equal	3	1	0	1	1	0	0	0	0	0	1	0	0

Table 9

A

	Gs. immaturus	Gs. sacculifer	Gs. obliquus	Dg. altispira	Gq. baroemoenensis	Gr. mayeri	Gr. siakensis	Gq. dehiscens	Gq. venezuelana	Gr. praemenardii	Gr. menardii	Gr. limbata	Gr. merotumida
Co-occur	0	10	12	27	31	19	13	5	30	4	10	7	6
Gs. immaturus higher	0	0	3	13	17	19	13	5	30	4	10	7	6
$\Delta\delta 13C > 0.2‰$	0	0	1	7	12	19	13	5	30	4	10	7	6
$\Delta\delta 13C > 0.4‰$	0	0	0	2	3	19	12	5	30	4	10	7	6
$\Delta\delta 13C > 0.5‰$	0	0	0	1	2	18	11	5	29	4	8	5	5
$\Delta\delta 13C > 1.0‰$	0	0	0	0	0	1	0	1	6	1	0	1	0
Gs. immaturus lower	0	10	5	13	10	0	0	0	0	0	0	0	0
$\Delta\delta 13C > 0.2‰$	0	6	0	4	5	0	0	0	0	0	0	0	0
$\Delta\delta 13C > 0.4‰$	0	3	0	0	3	0	0	0	0	0	0	0	0
$\Delta\delta 13C > 0.5‰$	0	1	0	0	1	0	0	0	0	0	0	0	0
$\Delta\delta 13C > 1.0‰$	0	0	0	0	0	0	0	0	0	0	0	0	0
Equal	0	0	4	1	4	0	0	0	0	0	0	0	0

B

	Gs. immaturus	Gs. sacculifer	Gs. obliquus	Dg. altispira	Gq. baroemoenensis	Gr. mayeri	Gr. siakensis	Gq. dehiscens	Gq. venezuelana	Gr. praemenardii	Gr. menardii	Gr. limbata	Gr. merotumida
Co-occur	10	0	9	11	11	1	1	1	10	0	7	7	6
Gs. sacculifer higher	10	0	9	11	11	1	1	1	10	0	7	7	6
$\Delta\delta 13C > 0.2‰$	6	0	8	6	8	1	1	1	10	0	7	7	6
$\Delta\delta 13C > 0.4‰$	3	0	2	1	6	1	1	1	10	0	7	7	6
$\Delta\delta 13C > 0.5‰$	1	0	1	0	2	1	1	1	10	0	7	7	6
$\Delta\delta 13C > 1.0‰$	0	0	0	0	0	0	0	0	8	0	3	3	2
Gs. sacculifer lower	0	0	0	0	0	0	0	0	0	0	0	0	0
$\Delta\delta 13C > 0.2‰$	0	0	0	0	0	0	0	0	0	0	0	0	0
$\Delta\delta 13C > 0.4‰$	0	0	0	0	0	0	0	0	0	0	0	0	0
$\Delta\delta 13C > 0.5‰$	0	0	0	0	0	0	0	0	0	0	0	0	0
$\Delta\delta 13C > 1.0‰$	0	0	0	0	0	0	0	0	0	0	0	0	0
Equal	0	0	0	0	0	0	0	0	0	0	0	0	0

C

	Gs. immaturus	Gs. sacculifer	Gs. obliquus	Dg. altispira	Gq. baroemoenensis	Gr. mayeri	Gr. siakensis	Gq. dehiscens	Gq. venezuelana	Gr. praemenardii	Gr. menardii	Gr. limbata	Gr. merotumida
Co-occur	12	9	0	14	13	3	3	1	14	1	11	7	6
Gs. obliquus higher	5	0	0	4	9	3	3	1	14	1	11	7	6
$\Delta\delta 13C > 0.2‰$	0	0	0	2	4	3	3	1	14	1	11	7	6
$\Delta\delta 13C > 0.4‰$	0	0	0	1	1	3	3	1	13	1	11	7	6
$\Delta\delta 13C > 0.5‰$	0	0	0	1	1	3	2	1	13	1	9	7	5
$\Delta\delta 13C > 1.0‰$	0	0	0	0	1	0	0	0	5	0	0	0	0
Gs. obliquus lower	3	9	0	9	4	0	0	0	0	0	0	0	0
$\Delta\delta 13C > 0.2‰$	1	8	0	3	0	0	0	0	0	0	0	0	0
$\Delta\delta 13C > 0.4‰$	0	2	0	0	0	0	0	0	0	0	0	0	0
$\Delta\delta 13C > 0.5‰$	0	1	0	0	0	0	0	0	0	0	0	0	0
$\Delta\delta 13C > 1.0‰$	0	0	0	0	0	0	0	0	0	0	0	0	0
Equal	4	0	0	1	0	0	0	0	0	0	0	0	0

D

	Gs. immaturus	Gs. sacculifer	Gs. obliquus	Dg. altispira	Gq. baroemoenensis	Gr. mayeri	Gr. siakensis	Gq. dehiscens	Gq. venezuelana	Gr. praemenardii	Gr. menardii	Gr. limbata	Gr. merotumida
Co-occur	27	11	14	0	38	22	15	8	37	5	14	8	6
Dg. altispira higher	13	0	9	0	22	21	14	7	36	5	14	8	6
$\Delta\delta 13C > 0.2‰$	4	0	3	0	13	20	13	7	36	5	13	8	6
$\Delta\delta 13C > 0.4‰$	0	0	0	0	4	19	12	7	34	4	13	8	6
$\Delta\delta 13C > 0.5‰$	0	0	0	0	2	16	11	7	34	3	10	8	6
$\Delta\delta 13C > 1.0‰$	0	0	0	0	0	2	0	2	9	1	0	0	0
Dg. altispira lower	13	11	4	0	7	0	1	0	1	0	0	0	0
$\Delta\delta 13C > 0.2‰$	7	6	2	0	4	0	0	0	0	0	0	0	0
$\Delta\delta 13C > 0.4‰$	2	1	1	0	0	0	0	0	0	0	0	0	0
$\Delta\delta 13C > 0.5‰$	1	0	1	0	0	0	0	0	0	0	0	0	0
$\Delta\delta 13C > 1.0‰$	0	0	0	0	0	0	0	0	0	0	0	0	0
Equal	1	0	1	0	9	1	0	1	0	0	0	0	0

E

	Gs. immaturus	Gs. sacculifer	Gs. obliquus	Dg. altispira	Gq. baroemoenensis	Gr. mayeri	Gr. siakensis	Gq. dehiscens	Gq. venezuelana	Gr. praemenardii	Gr. menardii	Gr. limbata	Gr. merotumida
Co-occur	31	11	13	38	0	32	17	11	43	5	13	8	6
Gq. baroemoenensis higher	10	0	4	7	0	29	15	10	40	4	12	8	6
$\Delta\delta 13C > 0.2‰$	5	0	0	4	0	26	13	10	38	4	12	8	6
$\Delta\delta 13C > 0.4‰$	3	0	0	0	0	24	12	10	35	2	11	8	6
$\Delta\delta 13C > 0.5‰$	1	0	0	0	0	19	11	8	31	2	8	8	5
$\Delta\delta 13C > 1.0‰$	0	0	0	0	0	2	0	1	8	1	0	0	0
Gq. baroemoenensis lower	17	11	9	22	0	1	2	0	1	1	1	0	0
$\Delta\delta 13C > 0.2‰$	12	8	4	13	0	1	0	0	0	0	0	0	0
$\Delta\delta 13C > 0.4‰$	3	6	1	4	0	0	0	0	0	0	0	0	0
$\Delta\delta 13C > 0.5‰$	2	2	1	2	0	0	0	0	0	0	0	0	0
$\Delta\delta 13C > 1.0‰$	0	0	1	0	0	0	0	0	0	0	0	0	0
Equal	4	0	0	9	0	2	0	1	2	0	0	0	0

Carbon Isotopes

Carbon isotopic analyses of the thirteen planktonic foraminiferal species reveals the presence of two groups separated by a minimum carbon isotopic difference of 0.4 ‰. Species with high δ^{13}C values include *Gs. immaturus, Gs. sacculifer, Gs. obliquus, Dg. altispira,* and *Gq. baroemoenensis.* These species are either known from modern forms or are inferred from morphological features to have been spinose (Table 12). Species with low δ^{13}C values include *Gr (J.) mayeri, Gr. (J.) siakensis, Gq. dehiscens, Gq. venezuelana, Gr. (M.) praemenardii, Gr. (M.) menardii, Gr. (M.) limbata,* and *Gr. (Gr.) merotumida.* These species are either known or inferred to have been non-spinose (Table 12).

Inter-species carbon isotopic comparisons for each of the thirteen species are shown in Tables 9 and 10. As with the treatment of the oxygen isotopic data, the taxa are compared at the 0.2 ‰ and 0.4 ‰ levels to identify inter-species differences. Furthermore, because of large inter-species differences in the δ^{13}C, cut-off levels of 0.5 ‰ and 1.0 ‰ were added for more effective evaluation of the groupings.

Taxa with High δ^{13}C Values

Among the species with high δ^{13}C values (Table 9), *Gs. sacculifera* in all cases records the highest δ^{13}C values, followed by *Gs. immaturus, Dg. altispira,* and *Gs. obliquus,* with similar carbon isotopic values, and *Gq. baroemoenensis,* with generally the lowest relative carbon isotopic values (Table 11).

Taxa with Low δ^{13}C Values

Differentiation of the low δ^{13}C species (Table 10) based upon their carbon isotopic values is more difficult because of the narrow range of values separating these species. From ~18.7 to 7.5 Ma, *Gr. mayeri, Gr. siakensis, Gr. praemenardii, Gr. menardii,* and *Gr. limbata* record nearly identical δ^{13}C values. *Gq. venezuelana* is more likely to have slightly lower values than the five species above, but the difference in carbon isotopic values is still generally less than 0.2 ‰. Of the eight species with data from this interval, *Gq. dehiscens* is the most likely to have the lowest δ^{13}C values (Table 11).

From ~7.5 to 5.8 Ma, *Gr. menardii, Gr. limbata,* and *Gr. merotumida* have δ^{13}C values that are essentially identical. *Gq. venezuelana* records carbon isotopic values that are ~0.2 ‰ lower than the other three species exhibiting low δ^{13}C values (Table 11).

Table 9. Inter-species δ^{13}C comparisons of planktonic foraminiferal species with high δ^{13}C values (*Gs. immaturus, Gs. sacculifer, Gs. obliquus, Dg. altispira,* and *Gq. baroemoenensis*). $\Delta\delta^{13}$O values <0.2 ‰ are not significant, $\Delta\delta^{13}$C values >0.2 ‰, but <0.4 ‰ represent slight differences, and $\Delta\delta^{13}$C values >0.4 ‰ are significantly different. $\Delta\delta^{13}$C values >0.5 ‰ and >1.0 ‰ are included to demonstrate the range of carbon isotopic differences. The shaded area shows that there is little carbon isotopic similarity between planktonic foraminiferal species with high δ^{13}C values and those with low δ^{13}C values (*Gr. mayeri, Gr. siakensis, Gq. dehiscens, Gq. venezuelana, Gr. praemenardii, Gr. menardii, Gr. limbata,* and *Gr. merotumida*).

Table 10

A

	Gs. immaturus	Gs. sacculifer	Gs. obliquus	Dg. altispira	Gq. baroemoenensis	Gr. mayeri	Gr. siakensis	Gq. dehiscens	Gq. venezuelana	Gr. praemenardii	Gr. menardii	Gr. limbata	Gr. merotumida
Co-occur	19	1	3	22	32	0	18	9	29	5	3	0	0
Gr. mayeri higher	0	0	0	0	1	0	3	5	14	3	1	0	0
$\Delta\delta^{13}C > 0.2‰$	0	0	0	0	0	0	0	3	4	1	0	0	0
$\Delta\delta^{13}C > 0.4‰$	0	0	0	0	0	0	0	1	0	1	0	0	0
$\Delta\delta^{13}C > 0.5‰$	0	0	0	0	0	0	0	0	0	0	0	0	0
$\Delta\delta^{13}C > 1.0‰$	0	0	0	0	0	0	0	0	0	0	0	0	0
Gr. mayeri lower	19	1	3	21	29	0	7	0	8	1	1	0	0
$\Delta\delta^{13}C > 0.2‰$	19	1	3	20	26	0	0	0	1	0	0	0	0
$\Delta\delta^{13}C > 0.4‰$	19	1	3	19	24	0	0	0	0	0	0	0	0
$\Delta\delta^{13}C > 0.5‰$	18	1	3	16	19	0	0	0	0	0	0	0	0
$\Delta\delta^{13}C > 1.0‰$	1	0	0	2	2	0	0	0	0	0	0	0	0
Equal	0	0	0	1	2	0	8	4	7	1	1	0	0

B

	Gs. immaturus	Gs. sacculifer	Gs. obliquus	Dg. altispira	Gq. baroemoenensis	Gr. mayeri	Gr. siakensis	Gq. dehiscens	Gq. venezuelana	Gr. praemenardii	Gr. menardii	Gr. limbata	Gr. merotumida
Co-occur	13	1	3	15	17	18	0	4	18	4	3	0	0
Gr. siakensis higher	0	0	0	1	2	7	0	3	12	1	1	0	0
$\Delta\delta^{13}C > 0.2‰$	0	0	0	0	0	0	0	2	1	1	0	0	0
$\Delta\delta^{13}C > 0.4‰$	0	0	0	0	0	0	0	0	0	1	0	0	0
$\Delta\delta^{13}C > 0.5‰$	0	0	0	0	0	0	0	0	0	1	0	0	0
$\Delta\delta^{13}C > 1.0‰$	0	0	0	0	0	0	0	0	0	0	0	0	0
Gr. siakensis lower	13	1	3	14	15	3	0	0	3	1	0	0	0
$\Delta\delta^{13}C > 0.2‰$	13	1	3	13	13	0	0	0	0	0	0	0	0
$\Delta\delta^{13}C > 0.4‰$	12	1	3	12	12	0	0	0	0	0	0	0	0
$\Delta\delta^{13}C > 0.5‰$	11	1	2	11	11	0	0	0	0	0	0	0	0
$\Delta\delta^{13}C > 1.0‰$	0	0	0	0	0	0	0	0	0	0	0	0	0
Equal	0	0	0	0	0	8	0	1	3	2	2	0	0

C

	Gs. immaturus	Gs. sacculifer	Gs. obliquus	Dg. altispira	Gq. baroemoenensis	Gr. mayeri	Gr. siakensis	Gq. dehiscens	Gq. venezuelana	Gr. praemenardii	Gr. menardii	Gr. limbata	Gr. merotumida
Co-occur	5	1	1	8	11	9	4	0	9	2	1	0	0
Gq. dehiscens higher	0	0	0	0	0	0	0	0	1	1	1	0	0
$\Delta\delta^{13}C > 0.2‰$	0	0	0	0	0	0	0	0	0	1	0	0	0
$\Delta\delta^{13}C > 0.4‰$	0	0	0	0	0	0	0	0	0	1	0	0	0
$\Delta\delta^{13}C > 0.5‰$	0	0	0	0	0	0	0	0	0	0	0	0	0
$\Delta\delta^{13}C > 1.0‰$	0	0	0	0	0	0	0	0	0	0	0	0	0
Gq. dehiscens lower	5	1	1	7	10	5	3	0	8	1	0	0	0
$\Delta\delta^{13}C > 0.2‰$	5	1	1	7	10	3	2	0	4	1	0	0	0
$\Delta\delta^{13}C > 0.4‰$	5	1	1	7	10	1	0	0	0	0	0	0	0
$\Delta\delta^{13}C > 0.5‰$	5	1	1	7	8	0	0	0	0	0	0	0	0
$\Delta\delta^{13}C > 1.0‰$	1	0	0	2	1	0	0	0	0	0	0	0	0
Equal	0	0	0	1	1	4	1	0	0	0	0	0	0

E

	Gs. immaturus	Gs. sacculifer	Gs. obliquus	Dg. altispira	Gq. baroemoenensis	Gr. mayeri	Gr. siakensis	Gq. dehiscens	Gq. venezuelana	Gr. praemenardii	Gr. menardii	Gr. limbata	Gr. merotumida
Co-occur	4	0	1	5	5	5	4	1	5	0	1	0	0
Gr. praemenardii higher	0	0	0	0	1	1	1	1	2	0	0	0	0
$\Delta\delta^{13}C > 0.2‰$	0	0	0	0	0	0	0	1	1	0	0	0	0
$\Delta\delta^{13}C > 0.4‰$	0	0	0	0	0	0	0	0	0	0	0	0	0
$\Delta\delta^{13}C > 0.5‰$	0	0	0	0	0	0	0	0	0	0	0	0	0
$\Delta\delta^{13}C > 1.0‰$	0	0	0	0	0	0	0	0	0	0	0	0	0
Gr. praemenardii lower	4	0	1	5	4	3	1	0	2	0	0	0	0
$\Delta\delta^{13}C > 0.2‰$	4	0	1	5	4	0	1	0	1	0	0	0	0
$\Delta\delta^{13}C > 0.4‰$	4	0	1	4	2	0	1	0	1	0	0	0	0
$\Delta\delta^{13}C > 0.5‰$	4	0	1	3	2	0	1	0	0	0	0	0	0
$\Delta\delta^{13}C > 1.0‰$	0	0	0	1	1	0	0	0	0	0	0	0	0
Equal	0	0	0	0	0	1	2	0	1	0	1	0	0

F

	Gs. immaturus	Gs. sacculifer	Gs. obliquus	Dg. altispira	Gq. baroemoenensis	Gr. mayeri	Gr. siakensis	Gq. dehiscens	Gq. venezuelana	Gr. praemenardii	Gr. menardii	Gr. limbata	Gr. merotumida
Co-occur	10	7	11	14	13	3	3	1	12	1	0	7	5
Gr. menardii higher	0	0	0	0	1	1	0	0	9	0	0	2	1
$\Delta\delta^{13}C > 0.2‰$	0	0	0	0	0	0	0	0	7	0	0	0	0
$\Delta\delta^{13}C > 0.4‰$	0	0	0	0	0	0	0	0	1	0	0	0	0
$\Delta\delta^{13}C > 0.5‰$	0	0	0	0	0	0	0	0	0	0	0	0	0
$\Delta\delta^{13}C > 1.0‰$	0	0	0	0	0	0	0	0	0	0	0	0	0
Gr. menardii lower	10	7	11	14	12	1	1	1	2	0	0	1	1
$\Delta\delta^{13}C > 0.2‰$	10	7	11	13	12	0	0	0	0	0	0	0	0
$\Delta\delta^{13}C > 0.4‰$	10	7	11	13	11	0	0	0	0	0	0	0	0
$\Delta\delta^{13}C > 0.5‰$	8	7	9	10	8	0	0	0	0	0	0	0	0
$\Delta\delta^{13}C > 1.0‰$	0	3	0	0	0	0	0	0	0	0	0	0	0
Equal	0	0	0	0	0	1	2	0	1	1	0	4	3

G

	Gs. immaturus	Gs. sacculifer	Gs. obliquus	Dg. altispira	Gq. baroemoenensis	Gr. mayeri	Gr. siakensis	Gq. dehiscens	Gq. venezuelana	Gr. praemenardii	Gr. menardii	Gr. limbata	Gr. merotumida
Co-occur	7	7	7	8	8	0	0	0	7	0	7	0	6
Gr. limbata higher	0	0	0	0	0	0	0	0	7	0	1	0	2
$\Delta\delta^{13}C > 0.2‰$	0	0	0	0	0	0	0	0	3	0	0	0	0
$\Delta\delta^{13}C > 0.4‰$	0	0	0	0	0	0	0	0	0	0	0	0	0
$\Delta\delta^{13}C > 0.5‰$	0	0	0	0	0	0	0	0	0	0	0	0	0
$\Delta\delta^{13}C > 1.0‰$	0	0	0	0	0	0	0	0	0	0	0	0	0
Gr. limbata lower	7	7	7	8	8	0	0	0	0	0	2	0	2
$\Delta\delta^{13}C > 0.2‰$	7	7	7	8	8	0	0	0	0	0	0	0	0
$\Delta\delta^{13}C > 0.4‰$	7	7	7	8	8	0	0	0	0	0	0	0	0
$\Delta\delta^{13}C > 0.5‰$	5	7	7	8	8	0	0	0	0	0	0	0	0
$\Delta\delta^{13}C > 1.0‰$	1	3	0	0	0	0	0	0	0	0	0	0	0
Equal	0	0	0	0	0	0	0	0	0	0	4	0	2

D

Co-occur	30	10	14	37	43	29	18	9	0	5	12	7	6
Gq. venezuelana higher	0	0	0	1	1	8	3	8	0	2	2	0	0
$\Delta \delta 13C > 0.2 \permil$	0	0	0	0	0	1	0	4	0	1	0	0	0
$\Delta \delta 13C > 0.4 \permil$	0	0	0	0	0	0	0	0	0	1	0	0	0
$\Delta \delta 13C > 0.5 \permil$	0	0	0	0	0	0	0	0	0	0	0	0	0
$\Delta \delta 13C > 1.0 \permil$	0	0	0	0	0	0	0	0	0	0	0	0	0
Gq. venezuelana lower	30	10	14	36	40	14	12	1	0	2	9	7	6
$\Delta \delta 13C > 0.2 \permil$	30	10	14	36	38	4	1	0	0	1	7	3	5
$\Delta \delta 13C > 0.4 \permil$	30	10	13	34	35	0	0	0	0	0	1	0	0
$\Delta \delta 13C > 0.5 \permil$	29	10	13	34	31	0	0	0	0	0	0	0	0
$\Delta \delta 13C > 1.0 \permil$	6	8	5	9	8	0	0	0	0	0	0	0	0
Equal	0	0	0	0	2	7	3	0	0	1	1	0	0

H

Co-occur	6	6	6	6	6	0	0	0	6	0	5	6	0
Gr. merotumida higher	0	0	0	0	0	0	0	0	6	0	1	2	0
$\Delta \delta 13C > 0.2 \permil$	0	0	0	0	0	0	0	0	5	0	0	0	0
$\Delta \delta 13C > 0.4 \permil$	0	0	0	0	0	0	0	0	0	0	0	0	0
$\Delta \delta 13C > 0.5 \permil$	0	0	0	0	0	0	0	0	0	0	0	0	0
$\Delta \delta 13C > 1.0 \permil$	0	0	0	0	0	0	0	0	0	0	0	0	0
Gr. merotumida lower	6	6	6	6	6	0	0	0	0	0	1	2	0
$\Delta \delta 13C > 0.2 \permil$	6	6	6	6	6	0	0	0	0	0	0	0	0
$\Delta \delta 13C > 0.4 \permil$	6	6	6	6	6	0	0	0	0	0	0	0	0
$\Delta \delta 13C > 0.5 \permil$	5	6	5	6	5	0	0	0	0	0	0	0	0
$\Delta \delta 13C > 1.0 \permil$	0	2	0	0	0	0	0	0	0	0	0	0	0
Equal	0	0	0	0	0	0	0	0	0	0	3	2	0

Table 10. (continued)

Discussion

Depth rankings inferred from the oxygen isotopic data do not agree with those inferred from the carbon isotopic data for all planktonic foraminiferal species or for all time intervals (Table 11). If calcification occurred in oxygen and carbon isotopic equilibrium, planktonic foraminiferal species living at similar depths should record similar oxygen and carbon isotopic values. However, two general groups exhibit inconsistencies between their oxygen and carbon isotopic values. First, during the interval from ~18.7 to 7.5 Ma, *Gr. (J.) mayeri* and *Gr. (J.) siakensis* record low $\delta^{18}O$ values typical of shallow-dwelling forms such as *Gs. sacculifer* and *Gs. obliquus,* yet low $\delta^{13}C$ values characteristic of the deep-dwelling forms *Gq. venezuelana* and *Gq. dehiscens.* Similarly, during the interval from ~7.5 to 5.8 Ma, *Gr. (M.) limbata, Gr. (M.) menardii* and *Gr. (Gr.) merotumida* exhibit low to intermediate $\delta^{18}O$ values, but also low $\delta^{13}C$ values typical of deep-dwelling forms.

Two possible explanations can account for the observed discrepancies in the inferred isotopic depth rankings. The first hypothesis is that the discrepancies resulted from differences in seasonal distribution among the species. The second is that the species lived under comparable oceanographic conditions, but that not all forms calcified their tests in isotopic equilibrium with respect to either oxygen or carbon, or both.

Table 10. Inter-species $\delta^{13}C$ comparisons of planktonic foraminiferal species with low $\delta^{13}C$ values (*Gr. mayeri, Gr. siakensis, Gq. dehiscens, Gq. venezuelana, Gr. praemenardii, Gr. menardii, Gr. limbata,* and *Gr. merotumida*). $\Delta\delta^{13}C$ values <0.2 ‰ are not significant, $\Delta\delta^{13}C$ values >0.2 ‰, but <0.4 ‰ represent slight differences, and $\Delta\delta^{13}C$ values >0.4 ‰ are significantly different. $\Delta\delta^{13}C$ values >0.5 ‰ and >1.0 ‰ are included to demonstrate the range of carbon isotopic differences. The shaded area shows that there is little carbon isotopic similarity between planktonic foraminiferal species with low $\delta^{13}C$ values and those with high $\delta^{13}C$ values (*Gs. immaturus, Gs. sacculifer, Gs. obliquus, Dg. altispira,* and *Gq. baroemoenensis*).

Table 11. Inferred oxygen isotopic depth rankings and carbon isotopic depth rankings from ~18.7 to 7.5 Ma and ~7.5 to 5.8 Ma.

Oxygen Isotopic Depth Rankings

from ~18.7 to 7.5 Ma

Shallow	Intermediate	Deep
Gs. obliquus		*Gq. dehiscens*
Gs. sacculifer		*Gr. (M.) limbata*
Gr. (J.) siakensis		*Gr. (M.) menardii*
Gr. (J.) mayeri		*Gq. venezuelana*
Dg. altispira		*Gr. (M.) praemenardii*
Gq. baroemoenensis		
Gs. immaturus		

from ~7.5 to 5.8 Ma

Shallow	Intermediate	Deep
Gs. obliquus	*Gr. (M.) menardii*	*Gq. venezuelana*
Gs. sacculifer	*Gr. (Gr.) merotumida*	
Dg. altispira		
Gq. baroemoenensis		
Gr. (M.) limbata		
Gs. immaturus		

Carbon Isotopic Depth Rankings

from ~18.7 to 7.5 Ma

Shallow	Intermediate	Deep
Gs. sacculifer		*Gr. (J.) siakensis*
Gs. immaturus		*Gr. (J.) mayeri*
Dg. altispira		*Gr. (M.) menardii*
Gs. obliquus		*Gr. (M.) limbata*
Gq. baroemoenensis		*Gr. (M.) praemenardii*
		Gq. venezuelana
		Gq. dehiscens

from ~7.5 to 5.8 Ma

Shallow	Intermediate	Deep
Gs. sacculifer		*Gr. (M.) menardii*
Gs. immaturus		*Gr. (M.) limbata*
Dg. altispira		*Gr. (Gr.) merotumida*
Gs. obliquus		*Gq. venezuelana*
Gq. baroemoenensis		

Table 12. Known or inferred spinose and non-spinose planktonic foraminiferal species.

Inferred spinose species
> *Globigerinoides obliquus* Bolli
> *Globigerinoides immaturus* Leroy
> *Globigerinoides sacculifer* (Brady)
> *Globoquadrina baroemoenensis* (Leroy)
> *Dentoglobigerina altispira altispira* (Cushman and Jarvis)

Inferred non-spinose species
> *Globorotalia (Menardella) praemenardii* Cushman and Stainforth
> *Globorotalia (Menardella) menardii* (Parker, Jones, and Brady)
> *Globorotalia (Menardella) limbata* (Fornasini)
> *Globorotalia (Globorotalia) merotumida* Blow and Banner
> *Globorotalia (Jenkinsella) siakensis* Leroy
> *Globorotalia (Jenkinsella) mayeri* Cushman and Ellisor
> *Globoquadrina venezuelana* (Hedberg)
> *Globoquadrina dehiscens* (Chapman, Parr, and Collins)

Discrepancies through Seasonal Variation

If the discrepancies are due to seasonal variation within the upper water column, then the oxygen and carbon isotopic gradients should co-vary in response to seasonal changes in vertical thermal gradients and associated surface water biological productivity. The carbon isotopic difference between *Gr. mayeri* and *Gr. siakenesis* (forms with low $\delta^{18}O$ values) and *Gq. venezuelana* (a form with high $\delta^{18}O$ values) is generally only ~0.2 ‰, while the $\delta^{18}O$ difference is large (~0.6 to 0.9 ‰). A weak carbon isotopic gradient coupled with a strong oxygen isotopic gradient suggests that these species calcified their tests under conditions of low surface water productivity and a well-developed vertical thermal gradient in surface waters. The carbon isotopic difference is large (~1.0 to 1.5 ‰) between *Gs. immaturus, Gs. sacculifer, Gs. obliquus, Dg. altispira* and *Gq. baroemoenensis* (forms with low $\delta^{18}O$ values), and *Gq. venezuelana* (a form with high $\delta^{18}O$ values). The $\delta^{18}O$ gradient is ~0.6 to 0.9 ‰. The strong $\delta^{13}C$ gradient between the inferred shallow and deeper-dwelling planktonic species could indicate calcification in highly productive surface waters during seasonal upwelling episodes that would produce higher $\delta^{13}C$ values in the shallow-dwelling species. However, if this was the case, vertical surface water temperature and oxygen isotopic gradients should also have been reduced. Instead, the $\delta^{18}O$ gradient between the shallow and deeper-dwelling forms remains essentially constant throughout, suggesting that no significant change occurred in the inferred temperature structure, nor was there any break-down of the thermocline.

Seasonal variability certainly must have occurred in vertical temperature gradients and surface-water productivity in this region of the equatorial Pacific during the Miocene. However, these were not large enough to account for the discrepancies in oxygen and carbon isotopic values exhibited by these particular groups of planktonic foraminifera.

Discrepancies through Isotopic Fractionation

The second hypothesis suggests that the apparent discrepancies between depth rankings based upon oxygen and carbon isotopic data may have resulted from non-equilibrium fractionation of either oxygen or carbon or both. Plankton tow and sediment trap studies of modern planktonic foraminifera strongly suggest that there is little departure of oxygen isotopic values from calculated $\delta^{18}O$ values based on sea water chemistry (Berger, 1969; Kahn and Williams, 1981; Sautter and Thunell, 1991). Furthermore, laboratory culture experiments with live forms indicates that shell calcite is secreted at or close to oxygen isotopic equilibrium (Erez and Luz, 1983; Bouvier-Soumagnak et al., 1986; Spero, in press). It is unlikely that the discrepancies in depth rankings in certain planktonic foraminifera resulted from oxygen isotopic fractionation during calcification. We assume that there was no significant fractionation among the Miocene forms investigated even though almost all are extinct.

In contrast, carbon isotopic fractionation is known to occur in planktonic foraminifera. The $\delta^{13}C$ values of modern planktonic foraminiferal species that host algal symbionts is a function of total dissolved CO_2 and the metabolic processes of respiration and algal symbiont photosynthesis, with the amount of photosynthetic activity dependent on irradiance levels. Cultures of *Orbulina universa,* a modern dinophyte symbiont-bearing spinose species, grown in varying irradiance levels (Spero and DeNiro, 1987; Spero and Williams, 1988) show that carbon isotopic values are ~0.5 to 1.5 ‰ higher in individuals grown in light compared with those grown in the absence of light. The possession of spines permits *O. universa* to host high concentrations of algal symbionts that photosynthesize in a sphere surrounding the individual. The metabolic process of algal symbiont photosynthesis preferentially takes up the light organic carbon, ^{12}C, from ambient water and increases the ratio of $^{13}C/^{12}C$ in the microenvironmental bicarbonate encompassing the foraminifera. During calcification, *O. universa* incorporates HCO_3-ions from a reservoir preferentially enriched in ^{13}C due to the algal symbiont photosynthesis, creating higher $\delta^{13}C$ values in the calcite than under equilibrium conditions. Modern species that host high concentrations of algal symbionts, such as *O. universa, Globigerinoides sacculifer* and *Gs. ruber,* record $\delta^{13}C$ values ~0.5 to 1.0 ‰ higher than in species not affected by algal symbiont photosynthesis, yet living at the same water depths (Reiss and Hottinger, 1984; Spero and DeNiro, 1987; Hemleben et al., 1989; Bijma et al., 1990).

The process of algal symbiont photosynthesis has no apparent effect upon oxygen isotopic values. Thus, it is possible to identify planktonic foraminiferal species having similar oxygen isotopic values and preferred water depths, that nevertheless calcified under carbon isotopic disequilibrium. Species with similar $\delta^{18}O$ values, but with $\delta^{13}C$ values offset by more than 0.5 ‰, almost certainly were influenced by the effects of symbiotic algal photosynthetic activity during calcification. This does not imply that forms with smaller $\delta^{13}C$ offsets (<0.2 ‰) did not contain symbionts, but suggests that certain species possessed algal symbionts in sufficiently large concentrations to alter the equilibrium ratio of $^{13}C/^{12}C$ of the microenvironmental bicarbonate.

Plots of $\delta^{18}O$ versus $\delta^{13}C$ values for each Miocene sample in DSDP Site 289 are

Table 13. Three groups of planktonic foraminifera in the interval from ~18.7 to 7.5 Ma based upon the similarity of their oxygen and carbon isotopic values and four groups of planktonic foraminifera in the interval from ~7.5 to 5.8 Ma based upon the similarity of their oxygen and carbon isotopic values.

Oxygen and Carbon Isotopic Groupings

from ~18.7 to 7.5 Ma

Low $\delta^{18}O$ values and high $\delta^{13}C$ values
Gs. obliquus
Gs. sacculifer
Gs. immaturus
Dg. altispira
Gq. baroemoenensis

Low $\delta^{18}O$ values and low $\delta^{13}C$ values
Gr. (J.) siakensis
Gr. (J.) mayeri

High $\delta^{18}O$ values and low $\delta^{13}C$ values
Gq. dehiscens
Gq. venezuelana
Gr. (M.) praemenardii
Gr. (M.) menardii
Gr. (M.) limbata

from ~7.5 to 5.8 Ma

Low $\delta^{18}O$ values and high $\delta^{13}C$ values
Gs. obliquus
Gs. sacculifer
Gs. immaturus
Dg. altispira
Gq. baroemoenensis

Low $\delta^{18}O$ values and low $\delta^{13}C$ values
Gr. (M.) limbata

Intermediate $\delta^{18}O$ values and low $\delta^{13}C$ values
Gr. (M.) menardii
Gr. (Gr.) merotumida

High $\delta^{18}O$ values and low $\delta^{13}C$ values
Gq. venezuelana

used to define three distinct groups of planktonic foraminifera: 1) shallow-dwelling forms with high $\delta^{13}C$ values (*Gs. obliquus, Gs. sacculifer, Gs. immaturus, Dg. altispira* and *Gq. baroemoenensis*); 2) shallow-dwelling forms with low $\delta^{13}C$ values (*Gr. siakensis* and *Gr. mayeri*); and 3) deep-dwelling forms with low $\delta^{13}C$ values (*Gq. dehiscens, Gq. venezuelana, Gr. praemenardii, Gr. menardii* and *Gr. limbata*). From ~7.5 to 5.8 Ma, a fourth group is added: 4) intermediate-dwelling forms with low $\delta^{13}C$ values (*Gr. menardii* and *Gr. merotumida*). Furthermore, a former deep-dwelling form, *Gr. limbata,* became shallow-dwelling with low $\delta^{13}C$ values at this time (Table 13).

Members of the two shallow-dwelling groups record similar $\delta^{18}O$ values, but the $\Delta\delta^{13}C$ values are ~0.5 to 1.0 ‰. Based on modern analogs, such a large carbon isotopic difference is possible between forms living in the same water mass as a result of the varying effects of algal symbiont photosynthesis. The five species exhibiting

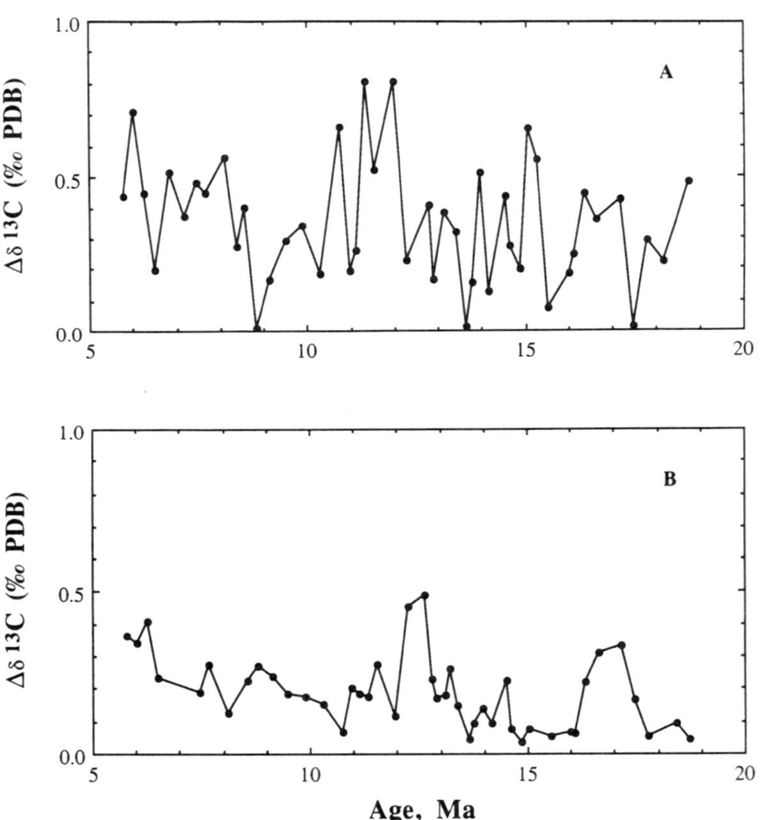

Fig. 5. A) Plot of $\Delta\delta^{13}C$ v. Age (Ma) showing the $\delta^{13}C$ differences (0.0 to 0.8 ‰) between planktonic foraminifera with high $\delta^{13}C$ values.

B) Plot of $\Delta\delta^{13}C$ v. Age (Ma) showing the $\delta^{13}C$ differences (0.0 to 0.5 ‰) between planktonic foraminifera with low $\delta^{13}C$ values.

high carbon isotopic values are known or inferred to have been spinose, while all of the species with low carbon isotopic values are known or inferred to have been non-spinose (Table 12). This difference in morphology between the two isotopically distinct groups is unlikely to be coincidental. The possession of spines, as in modern species such as *O. universa* and *Gs. sacculifer,* provides host species with the particular space needed to support high concentrations of algal symbionts.

A third observation of possible significance is that the group exhibiting the highest $\delta^{13}C$ exhibits a wider range of values (0.0 to 0.8 ‰) than the three groups exhibiting low $\delta^{13}C$ values (0.0 to 0.5 ‰) (Fig. 5). In contrast, a greater range of oxygen isotopic values is exhibited in the low $\delta^{13}C$ groups (0.0 to 1.3 ‰) than in the highest $\delta^{13}C$ group (0.0 to 0.6 ‰) (Fig. 6). The greater variability of carbon isotopic values in the high $\delta^{13}C$ group cannot be accounted for by a greater depth range experienced by these species, but is explained by different degrees of algal photosynthetic influ-

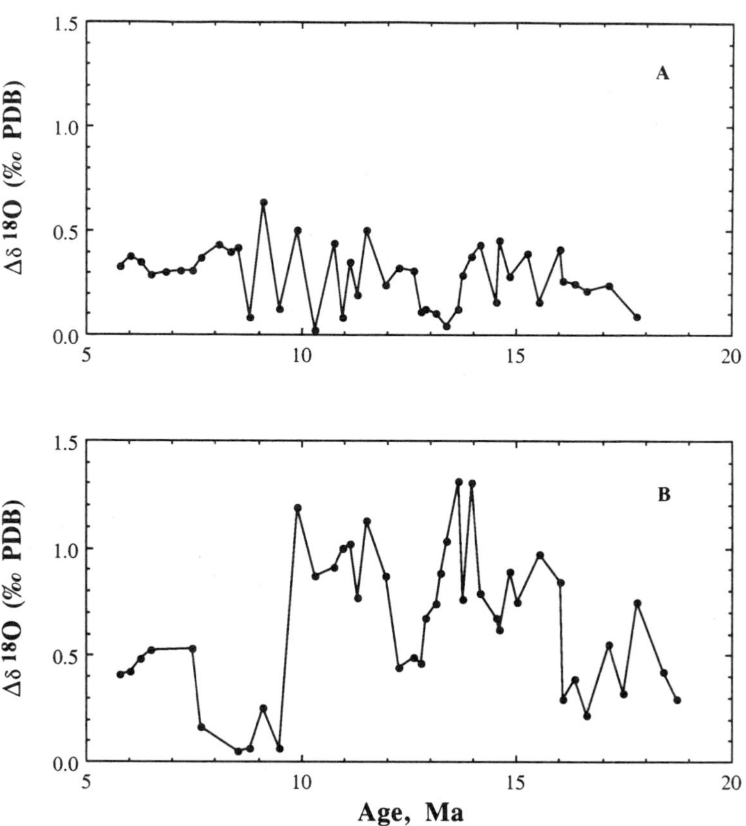

Fig. 6. A) Plot of $\Delta\delta^{18}O$ v. Age (Ma) showing the $\delta^{18}O$ differences (0.0 to 0.6 ‰) between planktonic foraminifera with high $\delta^{13}C$ values.

B) Plot of $\Delta\delta^{18}O$ v. Age (Ma) showing the $\delta^{18}O$ differences (0.0 to 1.3 ‰) between planktonic foraminifera with low $\delta^{13}C$ values.

ence amongst the species with high $\delta^{13}C$ values.

Further evidence supporting the hypothesis of algal symbiont-induced carbon isotopic fractionation in certain planktonic foraminifera is derived from modern forms of *Gs. sacculifer,* which are spinose, host high concentrations of algal symbionts and record carbon isotopic values that are fractionated from equilibrium in the same amount as observed in this study (Reiss and Hottinger, 1984; Hemleben et al., 1989; Bijma et al., 1990).

Thus, the discrepancies in the inferred depth rankings in the five species (*Gs. obliquus, Gs. sacculifer, Gs. immaturus, Dg. altispira,* and *Gq. baroemoenensis*), based upon oxygen and carbon isotopic data, can be readily explained by assuming that they hosted high concentrations of algal symbionts that influenced the carbon isotopic values. However, this hypothesis does not explain why there is no apparent carbon isotopic gradient between inferred shallow and deep-dwelling planktonic foraminifera with low $\delta^{13}C$ values.

The total dissolved CO_2 concentrations in surface water depth transects from the modern ocean have been documented using GEOSECS data (Kroopnick, 1985). In general, regions of strong upwelling and high biological productivity produce steep gradients of total dissolved CO_2. In contrast, the oceanic gyres and other areas with highly stratified surface waters have low surface water productivity and weak vertical

Table 14. Inferred depth rankings of planktonic foraminifera based upon agreement between oxygen and carbon isotopic depth rankings.

Justified Oxygen and Carbon Isotopic Depth Rankings		
from ~18.7 to 7.5 Ma		
Shallow	Intermediate	Deep
Gs. obliquus		*Gq. dehiscens*
Gs. sacculifer		*Gr. (M.) limbata*
Gr. (J.) siakensis		*Gr. (M.) menardii*
Gr. (J.) mayeri		*Gq. venezuelana*
Dg. altispira		*Gr. (M.) praemenardii*
Gq. baroemoenensis		
Gs. immaturus		
from ~7.5 to 5.8 Ma		
Shallow	Intermediate	Deep
Gs. obliquus	*Gr. (M.) menardii*	*Gq. venezuelana*
Gs. sacculifer	*Gr. (Gr.) merotumida*	
Dg. altispira		
Gq. baroemoenensis		
Gr. (M.) limbata		
Gs. immaturus		

gradients of total dissolved CO_2 in surface waters (Kroopnick, 1985). Other surface waters exhibit total dissolved CO_2 gradients of intermediate character. The oceanographic processes responsible for producing these varying vertical gradients in dissolved CO_2 were probably no different during the Miocene. Therefore, it is likely that the weak vertical $\delta^{13}C$ gradients we infer for the Miocene resulted from low surface water productivity in this region. The oxygen isotopic data suggests that this was never a region of strong seasonal upwelling during the Miocene from ~18.7 to 5.8 Ma. Indeed, the relatively low biogenic productivity and related increased competition for biotic resources in this region may be the reason why so many of the shallow-dwelling species in these assemblages are inferred to have been hosts to algal symbionts.

The inferred influence of algal symbionts on the carbon isotopic values of the spinose, shallow-dwelling planktonic foraminifera offers a new approach to establishing carbon isotopic depth rankings and an explanation for discrepancies observed between the inferred oxygen and carbon isotopic depth rankings. Although oxygen isotopic rankings reflect vertical thermal gradients and carbon isotopic rankings surface water productivity, the two approaches need not coincide because they record different oceanographic and paleobiological parameters, both of value for better understanding of paleoceanographic change (Table 14).

Conclusions

Studies of the oxygen and carbon isotopic composition of planktonic foraminifera provide a more comprehensive view of surface water conditions when used in combination than when used individually. The two approaches vary independently (oxygen isotopes are a function of temperature and carbon isotopes are a function of surface-water productivity), and therefore provide a check of the assumptions used to assign depth rankings.

The inferred shallow-dwelling, spinose species (*Globigerinoides immaturus, Globigerinoides sacculifer, Globigerinoides obliquus, Dentoglobigerina altispira,* and *Globoquadrina baroemoenensis*) were likely hosts to large concentrations of algal symbionts and preserve $\delta^{13}C$ values that are inferred to have been fractionated through the metabolic process of algal symbiont photosynthesis. There is no indication from shallow-dwelling, non-spinose species or from deeper-dwelling forms that the relatively high $\delta^{13}C$ values exhibited by these five species resulted from high biological productivity associated with enhanced seasonal upwelling. Instead the isotopic data suggests that the western equatorial Pacific was marked by low seasonality, a permanent well-developed thermocline and low surface-water biological productivity throughout the year. The inferred possession of large concentrations of algal symbionts among these five species may in fact have been an adaptation to living in this resource-limited environment.

From ~18.7 to 7.5 Ma, *Globigerinoides obliquus, Globigerinoides sacculifer, Globorotalia (Jenkinsella) siakensis, Globorotalia (Jenkinsella) mayeri, Dentoglobigerina altispira, Globoquadrina baroemoenensis,* and *Globigerinoides immaturus* were shallow-dwelling forms living within the same surficial water mass and indicating either a moderately weak depth stratification from shallow to deeper (as shown) or slight seasonal variations of temperature and species distribution. *Globoquadrina dehiscens, Globorotalia (Menardella) limbata, Globorotalia (Menardella) menardii, Globoquadrina venezuelana,* and *Globorotalia (Menardella) praemenardii* were deep-dwelling forms. During this time interval there were no intermediate-dwelling forms.

From ~7.5 to 5.8 Ma a marked change occurred in the depth distribution of the *Globorotalia (Menardella)* group, and the first intermediate-dwelling forms appeared. Shallow-dwelling forms at this time include *Globigerinoides obliquus, Globigerinoides sacculifer, Dentoglobigerina altispira, Globoquadrina baroemoenensis, Globorotalia (Menardella) limbata,* and *Globigerinoides immaturus.* Intermediate-dwelling forms were *Globorotalia (Menardella) menardii* and *Globorotalia (Globorotalia) merotumida,* while *Globoquadrina venezuelana* was a deep-dwelling form.

Evaluation of the isotopic data through the Miocene indicates that the relative depth rankings for most species did not change. However, between ~9.9 and 7.5 Ma the relative depth rankings of *Globorotalia (Menardella) menardii* and *Globorotalia (Menardella) limbata* changed from deep to intermediate and from deep to shallow, respectively.

Acknowledgements

We thank Dr. H. Spero for much valued advice during this investigation. We also thank Drs. B. Tiffney, S. Savin, and J. Wright for useful criticism of an early draft of this paper. Special thanks are extended to Professor Ryuichi Tsuchi for his enthusiastic and effective leadership of studies of Pacific Neogene stratigraphy through IGCP-246. This research was supported by NSF Grant DPP89-11554 (Division of Polar Programs) and OCE87-13391 (Submarine Geology and Geophysics).

References

Andrews, J.E., Packham, G. et al., 1975. Site 289, in, *Initial Reports of the Deep Sea Drilling Project,* v. 30, Washington, D.C., U.S. Government Printing Office, pp. 231–398.

Barrera, E., Keller, G., and Savin, S.M., 1985. Evolution of the Miocene ocean in the eastern North Pacific as inferred from oxygen and carbon isotope ratios of foraminifera. *Geol. Soc. Am. Memoir,* 163: 83–102.

Barron, J.A. et al., 1985. Synthesis of biostratigraphy; central equatorial Pacific, Deep Sea Drilling Project, Leg 85: Refinement of Oligocene to Quaternary biochronology, in, *Initial Reports of the Deep Sea Drilling Project,* v. 85, Washington D.C., U.S. Government Printing Office, Washington, D.C., pp. 905–934.

Bé, A.W.H., Harrison, S.M., and Lott, L., 1973. *Orbulina universa* d'Orbigny in the Indian Ocean. *Micropaleontology,* 19: 150–192.

Berger, W.H., 1969. Ecologic patterns of living planktonic foraminifera. *Deep-sea Research,* 16: 1–24.

Berger, W.H., Killingley, J.S., and Vincent, E., 1978. Stable isotopes in deep-sea carbonates: Box Core ERDC–92, West Equatorial Pacific. *Oceanologica Acta,* 1(2): 203–216.

Berger, W.H. and Vincent, E., 1986, Deep-sea carbonates: Reading the carbon isotope signal. *Geologische Rundschau,* 75(1): 249–269.

Berger, W.H. and Winterer, E.L., 1973. Plate stratigraphy and the fluctuating carbonate line, *In:* K.J. Hsü, and H.C. Jenkins, (Eds.), *Pelagic sediments on land and under the sea: International Association of Sedimentologists Special Publication,* 1: 11–48.

Berggren, W.A., Aubry, M.P., and Hamilton, N., 1983. Neogene magnetobiostratigraphy of DSDP Site 516 (Rio Grande Rise, South Atlantic), *In:* P.F. Barker, R.L. Carlson et al., *Initial Reports of the Deep Sea Drilling Project,* v. 72, U.S. Government Printing Office, Washington, D.C., pp. 675–713.

Berggren, W.A., Kent, D.V., and Van Couvering, J., 1985. Neogene geochronology and chronostratigraphy. *In:* N.J. Snelling (Ed.), *Geochronology and the Geologic Record,* Geol. Society London, Special Paper.

Bijima, J., Faber, W.W. Jr., and Hemleben, C., 1990. Temperature and salinity limits for the growth and survival of some planktonic foraminifers in laboratory cultures. *Journal of Foraminiferal Research,* 20(2): 95–116.

Bukry, D., 1975. Phytoplankton stratigraphy, southwest Pacific, Deep Sea Drilling Project, Leg 30, *In:* J.E. Andrews, G. Packham, et al., *Initial Reports of the Deep Sea Drilling Project,* Leg 30, U.S. Government Printing Office, Washington, D.C., pp. 539–548.

Bouvier-Soumagnak, Y., Duplessy, J.C., and Bé, A.W.H., 1986. Isotopic composition of a laboratory cultured planktonic foraminifer: Implications for paleotemperature reconstructions. *Oceanologica Acta,* 9: 519–522.

Craig, H., 1957. Isotopic standards for carbon and oxygen and correction factors for mass spectrometric analysis of carbon dioxide. *Geochimica et Cosmochimica Acta,* 12: 133–149.

Curry, W.B. and Matthews, R.K., 1981a. Paleo-Oceanographic utility of oxygen isotopic measurements on planktic foraminifera: Indian Ocean core-top evidence. *Paleo., Paleo., Paleo.,* 33: 173–191.

Curry, W.B. and Matthews, R.K., 1981b. Equilibrium ^{18}O fractionation in small size fraction planktic foraminifera: Evidence from recent Indian Ocean sediments. *Marine Micropaleontology,* 6: 327–337.

Curry, W.B., Thunell, R.C., and Honjo, S., 1983. Seasonal changes in the isotopic composition of planktonic foraminifera collected in Panama Basin sediment traps. *Earth and Planetary Science Letters,* 64: 33–43.

Deuser, W.G., 1987. Seasonal variations in isotopic composition and deep-water fluxes of the tests of perennially abundant planktonic foraminifera of the Sargasso Sea: Results from sediment-trap collections and their paleoceanographic significance. *Journal of Foraminiferal Research,* 17(1): 14–27.

Douglas, R.G. and Savin, S.M., 1978. Oxygen isotopic evidence for the depth stratification of Tertiary and Cretaceous planktic foraminifera. *Marine Micropaleontology,* 3: 175–196.

Elderfield, H., Gieskes, J.M., Baker, P.A., Oldfield, R.K., Hawkesworth, C.J., and Miller, R., 1982. $^{87}Sr/^{86}Sr$ and $^{18}O/^{16}O$ ratios, interstitial water chemistry and diagenesis in deep-sea carbonate sediments of the Ontong Java Plateau. *Geochim. Cosmochim. Acta,* 46: 2259–2268.

Erez, J. and Luz, B., 1983. Experimental paleotemperature equilibrium for planktonic foraminifera. *Geochim. Cosmochim. Acta,* 47: 1025–1031.

Fairbanks, R.G., Svendlove, M., Free, R., Wiebe, P.H., and Bé, A.W.H., 1982. Vertical distribution and isotopic fractionation of living planktonic foraminifera: Seasonal changes in species flux in the Panama Basin. *Nature,* 298: 841–844.

Haq, B.U., Worsley, T.R., Burckle, L.H., Douglas, R.G., Keigwin, L.D., Jr., Opdyke, N.D., Savin, S.M., Somner, M.A. III, Vincent, E., and Woodruff, F., 1980. Late Miocene marine carbon isotope shift and synchroneity of some phytoplanktic biostratigraphic events. *Geology,* 8: 427–431.

Hemleben, C., Spindler, M., and Anderson, O.R., 1989. *Modern Planktonic Foraminifera,* Springer-Verlag, New York, p. 327.

Johnson, D.A. and Nigrini, C.A., 1985. Synchronous and time-transgressive Neogene radiolarian datum levels in the equatorial Indian and Pacific Oceans. *Marine Micropaleontology,* 9: 489–523.

Kahn, M.I. and Williams, D.F., 1981. Oxygen and carbon isotopic composition of living planktonic foraminifera from the northeast Pacific Ocean. *Paleo., Paleo., Paleo.,* 33: 47–69.

Keigwin, L.D. Jr., 1982. Neogene planktonic foraminifers from Deep Sea Drilling Project Sites 502 and 503. *In:* W.L. Prell, J.V. Gardner et al., *Initial Reports of the Deep Sea Drilling Project,* Leg 68, U.S. Government Printing Office, Washington, D.C., pp. 269–288.

Keller, G., 1980. Middle to Late Miocene planktonic foraminiferal datum levels and paleoceanography of the north and southeastern Pacific Ocean. *Marine Micropaleontology,* 5: 249–281.

Keller, G., 1985. Depth stratification of planktonic foraminifers in the Miocene ocean. *Geol. Soc. Am. Memoir,* 163: 177–195.

Keller, G. and Barron, J.A., 1983. Paleoceanographic implications of Miocene deep-sea hiatuses. *Geol. Soc. Am. Bull.,* 94: 590–613.

Kennett, J.P. and Srinivasan, M.S., 1983. *Neogene Planktonic Foraminifera: A Phylogenetic Atlas,* Hutchinson Ross Publishing Company, Stroudsburg, Pennsylvania, 265p.

Kroopnick, P.M., 1985. The distribution of ^{13}C of ΣCO_2 in the world oceans. *Deep-sea Research,* 32(1): 57–84.

Loutit, T.S., Kennett, J.P., and Savin S.M., 1983. Miocene equatorial and southwest Pacific paleoceanography from stable isotope evidence. *Marine Micropaleontology,* 8: 215–233.

Miller, K.G. et al., 1985. Oligocene-Miocene biostratigraphy, magnetostratigraphy and isotope stratigraphy of the western North Atlantic. *Geology,* 13: 257–261.

Poore, R.Z. et al., 1983. Late Cretaceous-Cenozoic magnetostratigraphic and biostratigraphic correlations for the South Atlantic Ocean, Deep Sea Drilling Project Leg 73, *In:* K.J. Hsu, J.L. La Brecque et al., *Initial Reports of the Deep Sea Drilling Project,* Leg 73, U.S. Government Printing Office, Washington, D.C., pp. 645–655.

Reiss, Z. and Hottinger, L., 1984. *The Gulf of Aqaba, Ecological Micropaleontology, Ecological Studies,* v. *50,* Springer-Verlag. Berlin, Heidelberg, New York, Tokyo, 354p.

Ryan, W.B.F. et al., 1974. A paleomagnetic assignment of Neogene stage boundaries and the development of isochronous datum planes between the Mediterranean, the Pacific and Indian Oceans in order to investigate the response of the world ocean to the Mediterranean "salinity crisis". *Riv. Ital. Paleontol.,* 80: 631–688.

Saito, T., 1975. *In:* J.E. Andrews, G. Packham et al., *Initial Reports of the Deep Sea Drilling Project,* Leg 30, U.S. Government Printing Office, Washington, D.C., p. 252.

Sautter, L. and Thunell, R., 1991. Seasonal variability in the $\delta^{18}O$ and $\delta^{13}C$ of planktonic foraminifera from an upwelling environment: Sediment trap results from the San Pedro Basin, Southern California Bight. *Paleoceanography,* 6(3): 307–334.

Savin, S.M., Abel, L., Barrera, E., Hodell, D., Kella, G., Kennett, J.P., Killingley, J., Murphy, M., and Vincent, E., 1985. The evolution of Miocene surface and near-surface marine temperatures: Oxygen isotopic evidence. *Geol. Soc. Am. Memoir,* 163: 49–82.

Sclater. J.G., Meinke, L.M., Bennett, A., and Murphy, C., 1985. The depth of the ocean through the Neogene. *Geol. Soc. Am. Memoir,* 163: 1–19.

Shafik, S., 1975, Nannofossils biostratigraphy of the southwest Pacific, Deep Sea Drilling Project, Leg 30, *In:* J.E. Andrews, G. Packham et al., *Initial Reports of the Deep Sea Drilling Project,* Leg 30. U.S. Government Printing Office, Washington, D.C., pp. 549–598.

Spero, H., in press. Do planktonic foraminifera accurately record shifts in the carbon isotopic composition of sea water ΣCO_2? *Marine Micopaleontology.*

Spero, H.J. and De Niro, M.J., 1987. The influence of symbiont photosynthesis on the $\delta^{18}O$ and $\delta^{13}C$ values of planktonic foraminiferal shell calcite. *Symbiosis,* 4: 213–228.

Spero, H.J. and Williams, D.F., 1988. Extracting environmental information from planktonic foraminiferal del 13C data. *Nature,* 335: 717–719.

Srinivasan, M.S. and Kennett, J.P., 1981a. A review of Neogene planktonic foraminiferal biostratigraphy: Applications in the equatorial and south Pacific. *SEPM Special Publication,* 32: 395–432.

Srinivasan, M.S. and Kennett, J.P., 1981b. Neogene planktonic foraminiferal biostratigraphy and evolution: Equatorial to subantarctic. South Pacific, *Marine Micropaleontology,* 6: 499–533.

Thunell, R.C., Curry, W.B., and Honjo, S., 1983a. Seasonal changes in the isotopic composition of planktonic foraminifera collected in Panama Basin sediment traps. *Earth and Planetary Science Letters,* 64: 33–43.

Thunell, R.C., Curry, W.B., and Honjo, S., 1983b. Seasonal variation in the flux of planktonic foraminifera: Time series sediment trap results from the Panama Basin. *Earth and Planetary Science Letters,* 64: 44–55.

Thunell, R.C. and Reynolds, L.A., 1984. Sedimentation of planktonic foraminifera: Seasonal changes in species flux in the Panama Basin. *Micropaleontology,* 30(3): 243–262.

Williams, D.F., Sommer, M.A. II, and Bender, M.L., 1977. Carbon isotopic compositions of Recent planktonic foraminifera of the Indian Ocean. *Earth and Planetary Science Letters,* 36: 391–403.

Woodruff, F. and Douglas, R.G., 1981. Response of deep-sea benthic foraminifera to Miocene climatic events, DSDP Site 289. *Marine Micropaleontology,* 6: 617–632.

II
TECTONIC EVOLUTION OF THE PACIFIC

Paleomagnetic Evidence of the Deformation of Japan and Its Paleogeography during the Neogene

Kimio Hirooka

Department of Earth Sciences, Faculty of Science,
Toyama University, Toyama 930, Japan

Introduction

Previous paleomagnetic studies of the Japanese Islands concluded that the islands were bent at the central part of the Japanese main island of Honshu sometime between the end of Cretaceous and the early Neogene (Kawai et al., 1961, 1962, 1969, 1971). Recent studies revealed that the bending occurred during the early Miocene (Otofuji and Matsuda, 1983). Moreover, Otofuji et al. (1985a) proposed that the bending was completed within a very short time between 16 and 15 Ma. According to their results, the southwestern part of Honshu Island rotated clockwise while its northwestern part underwent a counterclockwise rotation, and they suggested that this deformation was caused by the opening of the Japan Sea.

If the bending of the island arc really corresponded in time with the opening of the Japan Sea, the pre-bending paleogeography around the Japanese islands would be very much different from that of the present, and a very different paleoenvironment would be expected in the region.

Paleomagnetic Evidence

As Otofuji et al. (1985b) and Hayashida et al. (1991) reported, paleomagnetic results obtained from Honshu Island show a clear contrast in that the declinations obtained from the southwestern part of the island are deflected to the east more than 45° and those of the northeastern part swing to the west.

Paleomagnetic data newly obtained from such as Echigoyuzawa (Hirooka et al., 1990) and Matsushima Bay Group (Yamazaki, 1989) show, however, easterly declinations, so that the northeastern part of the Honshu Island was not a single block, but consisted of many sub-blocks which rotated differently. The deformation of the island is not as simple as previously believed, i.e., deformation expressed by a clockwise rotation of the southwestern part of Honshu around an Euler pole and a counterclockwise rotation of its northeastern part around the another single pole.

Very recently, a paleomagnetic study was carried out at Tsushima Islands which are located to the northwest of Kyushu Island (Torii and Ishikawa, 1988; Ishikawa et al., 1989; Ishikawa and Tagami, 1991). Since the Euler pole for the rotation of southwestern Japan (129°E, 34°N) assumed by Otofuji and Matsuda (1983) locates to

"

INCLINATION
0°
30°
60°
44°N
40°N
36°N
32°N
132°E
136°E
140°E
144°E
17 Ma
(a)
INCLINATION
0°
30°
60°
44°N
40°N
36°N
32°N
132°E
136°E
140°E
144°E
16 Ma
(b)

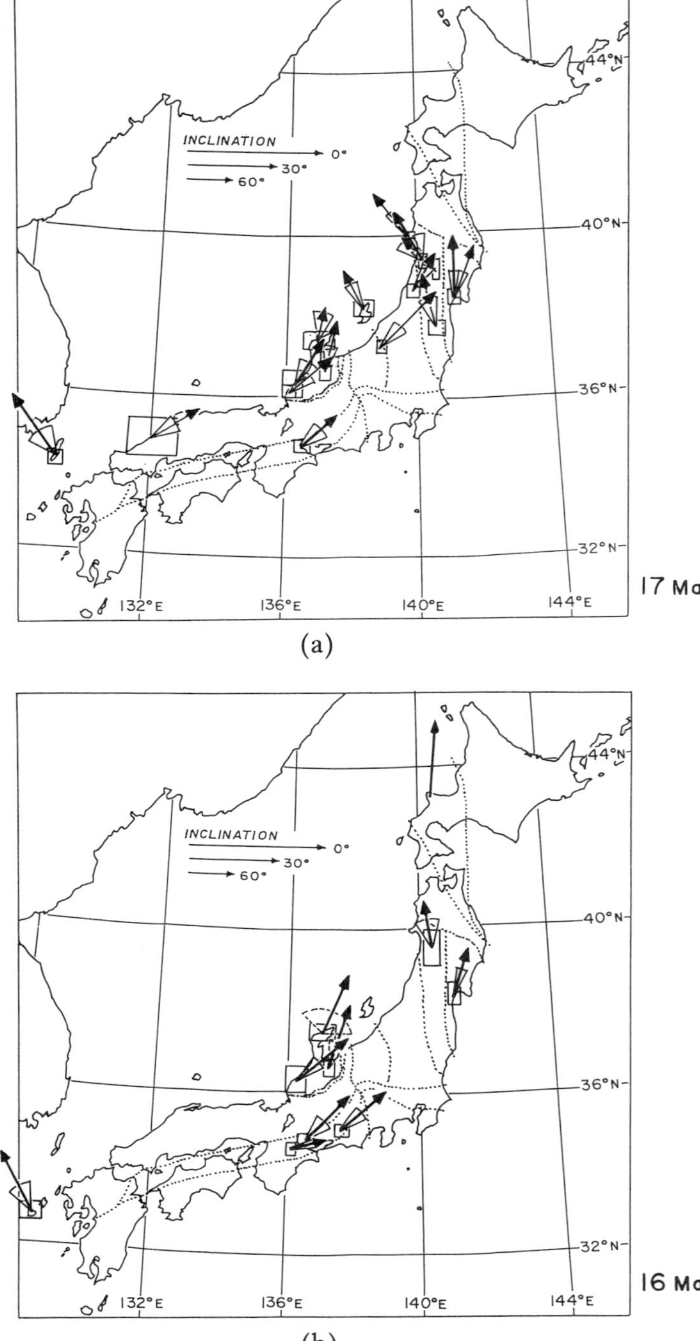

Fig. 1

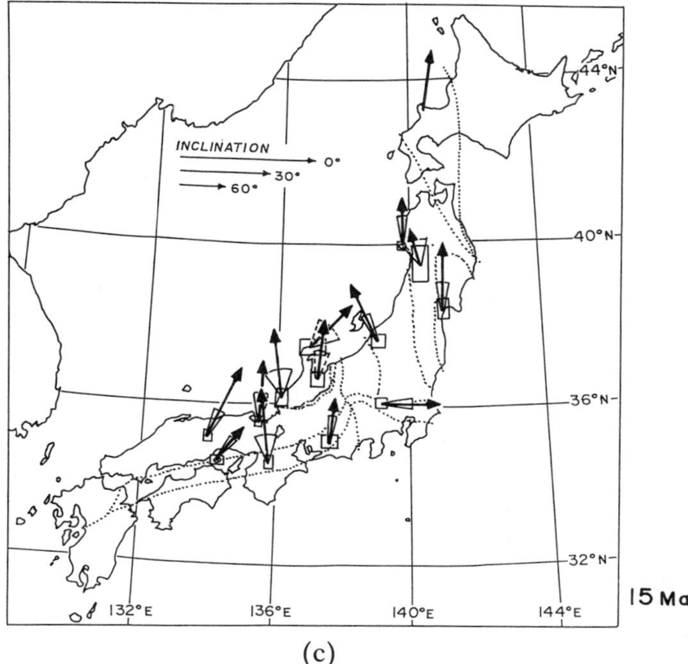

(c)

Fig. 1. Paleomagnetic directions of various localities on the Japanese Islands. The direction of arrows shows the paleomagnetic declination and the length of the arrows indicate the paleomagnetic inclination. (a), (b) and (c) are paleomagnetic directions for 17 Ma, 16 Ma, and 15 Ma, respectively.

southwest of Tsushima, the islands should have rotated clockwise according to their simple rotation model. But the results of paleomagnetism of the islands show that the islands rotated not clockwise, but counterclockwise. The newly obtained paleomagnetic data do not show any rotation in the Shakotan peninsula, western Hokkaido (Tanaka et al., 1991).

As we have very few paleomagnetic results from Hokkaido Island previous to the opening of the Japan Sea, we can draw no clear conclusion about the deformation of the Japanese Islands in the Neogene. The above-mentioned facts indicate that the deformation of the Japanese Islands in Neogene time is not so simple as previously supposed, but has a very complex history. To clarify the process of the deformation of the Japanese Islands, paleomagnetic directions so far obtained in the islands are summarized and divided into three stages of 17 Ma, 16 Ma and 15 Ma. Figure 1 shows paleomagnetic declinations of the stages. It is clearly seen that the easterly declinations are dominant in the southwestern part of Honshu Island at the time of 17 and 16 Ma. In the northeastern Honshu, however, are seen both easterly and westerly declinations at 17 Ma. Moreover, paleomagnetic direction of Tsushima Islands and western Hokkaido indicate that these regions belong to different blocks from the southwestern or northeastern parts of Honshu Island. It is very interesting

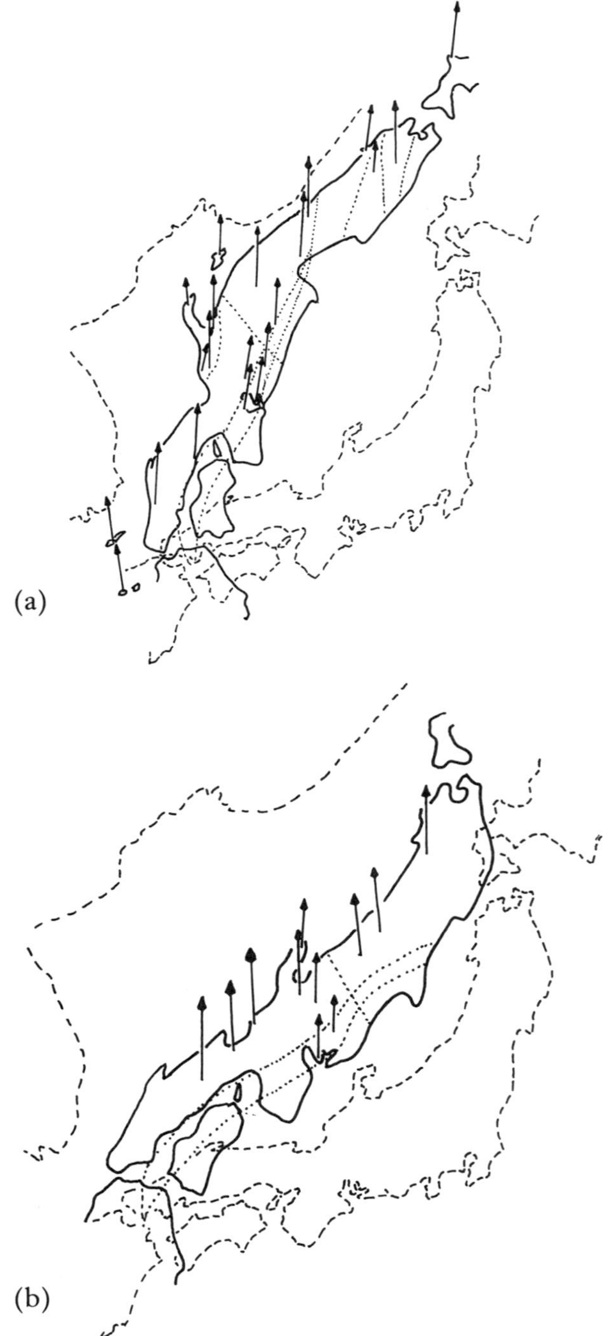

Fig. 2. Paleogeographic maps of Japan at 16 Ma and 15 Ma reconstructed from paleomagnetic directions. a and b are paleogeographic maps at 16 Ma and 15 Ma, respectively. The paleoposition (paleolatitude and paleolongitude) of Japanese Islands relative to the Asian Continent is arbitrary.

that no essential differences are found between paleomagnetic declinations of 17 Ma and those of 16 Ma. This fact suggests that no local relative deformation such as bending or rotation occurred within the Japanese Islands between 17 Ma and 16 Ma.

By arranging these paleomagnetic declinations parallel, the ancient shapes of the Japanese Islands at 16 Ma and 15 Ma are reconstructed as shown in Fig. 2.

It is necessary to obtain more paleomagnetic data from Kyushu and Hokkaido Islands as well as from the central part of Honshu Island where the boundary of southwestern and northeastern Japan is located, in order to establish the precise history of the geotectonism in the Japanese Islands.

References

Hayashida, A., Fukui, T., and Torii, M., 1991. Paleomagnetism of the early Miocene Kani Group in southwest Japan and its implication for the opening of the Japan Sea. *Geophys. Res. Lett.*, 18: 1095–1098.

Hirooka, K., Yamada, R., Yamashita, M., and Takeuchi, A., 1990. Paleomagnetic evidence of the rotation of central Japan and the paleoposition of Japan. *Palaeogeogr., Palaeoclimatol., Palaeoecol.*, 77: 345–354.

Ishikawa, N. and Tagami, T., 1991. Paleomagnetism and fissiontrack geochronology on the Goto and Tsushima Islands in the Tsushima Strait area: implications for the opening model of the Japan Sea. *J. Geomag. Geoelectr.*, 43: 229–253.

Ishikawa, N., Torii, M., and Koga, K., 1989. Paleomagnetic study of the Tsushima Islands, southern margin of the Japan Sea. *J. Geomag. Geoelectr.*, 41: 797–811.

Kawai, N., Ito, H., and Kume, S., 1961. Deformation of the Japanese Islands as inferred from rock magnetism. *Geophys. J. Roy. Astron. Soc.*, 6: 124–129.

Kawai, N., Kume, S., and Ito, H., 1962. Study on the magnetization of the Japanese rocks. *J. Geomag. Geoelectr.*, 13: 150–153.

Kawai, N., Hirooka, K., and Nakajima, T., 1969. Palaeomagnetic and potassium-argon age informations supporting Cretaceous-Tertiary hypothetic bend of the main island Japan. *Palaeogeogr., Palaeoclimatol., Palaeoecol.*, 6: 277–282.

Kawai, N., Nakajima, T., and Hirooka K., 1971. The evolution of the island arc of Japan and the formation of granites in the Circum-Pacific Belt. *J. Geomag. Geoelectr.*, 23: 267–293.

Otofuji, Y. and Matsuda, T., 1983. Paleomagnetic evidence for the clockwise rotation of Southwest Japan. *Earth Planet. Sci. Lett.*, 62: 349–359.

Otofuji, Y., Hayashida, A., and Torii, M., 1985a. When was the Japan Sea opened?: paleomagnetic evidence from Southwest Japan. *In:* N. Nasu, S. Uyeda, I. Kushiro, K. Kobayashi, and H. Kagami, (Eds.), *Formation of Active Ocean Margins*, 551–566, TERRAPUB, Tokyo.

Otofuji, Y., Matsuda, T. and Noda, S., 1985b. Paleomagnetic evidence for the Miocene counter-clockwise rotation of Northeast Japan — rifting process of the Japan Arc. *Earth and Planetary Science Letters*, 75: 265–277.

Tanaka, H., Tsunakawa, H., Yamagishi, H., and Kimura, G., 1991. Paleomagnetism of the Shakotan Peninsula, West Hokkaido, Japan. *J. Geomag. Geoelectr.*, 43: 277–294.

Torii, M. and Ishikawa, N., 1988. Paleomagnetic results from Tsushima Islands, southern margin of the Japan Sea basin (Extended Abstract). *J. Paleontol. Soc. Korea*, 4: 37–43.

Yamazaki, T., 1989. Paleomagnetism of Miocene sedimentary rocks around Matsushima Bay, Northeast Japan and its implication for the time of the rotation of Northeast Japan. *J. Geomag. Geoelectr.*, 41: 533–548.

Tectonic Approach to Changes in Surface-Water Circulation between the Tropical Pacific and Indian Oceans

Susumu NISHIMURA

Department of Geology and Mineralogy, Faculty of Science, Kyoto University, Kyoto 606, Japan

Abstract

Geophysical and geological data from Southeast Asia have been compiled and synthesized in order to constrain models of plate tectonic evolution which caused changes in surface water circulation within the Pacific Ocean and between the tropical Pacific and Indian Ocean.

Through the reconstruction of the Indonesian region back to 25 Ma with respect to the Philippine Sea and Australian-Indian Ocean plates and extrusion of Sundaland, we obtain the following results:

(1) 50 – 17 Ma, Sundaland extruded to the ESE.;

(2) 20 – 17 Ma, Borneo, Celebes Sea and western arm of Sulawesi rotated 50° counterclockwise;

(3) 20 Ma – Present, South China and Sula basins have undergone spreading;

(4) 17 – 15 Ma, Subduction jumped from the westside of the front of Sulawesi Island to present Java trench connecting to Savu Sea;

(5) 17 – 15 Ma, New-Guinea and Sulawesi collided;

(6) 10 – 3 Ma, Collision of the Banda Arc and the Australia continent;

(7) 3 Ma – Present, the Sunda Strait and the north part of Makassar Strait opened;

(8) 3 Ma – Present, the Andaman Sea underwent spreading.

The tectonic event show that the Indonesian Seaway would effectively have been closed around 17 – 16 Ma, preventing and further surface-water circulation between the tropical Pacific and Indonesian Oceans.

Introduction

A major objective of paleoceanography is to better understand patterns of surface-water circulation and the character of the upper part of the water column in ancient oceans. There are two principal approaches: Changes in biogeography and tectonic development between oceans.

The biogeographic approach is divided into two principal methods: determination of regional gradients in the oxygen isotopic compositions of planktonic foraminiferal tests which, in part, reflect changes in temperature and salinity related to paleocirculation; and changes in the distribution and character of planktonic microfossil assemblages. Modern planktonic microfossil group represent sensitive tracers of surface and near-surface water masses (Kennett, 1982).

Quantitative biogeographic maps of planktonic foraminifers in the Indo-Pacific region for three intervals of time during the Miocene have been present by Kennett and et al. (1985).

From these results, important changes occurred in bio-geographic patterns of planktonic foraminifera, especially between the early (16 Ma) and late Miocene (8 Ma), which are interpreted as reflecting major changes in surface-water circulation of the Indo-Pacific region: on the otherhand, paleoreconstructions have been made by some researchers (Hamilton, 1979; Silver et al., 1985; Nishimura and Suparka 1986, 1990) of the Banda Sea region between Indonesian and Australia — New Guinea. These show that the Indonesian seaway would effectively have been closed during the middle Miocene, preventing and further surface-water circulation between the tropical Pacific and Indian Oceans.

Geological and Geophysical Constraints

It must be stressed that the geological and geophysical data currently available from Indonesia and surrounding areas are insufficient to define a unique tectonic model. However, the geological and geophysical data available clearly restrict the possible interpretations.

Before discussing the models, we present an outline of the constrains which we consider to be important. Nishimura and Suparka (1986, 1990) have already presented some of the geological and geophysical constraints around East Indonesia and those data are omitted here.

Miocene Planktonic Foraminiferal Biogeography

Kennett et al. (1985) pointed out that biogeographic patterns have been quantitatively mapped for two time-slices in the early Miocene (22 and 16 Ma) and one in the late Miocene (8 Ma). Important differences are apparent between the early and late Miocene that resulted from changes in surface water circulation within the Pacific Ocean and between the tropical Pacific and Indian Oceans, as shown in Fig. 1.

The faunal changes are interpreted to reflect both the development, during middle Miocene, of the equatorial undercurrent system when the Indonesian seaway effectively closed and the general strengthening of the gyral circulation and the equatorial countercurrent that resulted from increased Antarctic glaciation and high-latitude cooling during the middle Miocene.

The distribution of faunas in the north Pacific indicates that the gyral circulation system was only weakly developed in the early Miocene, but was strong by the late Miocene. In the northwest Pacific, temperate faunas were displaced northward as the Kuroshio current intensified in the late Miocene along with a northward expansion of the polar-subpolar provinces and contraction of the tropical provinces.

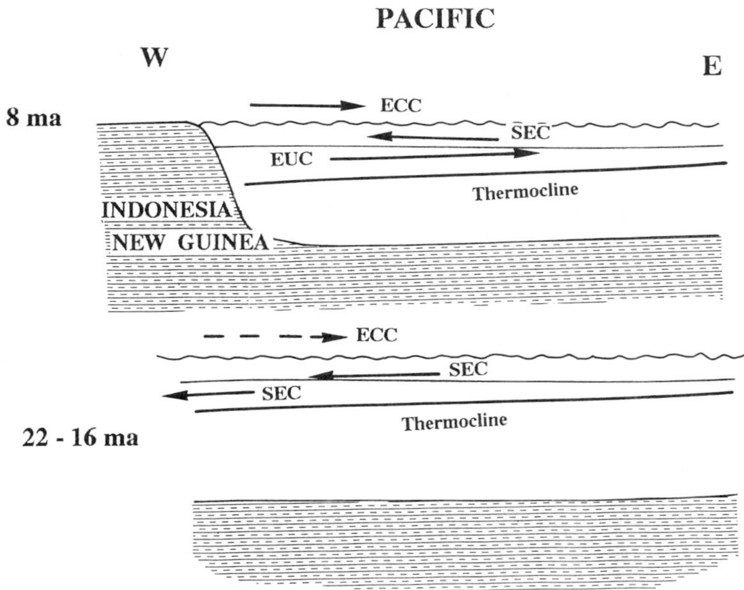

Fig. 1. Suggested Neogene changes in surface counter-mass structure in the equatorial Pacific Ocean between early Miocene (a) and late Miocene (modified from Kennett et al., 1985).
SEC: South Equatorial Current, **ECC:** Equatorial Counter Current, **EUC:** Equatorial Under Current.

Paleomagnetic Data

A significant paleomagnetic data set exists for Southeast Asia that has been reviewed by Sasajima (1984) and summarized by Nishimura and Suparka (1986, 1990). Since that review, a large number of data have been published, both supporting and contradicting data outlined by Sasajima. Hail et al. (1977) and Fuller (pers. com.) presented the paleomagnetic data, with radiometric ages of some localities in western and northern part of Borneo which indicate that Borneo has behaved as a unit since middle Cretaceous time, and has remained at about their present latitude, but has rotated counter-clockwise about 50° since then. Shibuya et al. (1990) reported that the Celebes Sea is trapped old ocean basin and has rotated counter-clockwise about 50° through Eocene and Oligocene but never experienced any latitude migration, and the Sula Sea basin has been sprending through Tertiary Period (Shibuya, 1989). From the data, Borneo (Haile et al., 1977; Mubroto, 1990), the southwestern arm of Sulawesi (Sasajima et al., 1980), and the Celebes sea basin (Shibuya et al., 1990) yield very similar late Mesozoic poles, suggesting little subsequent differential movement between these areas and indicating about 50° of counter-clockwise rotation and only little latitude shift during the Tertiary.

Tectonic Development around Luh-Ulo Area, Central Java

The area is located about 20 km just north of Kebumen, in central Java. It forms part of the south Serayu Mountain Range and is one of the two other localities in

Java, Ciletuh is west Java and Bayat in central Java where pre-Tertiary rocks crop-out.

The area is underlain by a complex Late Mesozoic to Paleocene tectono-stratigraphic terrains in the north and Tertiary sedimentary terrain in the south. The two regions are separated by a fault zone running approximately ENE-WSW.

This area is well known as a test field for training students, and has been deeply studied by many researchers, especially, Sukendar Asikin (LIPI, 1991). The general stratigraphy and relationships between the different formation of the region are shown in Fig. 2 and Table 1. The Luh-Ulo melange represents the subduction product of the Indian-Australian plate below the Southeast Asian Continental Plate during Late Cretaceous-Paleocene time (LIPI, 1991).

The Karangsambung formation is considered to be an olistostromal deposit, laid down directly on the floor of an accreing melange wedge associated with the subducting Indian Ocean Plate. The limestone blocks contain abundant large forams indicating late Eocene age (LIPI, 1991). The tectonic position of the sedimentary basin of Totogan formation is presumably still in the Fore-Arc with local uplifted melange basement as the source area. Faunal assemblages determined from the matrix indicate Late Oligocene age for the formation (LIPI, 1991).

The Watsuranda and Pensogan formations, which mainly consist of volcaniclastic and flysch turbidities, were deposited on top of olistrome basement with uncertain stratigraphic relationship. The Watsuranda formation is dominated by volcaniclastic turbidities and greywackes at the lower and uppermost part (LIPI, 1991). The age of the formation is constrained by the underlying Late Oligocene Totogan formation, and overlying Middle Miocene Penosogan formation, and hence an Early Miocene age has been assigned to the Watsuranda formation. The transition from Watsuranda formation to Penosogan formation is gradual.

The Halang formation is mainly made up of fine-grained tuffs interbeded with marls. Within the formation a number of volcanic breccia members are encountered. The Halang formation is characterized abundant slumping structure, indicating basin movements. The age of this formation is Late Miocene to Pliocene age.

From these occurrences of these formations, the following can be as follows;
(1) The Luh-Ulo Melange include greywacke, argillite, pillow basalt associated with red cherts and pink coloured limestones, dismembered ophiolites and high pressure, low temperature greenschists to bluishists facies metamorphite. This melange is similar to the Tonimbo shists of west Sulawesi and belongs to the same stratigraphic units as the ones of east Sulawesi (Helmers, 1991).

The Luh-Ulo Melange is not similar to the southeast Kalimantan metagabbro, especially in the character of geochemistry.
(2) The complex Late Mesozoic to Paleocene tectono-stratigraphic terrain in the north and the Tertiary sedimentary terrain in the south are separated by a fault zone.
(3) These formations could not traced to their strike directions, and even the uppermost formation, the Halang formation, contains characteristically abundant slumping structures, indicating many large earthquakes occurred during its deposition.

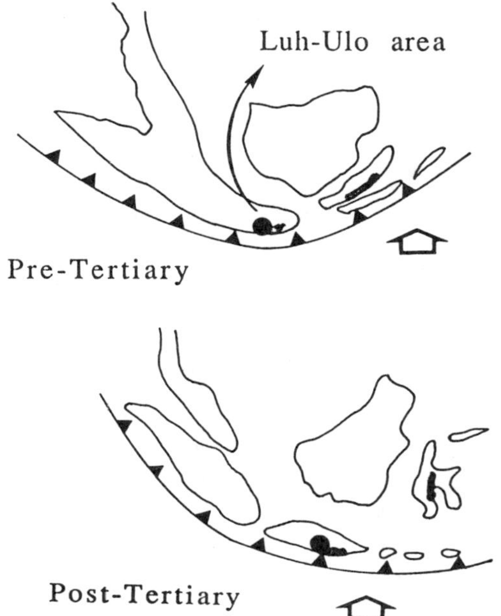

Fig. 2. The change of location of subduction during the Tertiary.
Luh-Ulo area: central Java.

Distribution of the Pre-Mesozoic Continental Basement

The southwest Pacific has the greatest concentration of marginal seas. It appears that many of these marginal seas are related to the breakup of the northern margin of Australia — New-Guinea or the southeast margin of China-Indochina. This breakup would have detached continental blocks, and seafloor spreading behind them would have carried the continental blocks into the South China Sea, Philippine-Indonesian region.

The microcontinental blocks that I have identified are shown on Fig. 3, and brief reasons are given for their identification. The meeting ground of these blocks coming from Australia with those coming earlier from China is in and around the Banda Sea. The east island arcs from middle Java is composed middle to late Miocene sediments. The basement of Celebes Sea is older than middle Miocene and Sula Sea basin has been spreading through Tertiary Period (Shibuya, 1989).

The Borneo is a compressed assemblage of continental blocks, ophiolite belts in southeastern part and intervening compressed and uplifted turbidite sequences.

Movement of the Major Plates

The marked decrease in convergence velocity beginning at 50 Ma is taken as the age of the continental collision between India and Eurasia (Patriat and Achache, 1984). After this collision, the movement of continent fragment has been interpreted to be the result of continued post collisional convergence rather than more fundamental

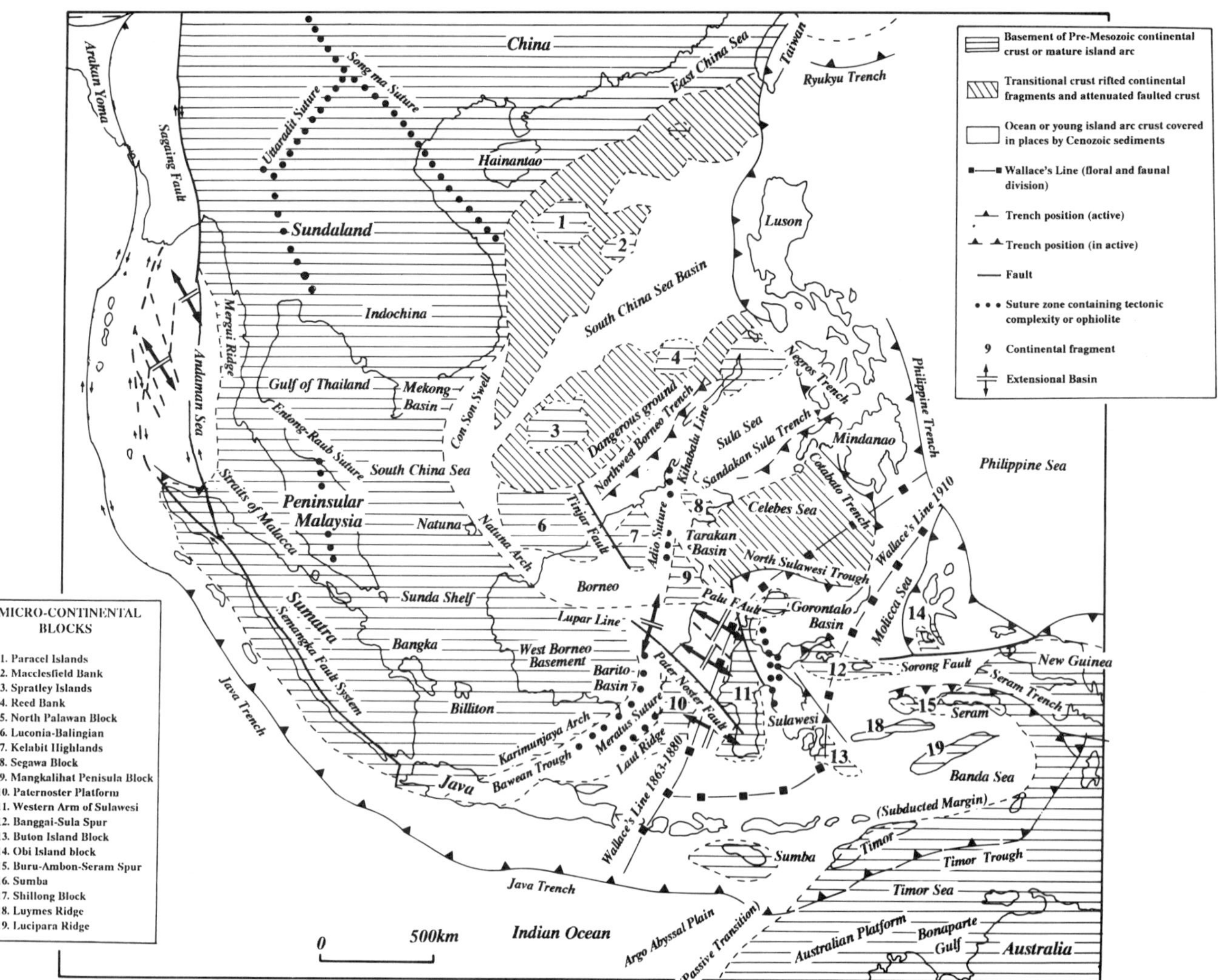

Fig. 3. Microcontinental blocks around Southeast Asia.

changes in global plate dynamics (Tapponnier et al., 1986).

The separation of Australia from Antarctica is estimated to be mid Cretaceous in age (Veevers, 1986), During the early period of break-up the motion of Australia was erratic, and in an absolute reference frame it was largely eastward. By 50 Ma, in the motion of Australia had changed from eastward to northward as rapid ocean-floor spreading episode is coincident with the onset of the continental collision between India and Southeast Asia. The Cenozoic northward motion of Australia has been discussed already by Nishimura and Suparka (1986).

The Philippine Sea Plate, which lies between the larger Eurasian and Pacific Plates, consists entirely of oceanic crust bordered by island arc systems. The evolution of the Philippine Sea plate is well constrained back to about 17 Ma, from magnetic lineations and paleomagnetic data (Seno and Maruyama, 1984). Prior to 17 Ma, the evolution is more speculative.

In the model presented here, the Philippine Sea plate developed as a result of a nearly 90° change in Pacific plate motion at 42 Ma. Prior to this, the oceanic crust that now forms the Philippine Sea plate was part of the Pacific plate. In general terms, the Philippine Sea plate has moved approximately northwest through time, and a large part of it has been subducted beneath the eastern Eurasian margin.

Tectonic development of these formation is considered as follows;

(1) Paleomagnetic data indicate Borneo, southwestern Sulawesi and the Celebes Sea are rotated counter-clockwise about 50° between 20 – 17 Ma.

(2) Sundaland was extruded during 25 – 17 Ma.

(3) Around 17 Ma, the subduction of Indo-Australia Plate jumped from the west side of the front of Sulawesi Island to the present Java trench connecting to Savu Sea.

The effect of the Collision between India and Asia

Tapponnier et al. (1986) present an extrusion model, based on data that suggests the Himalaya now absorb less than half of the total convergence between the Indian and Asian plates. The remainder of the convergence is taken up primarily by strike-slip faulting north part of the Himalaya belt. An echelon right-lateral, strike-slip faults in south Tibet allow the eastward displacement of the Tibetan Plateau with respect to India. The Tertiary geological record in Southeast Asia corroborates a polyphase extrusion model with displacements in excess of 1000–1500 km, in which India has successively pushed Sundland toward the ESE. Most of the Middle Tertiary movements may have occurred along the left-lateral Red River fault in south China, together with the counterclockwise rotation of Kalimantan (Borneo) with the Celebes-Sea floor and the opening of most of the South China Sea. Fast spreading (5 cm/yr) of the South China sea suggests that the Tibetan plateau formed mostly after 20 Ma.

Table 1. General stratigraphic column, Luh-Ulo region (after LIPI, 1991).

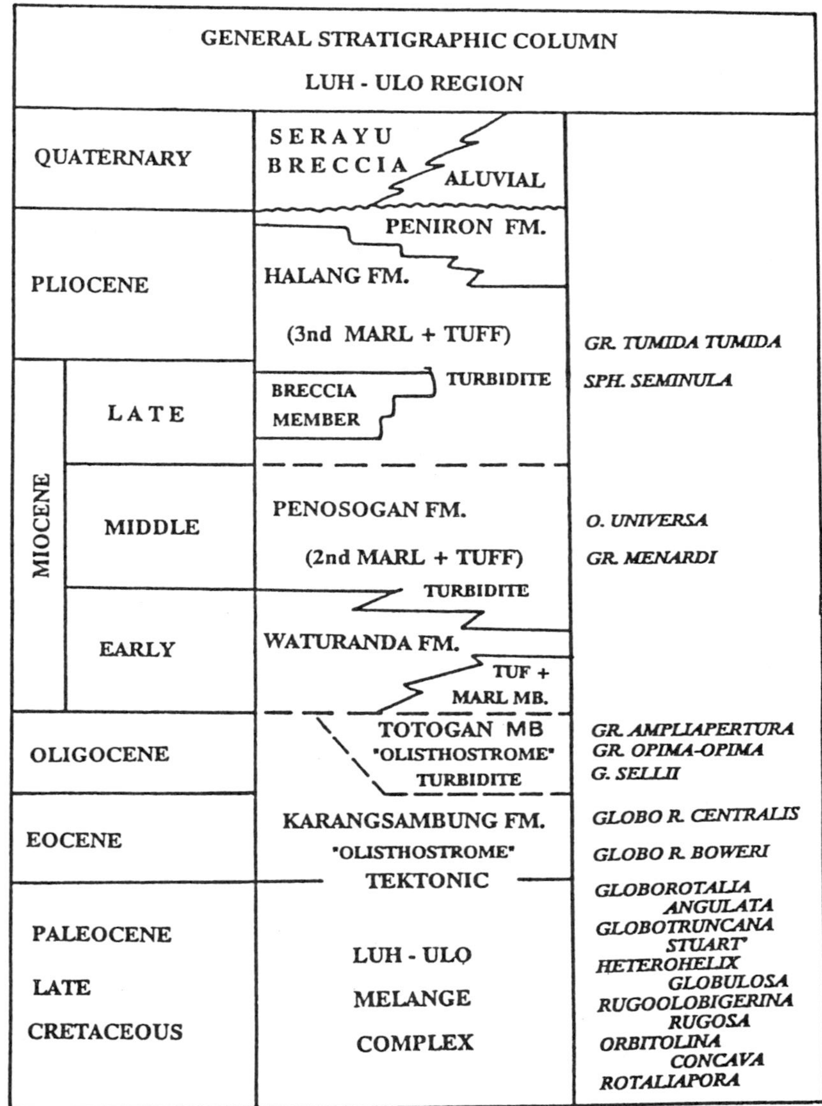

Reconstruction of Indonesian Region at 3, 17 and 25 Ma Age

For the reconstruction of the Indonesian region, we must consider the movement of the adjacent plates, i.e. the Philippine Sea and Australian-Indian Ocean plates and also the extrusion of Sundaland. A tectonic map of tihs region from late Miocene to Recent presented in Fig. 4, and includes the published interpretation on the tectonics of east Indonesia by Nishimura and Suparka (1986), and of Sumatra and Java by

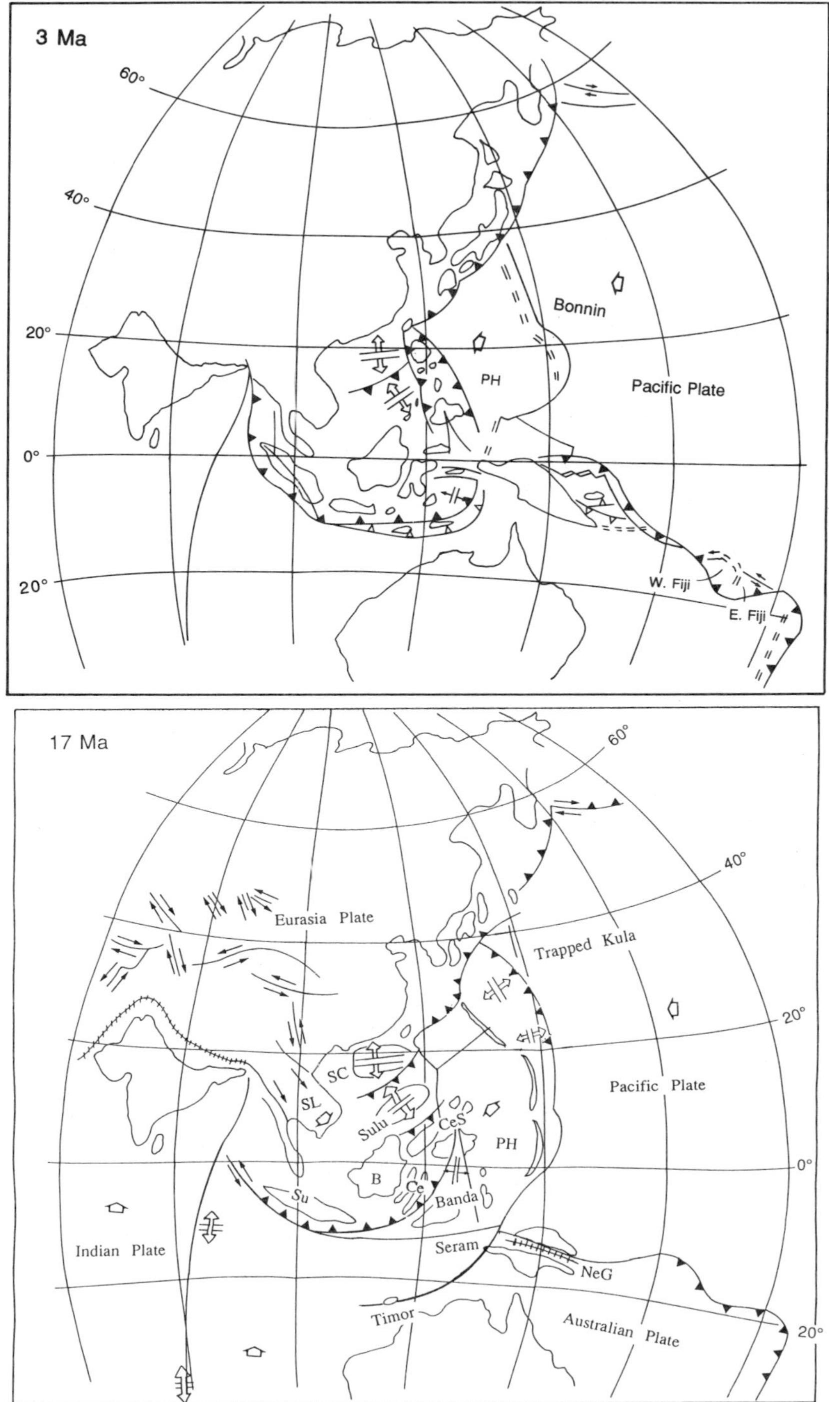

Fig. 4. The reconstruction of the Indonesian region.

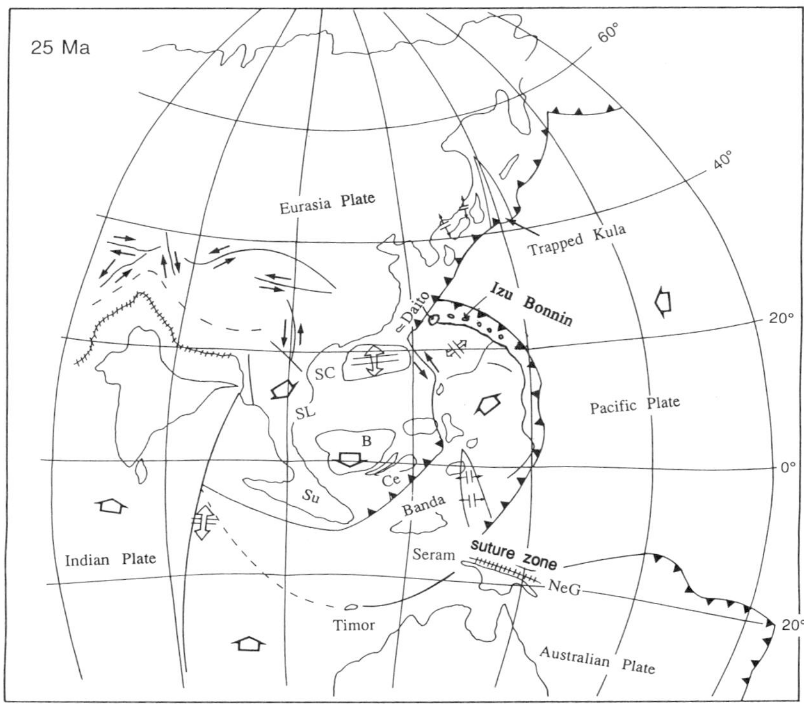

Fig. 4. (continued)

Nishimura and Suparka (1986).

In the reconstruction of the Indonesian region at 3 and 17 Ma ago, we consider that back-arc thinning occurred in the Banda and Sula basins and also a rotation of Sumatra, to fit the tectonic movement of Philippine Sea plate (Seno and Maruyama, 1984), as shown in Fig. 4. Up to 25 Ma, the subduction bounded the outer part of accreationaly microcontinent, which was the Proto Banda ridge, and Malay peninsula, Kalimantan and Celebes Sea underwent 50° counter clockwise rotation. We interpret volcanic arc from at 17 – 25 Ma stretched from west Java to the present northern arm of Sulawesi, as shown in Fig. 4. In this reconstruction, the Philippine Sea plate at 17 and 25 Ma ago (Seno and Maruyama, 1984) and Sundaland extrusion (Tapponnier et al., 1986) is fit into the Indonesian region.

Acknowledgement

This report summarizes part of the results of studies carried out jointly by Kyoto University and the National Institute of Geology and Mining*, Indonesian Institute of Science, on the physical geology of East Indonesia.

This study was also made possible by the assistance of the Center for Geotechnological Research and Development, Indonesian Institute of Sciences.

*Present Name: Center for Geotechnological Research and Development.

References

Fuller, M., 1990. Personal communication

Haile, N.S., McElhiny, M.W., and McDongall, I., 1977. Paleomagnetic data and radiometric ages from the Cretaceous of West Kalimantan (Borneo), and their significance in interpreting regional structure. *J. geol. Soc. Lond.*, 135: 133–144.

Hamilton, W., 1979. Tectonics of the Indonesian reigon. U.S. Geol. Surv., Prof. Paper 1078, 345 p.

Helmers, H., 1991. Sulawesi Bluschists and Subduction along the Sunda continent an alternative view, abstract, Symposium on the Dynamics of Subduction and its Products.

Kennett, J.P., 1982. Marine Geology: Englewood Cliffo, N.J.: Prentice-hall, 813 p.

Kennett, J.P., Keller, G., and Srinivasan, M.S., 1985. Miocene planktonic foraminiferal biogeography and paleoceanographic development of the Indo-Pacific region. *Geol. Soc. Am., Memsin.*, 163: 197–236.

LIPI, 1991. The geology of Luh-Ulo, Fieldtrup Guide Book, Symposium on the Dynamics of Subduction and its Products.

Mubroto, B., 1990. Paleomagnetic data on Sulawesi, Abstract, in Seminar on the first Progress Report of the Cooperation Program in the Field of Geodynamics, Mineral and Energy., Bandung, Indonesia.

Nishimura, S. and Suparka, S., 1986. Tectonic development of East Indonesia. *Jour. Southeast Asia Earth Sci.*, 1: 45–57.

Nishimura, S. and Suparka, S., 1990. Tectonics of East Indonesia. *Tectonophysics*, 181: 257–266.

Patriat, P. and Achache, J., 1984. India-Eurasia collision chronology has implications for critical shortening and driving mechanism of plates. *Nature*, 311: 615–621.

Sasajima, S., Nishimura, S., Hirooka, K., Otofuji, Y., Van Leeuwen, T., and Hehuwat, F., 1980. Paleomagnetic studies combined with Fission track dating of the western arc of Sulawesi. East Indonesia. *Tectonophys.*, 64: 163–172.

Sasajima, S., 1984. A hypothetical growth of the East Asian continent from the view point of Paleomagnetism-with special emphasis on the composite terrain of Southwest Japan. *In:* K. Huzita (Ed.): Tectonic Belts of Asia — between Himalayas and Japan Trench. pp. 239–256, Kaibundo Shuppan, Tokyo.

Seno, T. and Maruyama, S., 1984. Paleogeographic reconstruction and origin of the Philippine Sea. *Tectonophysics*, 102: 55–84.

Shibuya, H., 1989. Leg 124: Basins of the Southeast Asia, ODP News Lett., No. 7, 13–14.

Shibuya, H., Merrill, D., and Hsu, V., 1990. Core Orientation by Secondary Magnetization and the rotation of the Celebes Sea (in preparing).

Silver, E.A., Gill, J.B., Schwartz, D., Prasetyo, H., and Duncan, R.A., 1985. Evidence for a submerged and displaced continental borderland, North Banda Sea, Indonesia. *Geology*, 13: 687–691.

Tapponnier, P., Peltzen, G., and Armijo, R., 1986. On the mechanics of the collision between India and Asia. *In:* M.P. Coward, and A.C. Ries, (Eds.): Collision Tectonics. *Spec. Publ. Geol. Soc. Lond.*, 19: 115–157.

Veevers, J.J., 1986. Break up of Australia and Atractica estimated as mid-Cretaceous (95+25 Ma) from magnetic and seismic data at the continental margin. *Earth Planet. Sci. Letts.*, 77: 91–99.

III
PACIFIC NEOGENE EVENTS IN TIME AND SPACE

North Pacific: Neogene Biotic and Abiotic Events

Yu. B. Gladenkov

Geological Institute of the Russian Academy of Sciences, Moscow

Abstract

A zone transitional from the ocean to the continent was notable for its diverse geological history, with many abiotic and biotic events taking place during the Neogene. Development of these events was characterized by their cyclic character and certain trends. Peculiar to the ancient strata of the zone is an occurrence of facies wedges (transverse and longitudinal). A chronology of these geological events has been compiled on a new stratigraphic basis. Biotic assemblages evolved in the boreal belt and the adjacent subtropical and arctic regions in the changing paleogeographic and climatic environments. Their structure, areal distribution, rate of evolution, percentage of extinct species, and other factors changed as well.

Introduction

A transitional zone from the Pacific Ocean to the Asian continent is a very important area in which to study Cenozoic geological events, because different processes that occurred in the ocean and the continent have manifested themselves well there. The northern sector of the zone includes the sedimentary basins of Sakhalin, Kamchatka, and Chukotka, which developed on the Northeastern Asian margin.

During the last 15 years Russian geologists have been intensively studying this region within the framework of the national program and IGCP Projects 114 and 246. The investigations were carried out by a large number of specialists applying various methods and relying on fossils. This made the conclusions better grounded scientifically. As a result, a number of publications summarizing the data obtained have been issued (Gladenkov et al., 1987, 1988; Menner, 1984). Attention was concentrated on a systematic description of fauna and flora. A comparative analysis of Neogene data for the entire North Pacific area including Japan, the Sakhalin-Kamchatka region, and North America was given in a special publication (Gladenkov, 1988c).

Specific Features of Tectonics and Sedimentation of the Region

The studied territory stretches from Sakhalin in the south to Chukotka in the north, over more than 3 thousand km. Its structure is heterogenous. The southern part is an active margin of the Asian continent, while the northern one is related to a more stable area. Neogene strata occur on the continental crust. Specific geosynclinal formations were accumulated there: molassoids, flyshoids, and turbidites with a considerable share of volcanics. They are characterized by facies diversity, hiatuses, and great thickness (up to 2–3 km). Characteristic of the region is an avalanche of sedimentation (the rate of sedimentation is more than 100–1000 mm/th. years). Strata with a progradational structure are typical.

Facial Wedges Characterizing the Ancient Strata of the Transitional Zone

The paleogeographic setting of this zone was rather unstable. Along the strike of the basins there were continuous movements of the coastal line. When transgression occurred, relatively deep facies were deposited over the continental edge. In the period of regression they were replaced by shallow water facies. The deep-water facies formed wedges with the top indicated to the land, while the shallow-water ones formed wedges with the top to the sea (Gladenkov, 1988a).

These wedges are combined with the wedges of the second type ("climatic") that resulted from movements of warm and cold water masses along the coast. In some cases the two types of wedges coincided. One can see this when comparing sedimentary cycles with climatic optima and minima (Fig. 1).

While analyzing organic remains from the region one should take into consideration the phenomenon of the wedge structure proceeding from a close relation of fossils to certain types of strata.

The Neogene Stratigraphy

Neogene subdivision and correlations in the transitional zone present many difficulties, particularly due to the above-mentioned facies diversity and great thickness of sediments. There are two factors to add. First, the territory under discussion lies in different biogeographic belts (from subtropical to arctic). Second, across the strike of the region several different ecosystems are recognizable: from continental and shallow water (littoral-upper sublittoral) to relatively deep water (lower sublittoral-bathyal). Biotic correlations between these belts and ecosystems require all fossil remains to be studied. They form "a combined orchestra" in which the sounding of "violins" (diatoms) is as necessary as that of "pianos" (benthos) or "contrabass" (spore and pollen). It is their combination that provides the whole "musical composition" (subdivision and correlation): some yield long-distance correlations, others local zonations.

At the present time a new generation of stratigraphic schemes has been compiled for Kamchatka and Sakhalin on this basis, with 8–9 regiostages recognized in each region. Kamchatka: Miocene upper Viventekian, Kuluvian, Iljinian, Kakertian, Etolonian, Ermanovian, Yunyuvayamian; Pliocene-Enemtenian, Ust-Limimtevayamian, Tusatuvayamian. Each regiostage corresponds to a certain stage in the development of the ancient basins and their organic world and is characterized by peculiar faunal assemblages (benthonic, first of all). Paleontological descriptions of

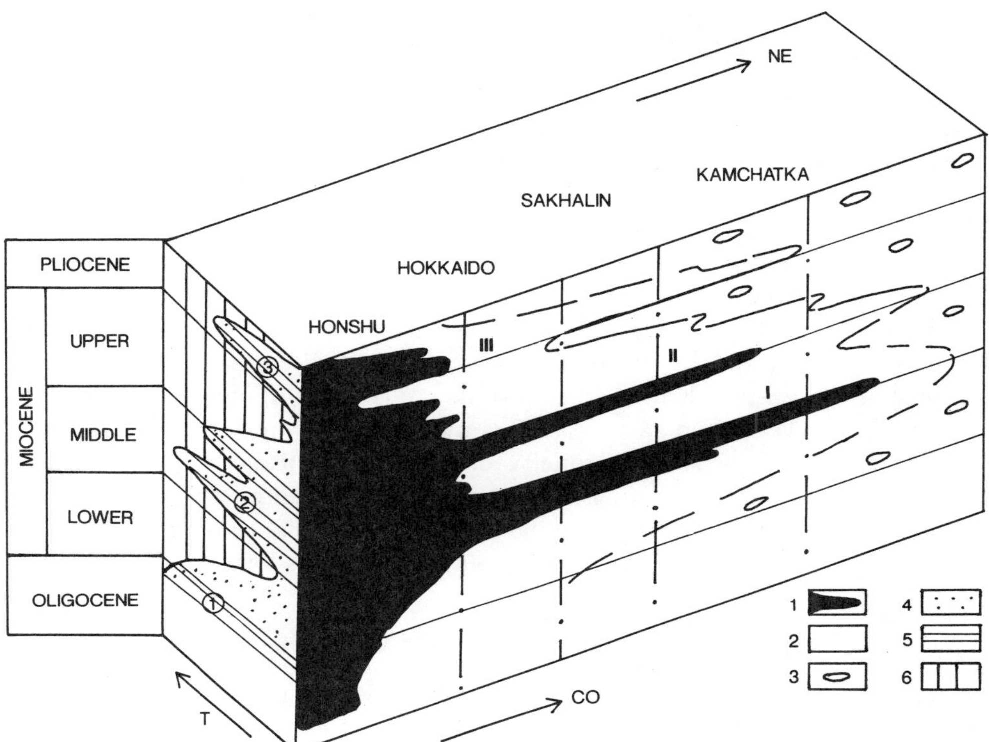

Fig. 1. Facies wedges in the Neogene strata of the ocean-to-continent transitional zone in the Far East region. 1–3. Fossil molluscan assemblages, 1. Tropical and subtropical, 2. South boreal, 3. North boreal, 4. Relatively shallow water marine formations, 5. Relatively deep-water formations, 6. Continental formations, I–III. Major climatic optima, 1–3. (Figures in circle) Major marine transgressions, T. Direction of transgressions from the Pacific to Asia, CO. Migration of warm water assemblages from southern latitudes to boreal regions during climatic optima.

the regiostages have already been published (Menner, 1984).

Correlation of the regiostages to the subdivisions of the International Standard Time Scale has been outlined reliably by means of diatoms, foraminifers, and mollusks. The correlation is based on zonal stratigraphy (first of all, using diatoms). There are 7–15 biotic zones recognizable in the Neogene strata now (Oreshkina, 1980).

Some progress was made in refining the regional schemes. In Kamchatka some smaller units — members and beds with fauna (about 20–25) — were distinguished. Many of them stretch over 700–800 km and have high affinity coefficients (65–95%). Some are treaceable in Sakhalin (Gladenkov and Sinelnikova, 1990).

The time range of these units in Eastern Kamchatka is 0.12–0.35 Ma (members) and 0.03–0.09 Ma (beds). The age was estimated by means of mollusks and foraminifers with application of paleomagnetic and fission track methods. Their duration corresponds to that of the Quaternary units.

Chronology of Geological Events, Their Correlations and Causal Relations

A chronology of Neogene geological events has been compiled on the basis of stratigraphic occurrences (Gladenkov and Shantser, 1990). We recorded as many as ten types of molluscan fauna, three significant climatic warmings (two in the Miocene, one in the Pliocene), and three intensive migrations of warm-water assemblages from south to north combined with some biotic migrations in the other direction. In addition, three cycles of volcanic activity took place there. Sedimentation processes are manifested in three sedimentation cycles and seismocomplexes which were accompanied by three epochs of hiatus, washout, and coalification. Other events occurred at the same time: four significant tectonic events (fault and block movements), appearance of the Eastern Kamchatka volcanic arc (in early Pliocene), destruction of the Bering Bridge, etc. You can see in this model how complex the geology of this zone was. We can suggest synchroneity of some geological events (for example, warmings and regressions, opening of the Bering Strait, and biotic migrations). It is often difficult to determine their regional or subglobal scope. This made us analyze comparative data on the adjacent regions and the ocean as well.

Characteristics of Biotic Evolution in the Shelf Zone

Up to recent time the peculiarities of the biotic evolution in the transitional zone were not an object of thorough study. We only noted changes in faunal and floral assemblages in the various stratigraphic sections. Meanwhile, it is very important to establish some evolutionary trends for better understanding the Neogene biotic events in the entire Pacific area. This can be illustrated by means of mollusks.

Composition and Distribution of Molluscan Assemblages

A significant part of the oceanic biota (157 of 160 thousand marine species) is concentrated in the shelf zone. Not without reason, this zone is often called the "warehouse" of the ocean. Mollusks were one of the dominant fossils of the Cenozoic; G. Thorson called them "index-fossils of our time". They compose the core of all shelf biocoenoses. Their good preservation in ancient sediments allows reconstruction of paleocoenoses of many different types.

Distribution of mollusks as well as other benthonic faunas is divided into some climatic and vertical (sea) zones and provinces. Climatic zonation is reflected by benthonic assemblages first of all. The difference between summer and winter temperatures reaches 7°C in the northern part of the region and 20°C or more in the southern part. Some biogeographic associations were formed there: tropical-subtropical (by origin), lower boreal, boreal, upper boreal, boreal-arctic and arctic. Table 1 shows abundances of Bivalvia spp. in these associations (Skarlato, 1981). Table 2 shows the distribution of species over the subareas and provinces. Two large subareas were distinguished there: lower boreal Northern Japanese and high boreal Beringian;

Table 1. Bivalvia species abundances in the Far East shelf zone in latitudes 43°–66°N (Skarlato, 1981).

Tropical-subtropical	3
Subtropical	45
Low boreal	50
Boreal	50
High boreal	46
Boreal-arctic	41
Arctic	5

Table 2. Bivalvia species abundances in various biogeographic regions (Skarlato, 1981).

Subarea: Northern Japanese low boreal	188
Province: Northern Japanese	185
Districts: Southern Primorie	135
Southern Kurils	149
Northern Primorie	114
Southern Sakhalin	117
Subarea: Bering Sea high boreal	150
Provinces: Sea of Okhotsk	110
Northern Kurils	89
Komandor Islands	68
Districts: Anadyr	43
Eastern Chukotka	68
Percentage of endemic species: in provinces:	20–50%
in subprovinces:	10–25%

each of them includes several provinces (with subprovinces and districts). Percentage of endemic species in assemblages was one of the factors used to distinguish them (20–50% for provinces and 10–25% for subprovinces).

Vertical zonation of the basin bottom also affected the benthonic distribution. Littoral, upper and lower sublittoral, and bathyal molluscan assemblages can be found (200–500 m in depth). Their bathymetric location is determined by temperature and dynamics of the water column, character of suspended sediments and substrates, trophic connections, and other factors. In upper horizons of the sea immobile, slow, and mobile suspension feeders are predominant while in lower horizons deposit-feeders and collecting detritus-feeders are dominant.

A general trend can be noted: the systematic diversity and biomass of mollusks decreases with depth. The greatest biomass is recorded in the range 0–50 m (with slight increase at 100–200 m). In addition, a change in ecological faunal groups takes place: immobile forms of the epifauna that predominated near the coasts are replaced first by mobile benthos and at 500 m and deeper by the infauna. The sublittoral assemblages are more changeable than relatively deep water ones, which had, as a rule, a wide distribution. Noteworthy is the predominance of some molluscan species in the biocoenoses. They comprises 80% of the whole biomass although their species percentage is small — 14%. The core of the biocoenoses consists of species of different genera and even families. This is a result of the incompatibility of species belonging to one genus and having similar modes of life.

Depending on latitudinal and vertical zonation, different biocoenoses occur on the shelf bottom. They replace each other along the coastal line and across the strike (in depth). It is noteworthy that on the steep shelf they had narrow linear or striped patterns and on the sloping shelf a mosaic one. The biocoenoses reflect a stable balance of different species in the biotopes. Even in changeable environments the assemblages tend to keep their structure. This phenomenon is called "homeostasis of associations".

Stages and Rates of Evolution

Rates of benthonic evolution on a species level are not much lower than these of other faunas. Changes in the assemblages by which the regiostages were distinguished occurred each 2.5 Ma on average. Bed-by-bed analysis combined with the quantitative methods suggests shorter evolutional steps — a maximum of 0.5 Ma.

Analysis of some molluscan genera shows phases of their flourishing and extinction (for example, *Yoldia* flourishing in the Oligocene). No large changes have been recorded in the Neogene molluscan genera: the succession of assemblages is similar from one stage to another.

Taxa of higher rank seem to have originated in the warm-water areas. Centers of origin of many species were apparently located on the boundary of tropical/boreal belts, along junctions of warm and cold water masses. New species appeared as a result of gradual changes. This process was apparently accelerated by warmings. The intensity or rate of transformation of thermotropic forms into cold-water ones is not clear now. But periods of rapid appearance of new species correspond in some cases to regressions or beginnings of transgressions (early Middle and Late Miocene, Early Pliocene).

Attention is attracted by parallelism (the homologous rows) in the evolution of species of many genera. Such homeomorphic faunas are very typical (*Yoldia, Nuculana*, etc.). Of interest also are short-term population bursts (waves) manifested by the burial of forms at particular age levels over enormous areas.

The above data show that the core of the recent boreal fauna started to form in the late Miocene and finished in the late Pliocene after the warming at 3.5–4 Ma. By this time a number of now-extinct forms had been reduced by several percent, and assemblage structure was not changed.

Change in Percentage of Extinct Species

This change can be considered a reflection of evolution and its linear character. During the Neogene in Kamchatka the percentage of now-living species increased: 10 to 40–50% in the Miocene, 60–65% to 96% in the Pliocene. This parameter has changed in the same way in the North Atlantic Neogene. It grew from 60% to 96–98% in the Pliocene of England and Iceland (Gladenkov, 1981).

Paleoclimatic and Paleogeographic Environments
and Benthonic Migrations

Biotic evolution in the Pacific went on in the background of various subglobal geological phenomena: climatic oscillations and eustatic changes in sea level first of all. In the Neogene a general climatic deterioration occurred with three significant warmings (in early Middle and Late Miocene and in Middle Pliocene). Duration of the temperature optima is estimated at 0.5–1 Ma, and that of the minima at 3.5 Ma and more. The optima are marked by mass occurrence of warm-water subtropical molluscan species in the boreal regions (about 60% in early Middle Miocene in Kamchatka) and by thermophilic flora (33% of *Fagus*) (Fig. 2). The warm-water assemblages migrated over 10–15° south-north (2,000 km and more). The most significant was the warming in early Middle Miocene; less significant was the warming in late Middle Miocene, and the least important was the warming in the Pliocene

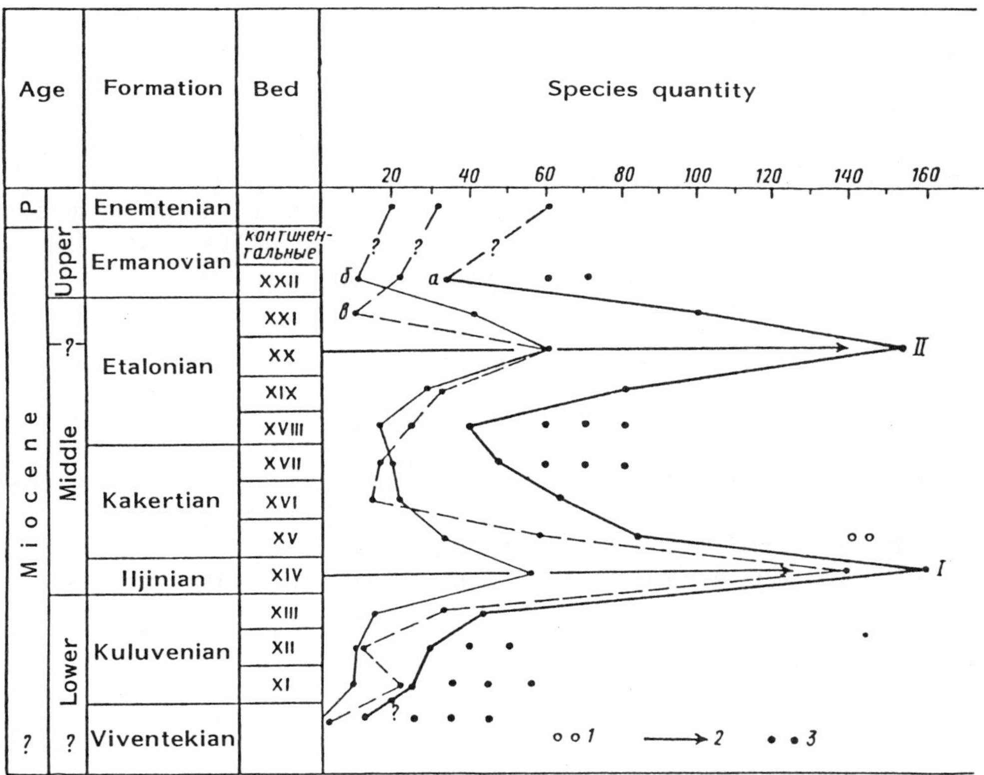

Fig. 2. Distribution of molluscan assemblages in the Miocene beds of western Kamchatka and their numerical characterstics.

I–II. Major climatic optima; Number of species: (a) Total, (b) Subtropical and south boreal, (c) Newly appearing in section; 1. Northward migrations of subtropical and south boreal assemblages, 2. Floating pebbles (possibly ice-rafted), 3. Beds with *Ammonia* and Fagaceae.

(Gladenkov, 1988b; Gladenkov and Sinelnikova, 1990). Sometimes periods of warming coincided with eustatic rises of the sea level.

Formation or destruction of physical barriers has also greatly influenced biocoenoses patterns. A clear illustration is an opening of the Bering Strait in the Middle Pliocene that resulted in the Pacific and arctic biotic exchange. Some Pacific migrants also appeared in the North Atlantic (they comprise about 20–25% of the *Serripes* zone in Iceland — Red Crag in England). Some species (*Cyrtodaria, Astarte, Portlandia*) of Atlantic or arctic origin appeared in Kamchatka. During the periods of low sea level (lasting 0.2–0.3 Ma) continental faunas migrated along the Beringian Bridge, and Asiatic species of mastodons, bears, deer, rodents, and others appeared in North America (Gladenkove et al., 1991).

Benthonic Migrations

We noted above the influence of climatic oscillations on the migrations of benthonic assemblages (their movements reached 2,000 km and more). Sea currents have affected the migrations as well, but one more factor still remains underestimated, and this is a biotic one. In some cases it has caused mass migrations, and, as a result, many ecological niches have been inhabited by individual species. In the Pliocene, when the Bering Strait was open, North American and arcto-atlantic faunal exchange was rather unilateral: the number of North Pacific species that migrated into the North Atlantic was eight times greater than the number of species that migrated the other way. It is noteworthy that the migrations were instrantaneous from a geological point of view. Some changes in molluscan distribution in the North Asian and North American shelves also can be explained by such biotic causes.

Conclusion

The North Pacific area had a complex geological history in the Neogene. The biota evolved under the influence of changeable paleoclimatic and paleogeographic events of regional and subregional scale. Sedimentary and volcanic formations were affected by active tectonic movements and sedimentation processes. Study of the abiotic and biotic events that occurred in the region and their cyclicity, stages, and in some cases synchroneity may help in deciphering the same trends in other regions of the Earth. Trends in biotic evolution allow us to infer a reliable prognosis for the future.

References

Gladenkov, Y.B., 1981. Marine Plio-Pleistocene of Iceland and problems of its correlation. *Quaternary Research*, 15: 18–23.

Gladenkov, Y.B., Bratseva, G.M., and Sinelnikova, V.N., 1987. Marine Cenozoic of the Bay of Korf in the Eastern Kamchatka. — Review of geology of the Northwestern Pacific tectonic belt. Moscow, "Nauka", pp. 5–73 (in Russian).

Gladenkov, Y.B., 1988a. Pacific wedges — a characteristic of the Far East Cenozoic strata (a transitional ocean-continental zone). — "Pacific Neogene Events", Abst. Vol. Oji International Seminar for IGCP-246. Shizuoka, p. 51.

Gladenkov, Y.B., 1988b. The North Pacific Holarctic Neogene from an ecostratigraphic point of view. Saito Ho-on Kai spec. publ. (Prof. T. Kotaka Commem. vol.), pp. 33–45.

Gladenkov, Y.B., 1988c. Marine Neogene stratigraphy of the Northern Pacific belt. Moscow, "Nauka", 212 p. (Trans. GIN, vol.428) (in Russian).

Gladenkov, Y.B., Sinelnikova, V.N., and Titova, L.V., 1988. Stages in faunal evolution in the Noeogen shelf bassins of Kamchatka (exemplifie by Buccinidae). — Mesozoic and Cenozoic lithology and stratigraphy of the Eastern USSR. Moscow, "Nauka", pp. 38–134 (in Russian).

Gladenkov, Y.B. and Shantser, A.E., 1990. Neogene of Kamchatka: stratigraphy and correlation of geological Events. — Pacific Neogene Events. Univ. Tokyo Press, pp. 173–182.

Gladenkov, Y.B. and Sinelnikova, V.N., 1990. Miocene mollusks and climatic optima in Kamchatka. Moscow, "Nauka", 174 p. (Trans. GIN, vol.453).

Gladenkov, Y.B., Barinov, K.B., Basilian, A.E., and Cronin, T.M., 1991. Stratigraphy and Paleoceanography of Pliocene deposits of Karaginski island, Eastern Kamchatka, USSR. *Quaternary Science Reviews*, 10: 239–245.

Menner, V.V. (ed.), 1984. Atlas of fauna and flora of the Far East Paleogene and Neogene deposits (the Tochilian key section). Moscow, "Nauka", 332 p. (Trans. GIN, vol.385) (in Russian).

Oreshkina, T.V., 1980. Diatom assemblages from the marine Neogene deposits of Karaginsky Island (Eastern Kamchatka) and their stratigraphic significance. Izvestija AN SSSR, ser. geol. N 11, pp. 57–66.

Skarlato, O.A., 1981. Bivalvia of the Western Pacific boreal regions. Leningrad, "Nauka", 479 p.

Evolution of the Planktic Foraminiferal Clade *Globoconella* during the Late Neogene: Paleoceanographic Modulation

Kuo-Yen WEI

Department of Geology and Geophysics, Yale University, New Haven, CT 06511-8130, U.S.A.

Abstract

The *Globorotalia (Globoconella) conomiozea-inflata* clade has been a dominant planktic foraminiferal group in the temperate and cool subtropical southwest Pacific during the late Neogene. The evolutionary responses of the *Globoconella* species to the changing paleoceanographic conditions during the last 7 million years have been investigated using morphometric, stable isotopic, and faunal data collected from a suite of DSDP sites (206, 207A, 284, 588, 590A, and 593) along a meridional transect in this area.

A chronicle of the paleoceanographic control of the evolution of the *Globoconella* clade is presented:

The intensification of the Tasman Front at the Miocene/Pliocene boundary (5.05 Ma) isolated the northern peripheral demes of *Globoconella* in the cool subtropics from the main populations in the temperate areas and triggered the rapid speciation of *G. pliozea* in the cool subtropics at 5.05 Ma.

The southward migration of the Tasman Front at about 4.4 Ma allows the dispersal of the main stock species, *G. puncticulata*, into the northern sites (DSDP 208 and 588). The biotic competition between two coexisting sister species, *G. pliozea* and *G. puncticulata*, may have caused *G. pliozea* to become extinct at 3.97 Ma. The extinction of *G. pliozea* seems to be a biotic rather than a paleoceanographic event.

The origination of *G. inflata* at about 3.50 Ma is characterized by a gigantism trend as it branched off the *G. puncticulata* lineage. Further divergence between these two lineages is accomplished by a delay (post-displacement) of the ontogenetic development in *G. inflata*. The enhancement of the Antarctic Intermediate Water mass (AAIW) during this period might have created new niches and driven the evolution of *G. inflata*. The extinction of *G. puncticulata* at 2.35 Ma is coincident with the onset of significant glaciation in the Northern Hemisphere. This middle Pliocene event also caused a change in heterochronic mode in the *G. inflata* lineage from a post-displacement trend to a pre-displacement trend. The high-frequency glacial-interglacial oscillations during the late Pleistocene appear to have stabilized the allometric trajectories in *G. inflata* during the periods from 0.9 to 0.26 Ma.

Introduction

Several studies have demonstrated a close relationship between the evolutionary changes of planktic foraminifera and major paleoceanographic events during the last 100 million years (Wei and Kennett, 1986; Corfield and Shackleton, 1988; Leckie, 1989). The recognition of such a coincidence between biotic evolution and paleoceanographic changes has been mostly based upon large-scale evolutionary patterns documented in terms of taxonomic analysis (e.g. Hoffman and Kitchell, 1984; Wei and Kennett, 1986; Corfield and Shackleton, 1988). However, only a few studies have investigated the processes of evolutionary changes in relation to oceanographic changes (e.g. Malmgren and Berggren, 1987; Wei and Kennett, 1988). To explore the cause-effect relationships between evolutionary and environmental changes, it is essential to gather and analyze morphological, genealogical, and paleoenvironmental data. This contribution presents a summary of morphometric studies on the evolutionary patterns in the late Neogene planktic foraminiferal clade, *Globorotalia (Globoconella) conomiozea-puncticulata-inflata*, during the last 7 million years in the southwest Pacific, and discusses possible causal effects of paleoceanographic events.

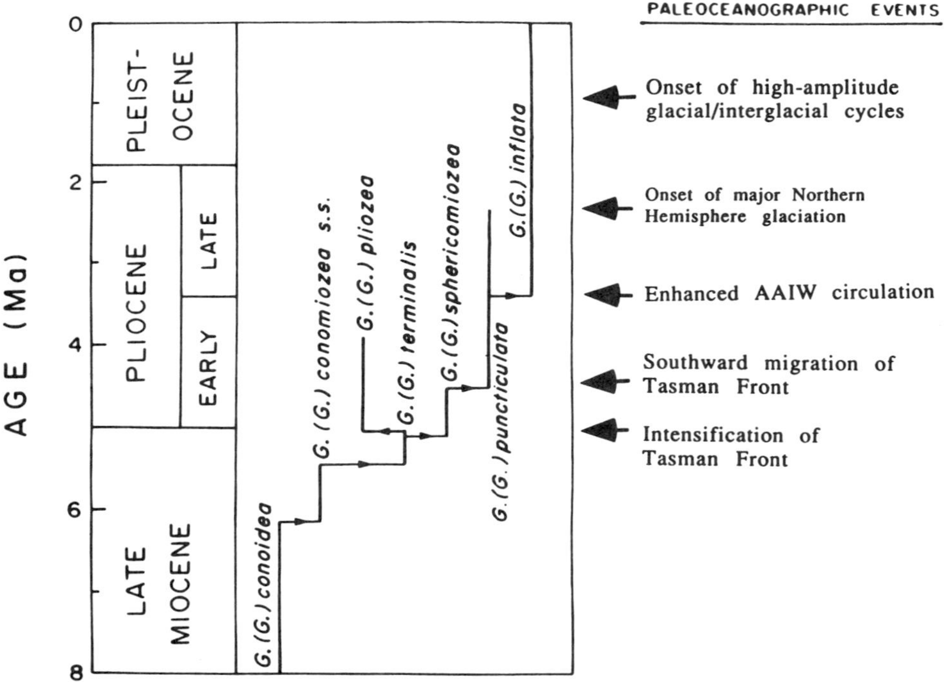

Fig. 1. Phylogenetic tree of the *Globorotalia (Globoconella)* clade and timing of the late Neogene paleoceanographic events.

Background

Globoconella clade

The *Globorotalia (Globoconella) conomiozea-puncticulata-inflata* bioseries (hereafter referred as *Globoconella* clade) has been a dominant member of planktic foraminiferal assemblages in the temperate and cool subtropical oceans during the late Neogene. The extant species, *G. inflata*, occurs in middle latitudes from 10° to 40° in both hemispheres (Bé, 1977). The presence of warm pool water and the warm western boundary current in the western equatorial Pacific has limited the northern distribution extent of this species in the studied region at about 20°S (Eade, 1973). The succession of members in this clade has been well documented throughout the temperate region of the southwest Pacific (e.g. Kennett, 1966, 1973; Scott, 1979, 1982; Hornibrook, 1981, 1982). Opinions on the phylogeny of this group vary among workers (Maiya et al., 1976; Kennett and Srinivasan, 1983; Cifelli and Scott, 1986; Hornibrook et al., 1989; Hoskins, 1991). Figure 1 represents a phylogenetic scheme of the *Globoconella* clade as summarized from a series of recent studies (Wei, 1987a,b; Wei and Kennett, 1988; Wei, in press.).

Paleobiogeographic Setting

Materials used for this investigation are from six Deep Sea Drilling Project (DSDP) sites located along a meridional transect from 26°S to 40°S in the southwest Pacific (Fig. 2). The circulation pattern in the region is dominated by the eastward moving subtropical water derived from the East Australian Current and Tasman Current (Eade, 1973; Heath, 1985), except that the northern site, DSDP 588, located at 26°S, is under the influence of the Trade Wind Drift during the summer months (Stanton, 1969). The conjoining of waters from the Tasman Current and the East Australian Current forms an eastward zonal jet, known as the Tasman Front (Stanton, 1979; Andrews et al., 1980; Heath, 1985). The Tasman Front takes a circuitous route, meandering between latitudes 31°S and 35°S, to cross over the northern part of the Tasman Sea (Mulhearn, 1987). It is believed that the front causes upwelling when it encounters the bathymetrically shallower topography of the Lord Howe Rise and the Norfolk Ridge (where the studied DSDP cores were drilled; see Fig. 2) (Heath, 1985).

Geographic distribution of living planktic foraminifera in this region is closely related to the surface circulation pattern (Eade, 1973). The Tasman Front separates the Subtropical fauna from the Transitional fauna (Bé and Toderlund, 1971; Eade, 1973). Relative abundance of the *Globoconella* in the studied sites indicates that the Tasman Sea has been the central biogeographic province for the *Globoconella*, whereas the subtropical areas to the north of the Tasman Front have remained a peripheral biogeographic province (Wei and Kennett, 1988).

With the belief that the DSDP sites located in the same water mass should contain relatively similar faunas compared to those situated in different water masses, the difference in foraminiferal faunal compositions among three sites, DSDP 284, 590A,

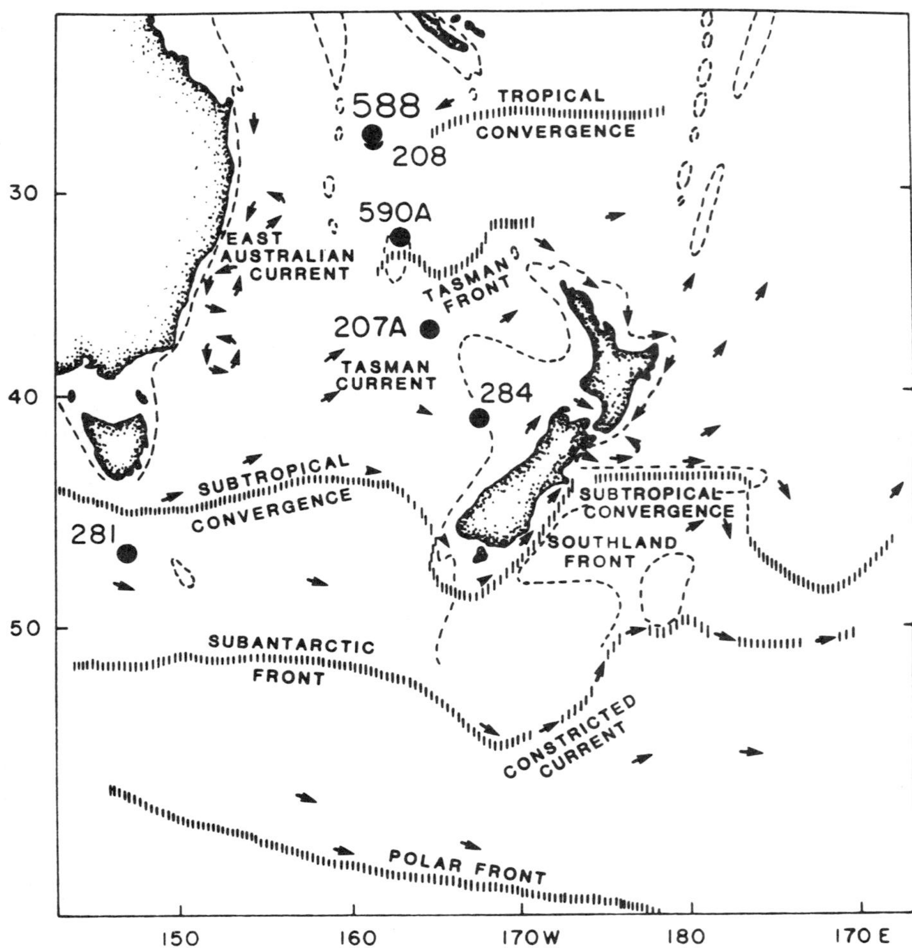

Fig. 2. Location of studied DSDP Sites and the oceanic circulation in the southwest Pacific (modified after Heath, 1985).

and 588, has been monitored using a dissimilarity index called chord distance. (It is done by projecting the faunal compositions of two samples onto to a circle of unit radius through the use of direction cosines and the measure is the distance of the chord after such a projection, see Pielou, 1984, p.48; Ludwig and Reynolds, 1988, p. 170) The chord distances have been calculated for 26 time-slices ranging from 5.34 to 3.46 Ma, and depicted as the three sides of a triangle to indicate the dissimilarity among the three sites (Fig. 3). The area of each triangle indicates the overall biogeographic differentiation among these sites during each time interval. The faunal data were obtained from Elmstrom (1985) for DSDP 588 and 590A, and from Kennett and Vella (1975) for DSDP 284.

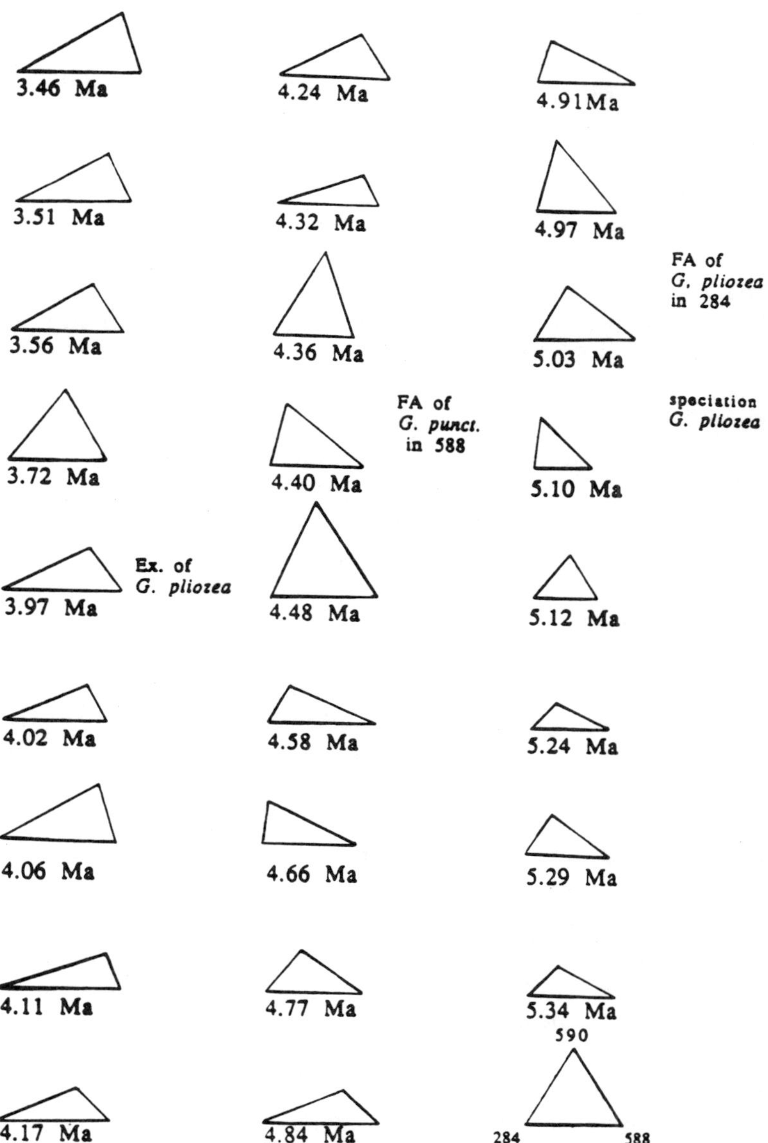

Fig. 3. Dissimilarity triangles showing the planktic foraminiferal faunal difference among DSDP Sites 284, 588, and 590 during the latest Miocene (5.34 to 5.0 Ma) and early Pliocene (5.0 to 3.5 Ma). FA: first appearance, Ex: extinction.

Prior to 5.10 Ma, the triangles are relatively small, indicating a more or less homogeneous distribution of planktic foraminifera in the studied region. The increase in size shown by the 5.03 Ma triangle suggests a major biogeographic differentiation at the Miocene/Pliocene boundary.

The faunas in DSDP Site 590A were more similar to those in DSDP 284 (temperate) before 4.48 Ma, but became more similar to that of DSDP 588 (subtropical) after 4.36 Ma (Fig. 3). The intervening time interval, from 4.48 Ma to 4.36 Ma, is marked by an increase in dissimilarity among these sites, as indicated by the relatively large sizes of the triangles (Fig. 3). It appears that a major biogeographic reorganization occurred from 4.48 to 4.36 Ma, and caused DSDP Sites 588 and 590A to have common faunal composition afterwards. The early Pliocene climatic optimum during the early Pliocene (Keany, 1978; Abelmann et al., 1990; Hodell and Warnke, 1991) might have pushed the Tropical Convergence and the Tasman Front southward, especially during the summer months, and introduced more subtropical elements to the DSDP 590A assemblage. The interpretation is supported by the fact that a major reduction in the relative abundance of the cold-water variant of *Neogloboquadrina pachyderma* (sinistral forms) and an increase of the warm-water variant, right-coiling forms of the same species in Site 590A occurred after 4.32 Ma (Elmstrom, 1985), but this phenomenon is not seen in Site 284 (Kennett and Vella, 1975).

Chronological Framework

Magnetic polarity sequences were established for the last 13 million years for DSDP 588 and 3.5 million years for DSDP 590A (Barton and Bloemendal, 1986). The paleomagnetic stratigraphy of DSDP 588 has been used to construct a chronological framework for all the studied sites. The magnetostratigraphic scheme of DSDP 588 has been modified slightly from that reported by Barton and Bloemendal (1986) following the suggestion made by Hodell and Kennett (1986). The ages of polarity boundaries and various nannofossil and planktic foraminiferal datum levels are from Berggren et al. (1985). The datum levels of *Globoconella* were not employed for correlation in order to allow the dating of the various evolutionary events in the *Globoconella* to be derived from independent stratigraphic markers. The resulting correlation scheme of the studied sites is shown in Fig. 4. A detailed account of the correlation was presented elsewhere (Wei and Kennett, 1988).

Morphometric Measurements

A total of 116 samples were obtained from DSDP Sites 207A, 208, and 588. For each sample at least 25 specimens were randomly picked and measured from the size fraction larger than 150μ. In addition, the morphometric data of *Globoconella* obtained from DSDP Site 284 (Malmgren and Kennett, 1981) were included for comparison. Figure 5 shows the definitions of the various morphometric variables used in this study.

Speciation of *Globorotalia Pliozea* at the Miocene/Pliocene Boundary

Globoconella formed a geographic cline during the late Miocene (7.0 to 5.05 Ma). Forms in the temperate areas differ from those in the cool subtropical sites in having larger conical angles and fewer chambers in the last whorl (Fig. 6). During the late Miocene all the forms in the cline showed a common evolutionary trend marked by an increase in conical angle and a decrease in chamber numbers. The increase in conical angle culminated eventually in an architectural threshold when the conical angle approached 90° and the angular periphery can no longer be maintained. The central and peripheral populations responded differently when they approached this threshold during the latest Miocene (5.15–5.0 Ma): the central populations transformed gradually into a spherical form, *Globorotalia sphericomiozea*, whereas the peripheral populations split into two morphotypes, *G. conomiozea terminalis* and *G. pliozea*. The splitting manifested most clearly by a disparity in their conical angles: one morphotype (*Globorotalia conomiozea terminalis* Wei; Wei, 1987a) continued the previous trend of increasing degrees in conical angle while the other (*G. pliozea*) reversed the trend and possessed much smaller conical angles (Fig. 7). *G. conomiozea terminalis* became locally extinct at 5.03 Ma, whereas the new species, *G. pliozea*, originated at 5.05 Ma and persisted through the Miocene/Pliocene boundary in the cool subtropical areas (DSDP 588 and 208).

The origination of *G. pliozea* coincided with a decrease in the $\delta^{18}O$ gradient between the planktic foraminifera and benthic foraminifera at DSDP 590A (Fig. 6). The decrease in $\delta^{18}O$ gradient reflects a reduction of vertical temperature gradient, which, in turn, indicates intensification of upwelling associated with the Tasman Front. The intensification of the upwelling resulted from a strengthening of the Tasman Front (Elmstrom and Kennett, 1986). The faunal dissimilarity among the three DSDP sites (Fig. 3) increased concomitantly at the Miocene/Pliocene boundary. The intensified Tasman Front might have restricted the genetic flow between the central and peripheral populations of *Globoconella* and triggered the punctuated speciation of *G. pliozea* in the northern isolated peripheries (Wei and Kennett, 1988).

Migration of *Globorotalia puncticulata* at 4.4 Ma

Following its origination at 5.05 Ma, the new daughter species, *Globorotalia pliozea*, remained in morphological stasis for 0.65 million years until 4.4 Ma (Figs. 7 and 8). Meanwhile, the central populations in the Tasman Sea (DSDP 207A and 284) continued their previous gradual evolutionary trend; the forms became more rounded in shape, and further reduced chambers in the final whorl (Malmgren and Kennett, 1981). This morphological trend resulted in a transformation from *G. sphericomiozea* to *G. puncticulata* at 4.6 Ma (Malmgren and Kennett, 1982; Wei, 1987a). The descendant species, *G. puncticulata*, made its first appearance in the cool subtropical sites (DSDP 588, 208) at about 4.4 Ma (Fig. 8). The dispersal of *G. puncticulata* into the cool subtropical regions appears to coincide with the bio-

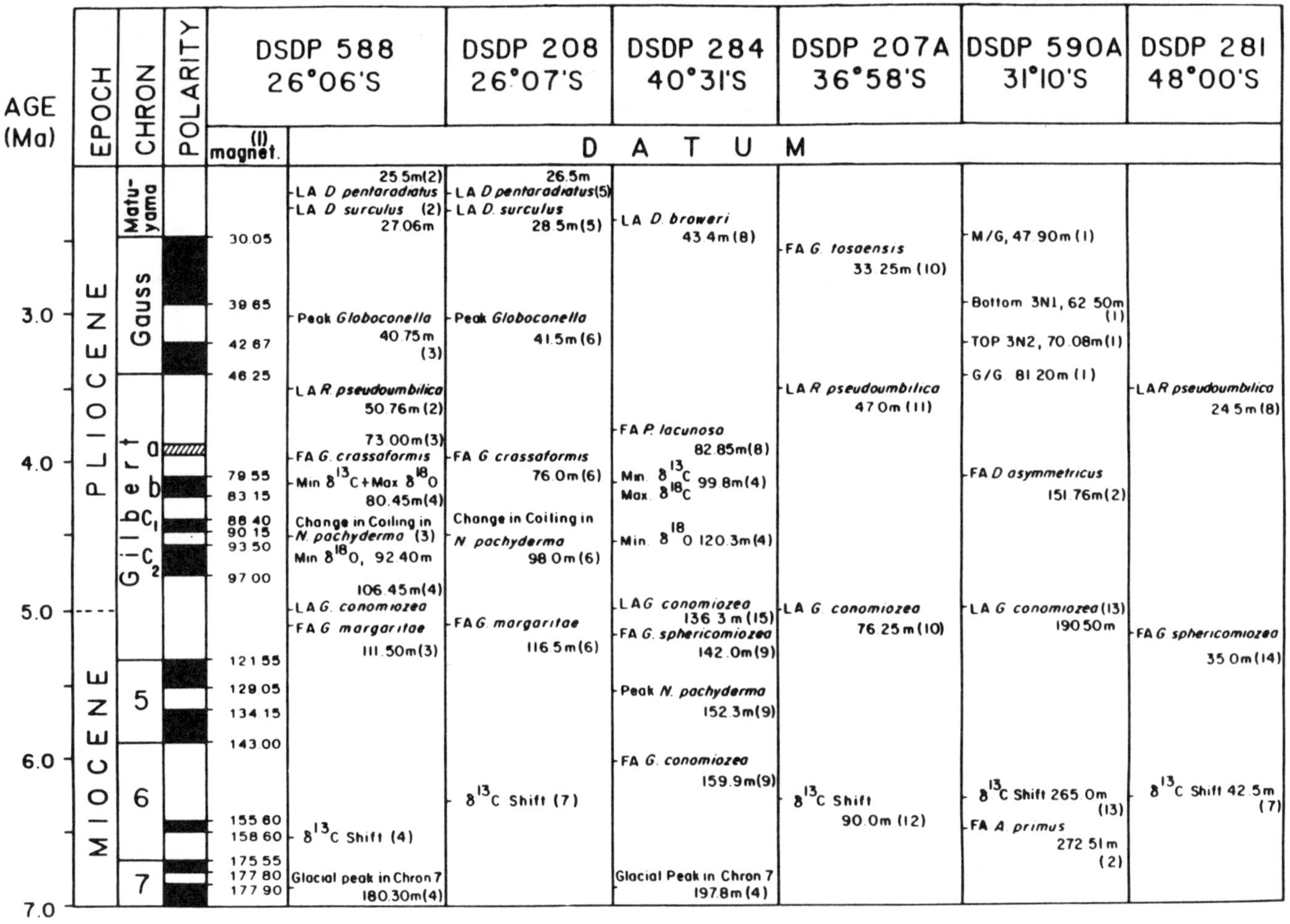

K.Y. Wei

Fig. 4. Correlation datum levels used in constructing the chronological framwork of studied sites. The references for each datum level are indicated by the numbers shown in parentheses. Check Wei and Kennett (1988) for details.

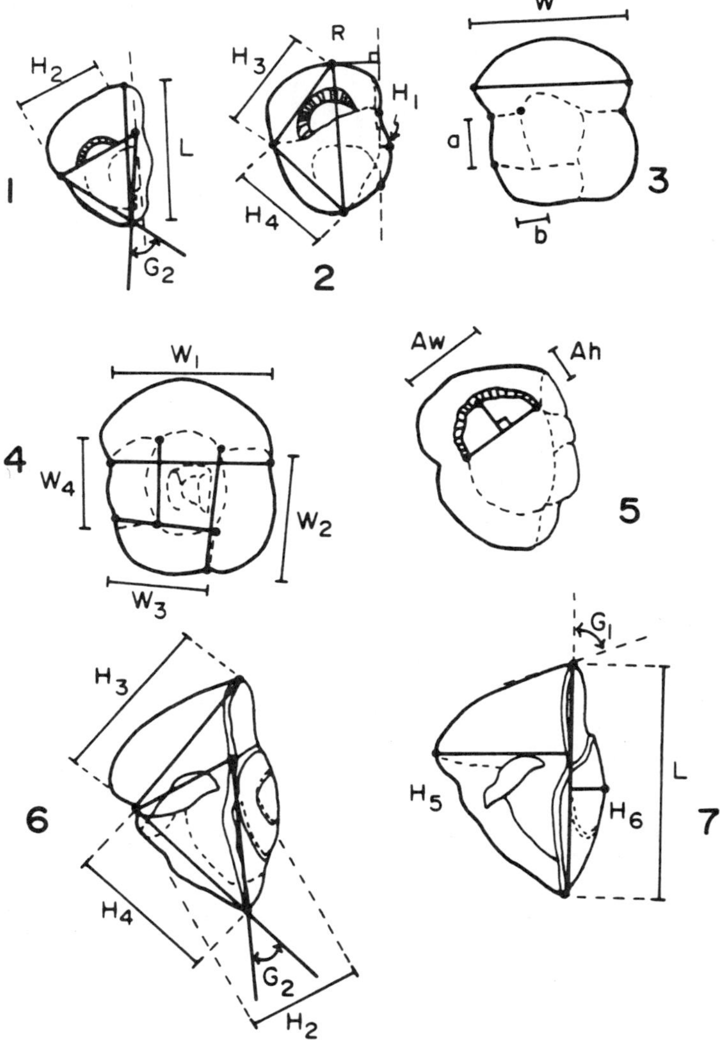

Fig. 5. Definitions of morphometric variables. L is the length of the test measured in edge view as shown by specimens 1 and 7. W is the maximum width of the final chamber as shown by specimen 3. W1, W2, W3, and W4 are widths of the last (N), N-1, N-2, and N-3 chamber, respectively, defined in spiral view. H1, H2, H3, H4, H5, and H6 are defined in edge view. The number of chambers in the final whorl is measured as N + a/(a+b), where N is the integer number of whole chambers and a and b are defined as shown by specimen 3. R is the peripheral roundness, defined in edge view as the distance from the point of the maximum inflation of the final chamber along the normal to the reference line defined based upon two points as shown by specimen 2. G1 is the conical angle, defined in edge view as the angle between the apices (keel) of the final chamber and the antepenultimate chamber as shown by specimen 7. G2 is the peripheral-umbilical angle (P–U angle) as defined by specimens 1 and 6. Aw and Ah are the maximum width and height of the aperture as defined by specimen 5.

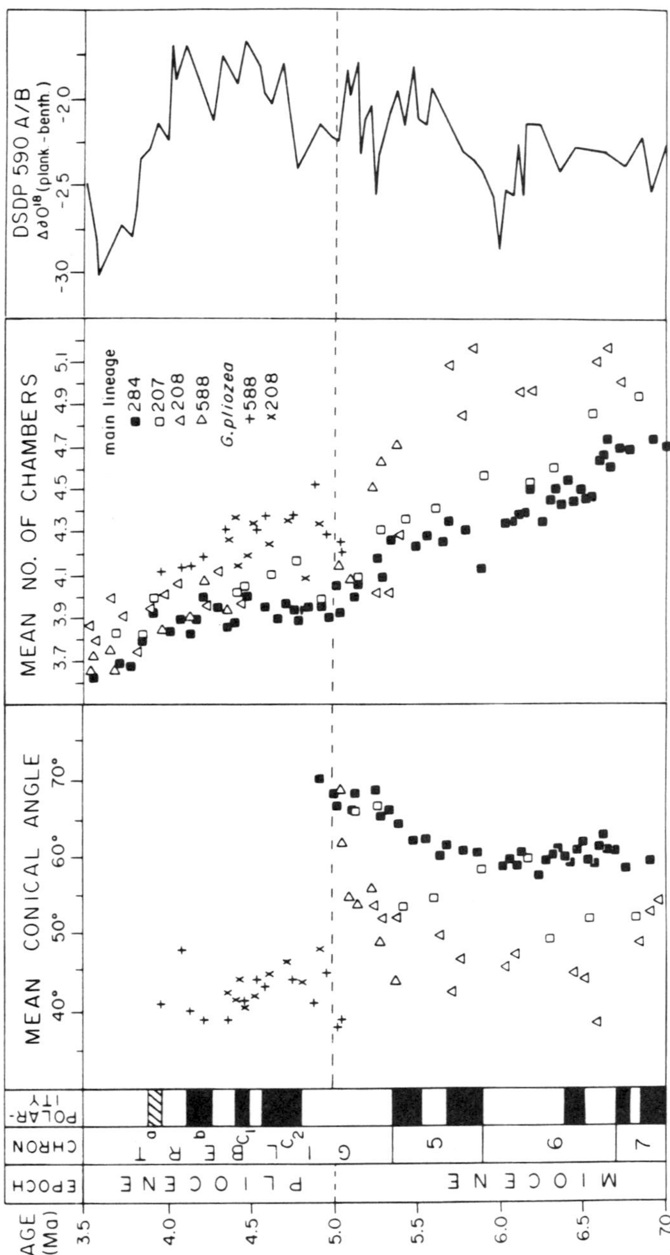

Fig. 6. Clinal variation shown in average conical angle and average number of chambers in the last whorl of the *Globoconella* specimens from the southwest Pacific DSDP Sites. Also shown in the rightmost column are the oxygen isotopic gradients between the surface and the bottom waters at DSDP Site 590.

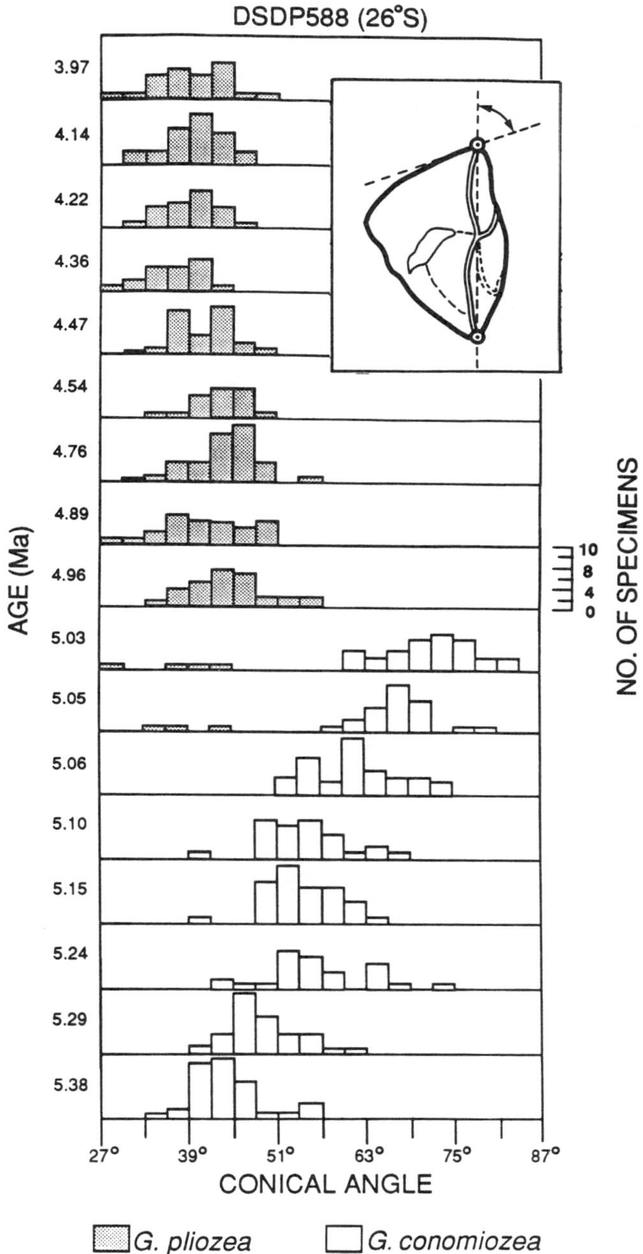

Fig. 7. Stacked histograms showing the distribution pattern of conical angles of the ancestral species, *Globorotalia (Globoconella) conomiozea terminalis*, and the descendant species, *G. (G.) pliozea*, at DSDP Site 588 in the intervals prior and after the speciation event at about 5.05 Ma.

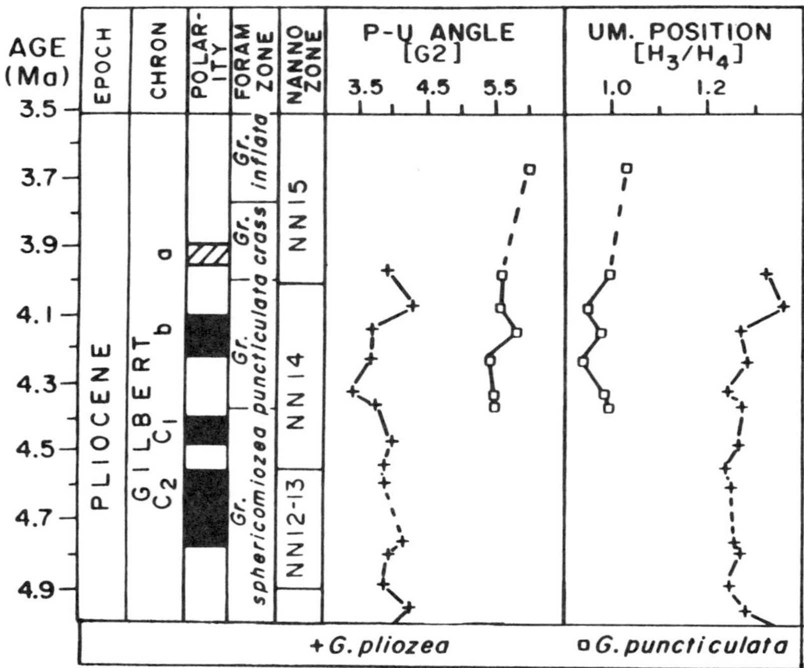

Fig. 8. Morphological variation of two lineages in DSDP Site 588 during the Pliocene: the indigenous cool subtropical species, *G. (G.) pliozea*, and the migrated species, *G. (G.) puncticulata*, which is the descendant species of the central main stock, *G. (G.) sphericomiozea*.

geographic reorganization during 4.48 Ma to 4.36 Ma, and prior to the southward movement of the Tasman Front. The coexistence of *G. puncticulata* and *G. pliozea* in the cool subtropics may have caused competition between these two species for common habitat and resources. The morphology of *G. pliozea* became more variable, as shown by several shape variable (Fig. 8). Following the next 0.45 m.y. of morphological fluctuation, *G. pliozea* became extinct at 3.97 Ma (Fig. 8). The extinction of *G. pliozea* is apparently a case of biotic exclusion resulting from the competition between the two daughter species that originated from the same ancestral stocks. No direct, immediate paleoceanographic cause can be identified.

Heterochrony in the *G. puncticulata–G. inflata* Clade

The extant species of the *Globoconella* clade, *G. inflata*, originated from its ancestral species, *G. puncticulata*, at about 3.5 Ma (Wei, 1987b; Wei, in press.). Compared to *G. puncticulata*, *G. inflata* is characterized by relatively larger size, more globular form, and fewer chambers in the last whorl. According to a preliminary analysis of various morphometric parameters (Wei, 1987b), the width of the last chamber (W) is regarded to be the best indicator of test size, and the peripheral roundness (R) the best shape parameter for differentiating *G. inflata* from *G. puncticulata*. A combination of these two parameters illustrates the different ontogenetic trajectories of the two species (Fig. 9). A comparison of such ontogenetic trajectories obtained from successive stratigraphic intervals reveals the heterochronic trends in the lineages. *G. puncticulata* exhibits a heterochronic acceleration in which, relative to the ancestral forms, the descendants developed faster and acquired the same shape when the sizes of the specimens are smaller (Fig. 10). The acceleration is indicated by the increase in the slope (allometric coefficient) of the average growth trajectories from 0.7 to 1.3. In other words, there is a corresponding increase in peripheral roundness through geologic time. In contrast, *G. inflata* proceeds with a post-displacement mode (McNamara, 1986; McKinney, 1986), in which the development is delayed, occurring at a later stage marked by larger sizes (Figs. 10 and 11). The different developmental strategies—namely, acceleration in *G. puncticulata* and post-displacement in *G. inflata*—account for the divergence between the two sister lineages. A comparison of the allometric trajectories of *G. inflata* with that of *G. puncticulata* suggests that the divergence began at about 3.5 Ma and proceeded gradually during the interval from 3.5 to 2.4 Ma.

At about 2.35 Ma, *G. puncticulata* became extinct. At the same time, the post-displacement trend in *G. inflata* lineage (Fig. 11) reversed: the y-intercepts of the allometric trajectories exhibit a trend of increasing values, which indicates a pre-displacement trend of heterochrony (i.e., shape development begins ontogenetically earlier, Fig. 12). The pre-displacement trend ceased at about 0.9 Ma and was followed by a stasis that lasted until 0.26 Ma, whereas the last 0.26 Ma witnessed a neoteny mode (Fig. 13).

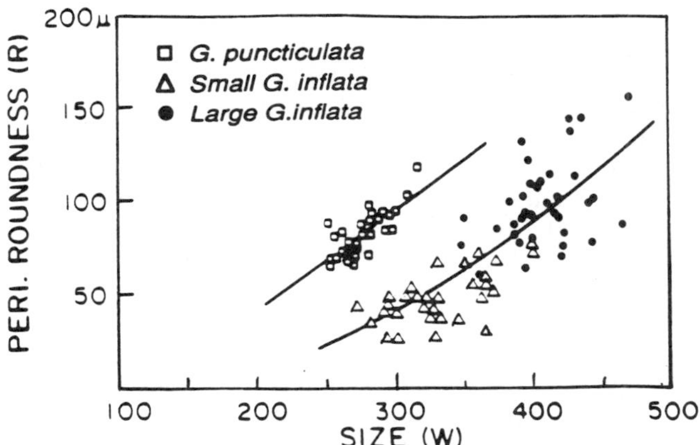

Fig. 9. Scatter plot of the peripheral roundness (R) versus the width of the last chamber (W) of *Globoconella* specimens in sample 2.76 Ma, in DSDP Site 588. The solid lines are best-fitted regression lines representing the mean growth trajectories. The regression lines were derived using Huxley's (1932) allometry equation (R = bW^a, where a and b are two coefficients derived from the least square regression. a is the slope, and b is the intercept shown in Figs. 10 to 13).

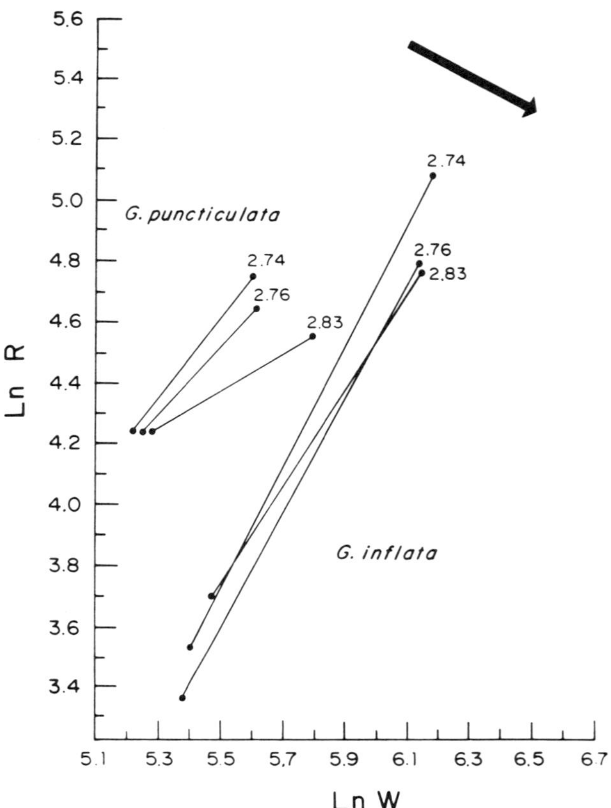

Fig. 10. Allometric trajectories of *G. puncticulata* and *G. inflata* in three consecutive stratigraphic samples obtained from the lower Pliocene of the DSDP Site 588. The X axis is the width (W) of the final chamber at natural logarithmic scale, Y axis the peripheral roundness (R). The number associated with each line indicates the age (in myrs) of the sample. The endpoints of each line mark the minimum and maximum chamber sizes, respectively, in each sample. The solid arrow on the upper right corner stands for the post-displacement trend which characterizes the heterochronic mode of *G. inflata*, whereas the evolution in the ancestral species, *G. puncticulata*, is marked by acceleration, a different peramorphosis mode. The difference in their heterochronic modes accounts for the morphological divergence between the two species.

One obvious consequence of post-displacement is, *ceteris paribus*, the occurrence of large forms in the descendant lineage, such that the first appearance of *G. inflata* is marked by the sudden appearance of large forms at 3.5 Ma. The turnover from *G. puncticulata* to *G. inflata* during the late Pliocene suggests that the new developmental strategy adapted the prevailing paleoceanographic conditions better than acceleration mode. Two questions arise: (1) what would be the advantage of becoming larger? and (2) what were the paleoecological factors which triggered the post-displacement? To

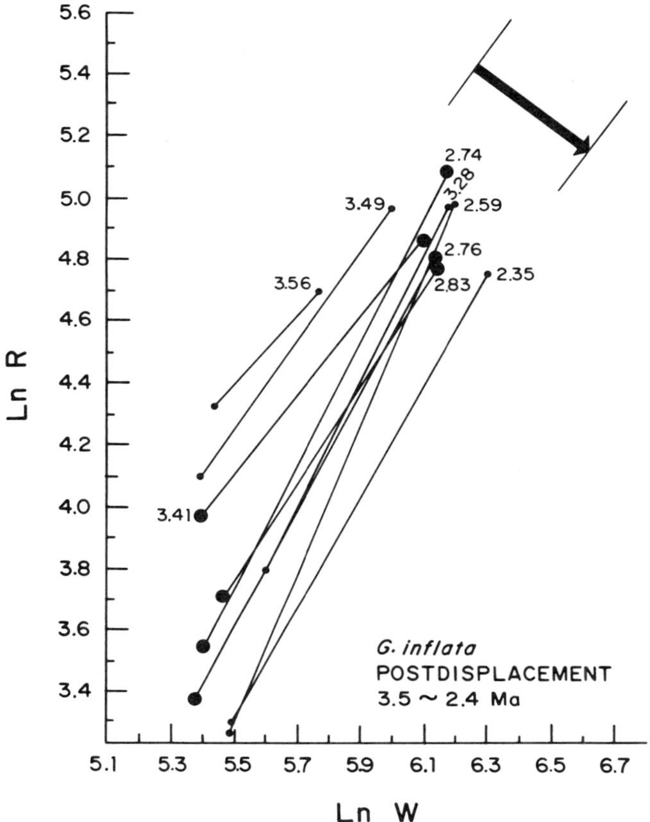

Fig. 11. Allometric trajectories of *G. inflata* in DSDP Site 588 during the time interval from 3.5 to 2.35 Ma. The shift of trajectories from left upper corner to lower right suggests a post-displacement trend of heterochrony. The regression lines denoted by larger dots at the two ends have higher values (r > 0.80) of correlation coefficient than those with smaller dots. Detailed caption is referable to Fig. 10.

assist in answering these two questions, a preliminary isotopic analysis was conducted for three morphotypes: *G. puncticulata*, small forms of *G. inflata* (W < 330μ), and large forms of *G. inflata* (W > 330μ). The analysis was done for two selected samples, dated 3.28 Ma and 2.74 Ma, respectively. The results are shown in Fig. 14. The adult (large) *G. inflata* have the highest $\delta^{18}O$ values (0.38 and 0.69 per mil) at both levels, indicating a deeper (colder) habitat in the water column, whereas the young (small) forms of *G. inflata* exhibit the lowest $\delta^{18}O$ values (0.17 and −0.09 per mil) among the three morphotypes. The differences in $\delta^{18}O$ among the three morphotypes appear to increase through time. The $\delta^{13}C$ values of the large forms of *G. inflata* are similar to that of benthic foraminifera, and the enrichment trend of the $\delta^{18}O$ values are parallel to the trend shown by the benthic foraminifera. The benthic isotopic composition at this site is representative of the Antarctic Intermediate Water (AAIW)

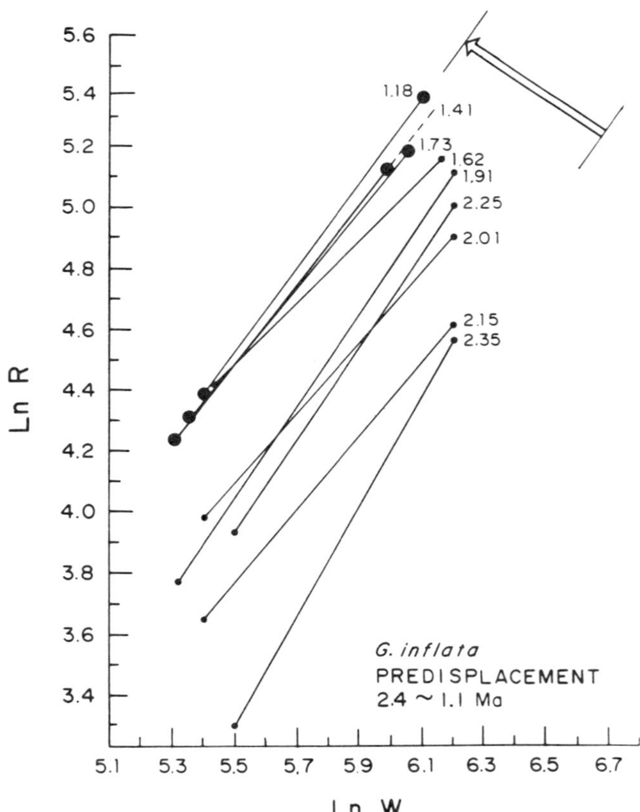

Fig. 12. Allometric trajectories of *G. inflata* in DSDP Site 588 during the time interval from 2.4 to 1.1 Ma. The shift of trajectories from lower right corner toward upper left exhibits a trend of pre-displacement peramorphosis.

(Kennett, 1986). Because large forms of *G. inflata* exhibit characteristic $\delta^{13}C$ values of the AAIW and parallelism between *G. inflata* and the AAIW in their $\delta^{18}O$ values, I infer that the adult (large) *G. inflata* were inhabiting deeper waters, probably close to the AAIW, during the middle Pliocene. Apparently, the gigantism associated with the post-displacement enabled *G. inflata* to exploit a deeper habitat. Sedimentological evidence obtained from the southwest Pacific suggests that the intensification of the intermediate water circulation was initiated at the same time, 3.5 Ma (Stein and Robert, 1986; Stein, 1986). The enhancement of the AAIW is believed to respond to a cooling in the high latitudes (Kennett and von der Borch, 1986). This intensification of the AAIW might have enhanced the vertical stratification of the surface water column, creating new ecological niches and triggering post-displacement change in the *G. inflata*.

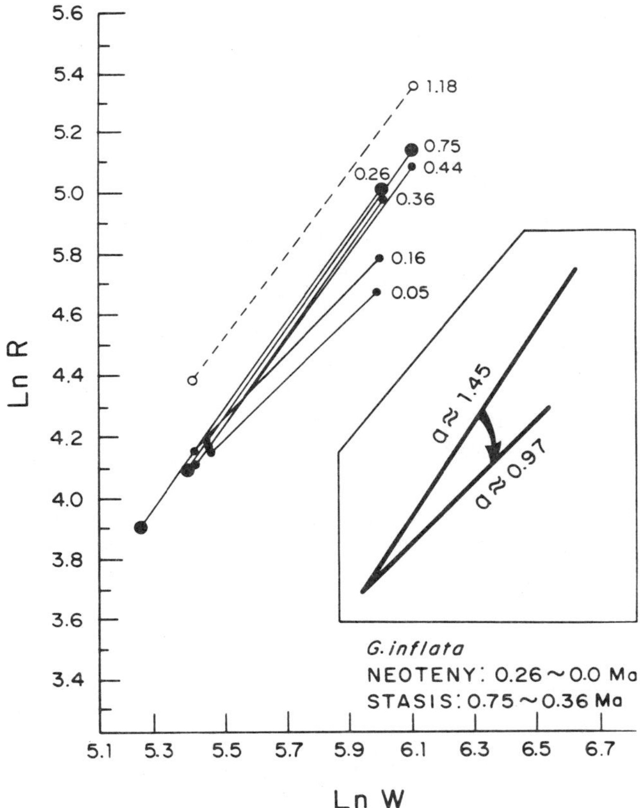

Fig. 13. Allometric trajectories of *G. inflata* during the late Quaternary. The inlet box presents the inferred neoteny based upon trajectories of samples of 0.26 Ma to 0.05 Ma.

The coincidence between the timing of paleoceanographic events and the shifts in heterochronic mode is evident in the evolution of the *G. inflata* lineage: (1) the change from the post-displacement to the pre-displacement trend at about 2.4 Ma is coincident with the onset of the major continental glaciations in the northern hemisphere (Shackleton et al., 1984; Rea and Schrader, 1985; Ruddiman and Raymo, 1988; Raymo et al., 1989); (2) the cessation of the pre-displacement and the beginning of stasis at about 0.9 Ma is synchronous with the beginning of the large-scale, high-amplitude glacial/interglacial cycles (Nelson et al., 1986; Jansen and Seirup, 1986; Maasch, 1988). Clearly paleoenvironmental perturbations have caused developmental perturbations in *Globoconella*. Allometric analysis proves to be an insightful approach to revealing how paleoceanographic conditions modulate the size, shape, and developmental timing of planktic foraminifera.

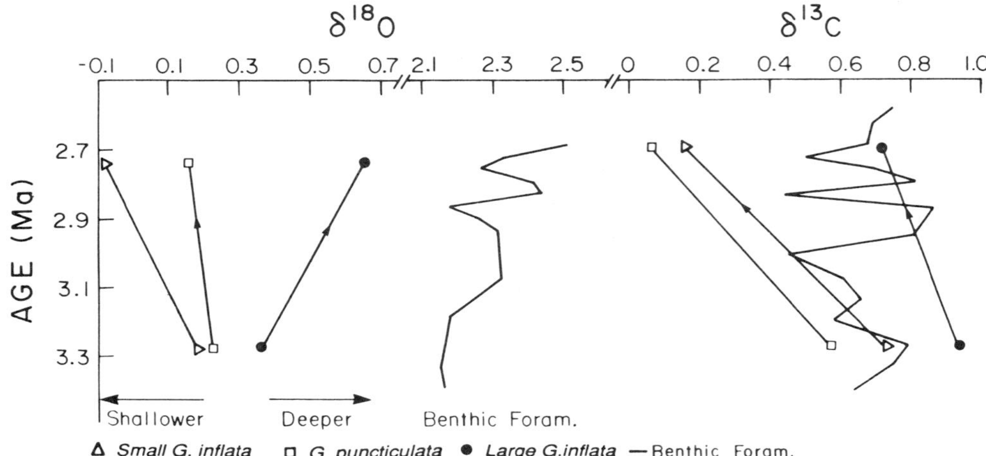

Fig. 14. δ¹⁸O and δ¹³C values of *G. puncticulata* and *G. inflata* from two stratigraphic levels (3.28 and 2.74 Ma). Also shown are isotopic compositions of the benthic foraminifera obtained from the interval between 2.65 and 3.35 Ma. The benthic isotopic values are adopted from Elmstrom (1985).

Summary

(1) The intensification of the Tasman Front at the Miocene/Pliocene boundary isolated genetically the northern peripheral populations from the central populations of the *Globoconella* cline in the Tasman Sea and triggered the punctuated speciation of *G. pliozea*.

(2) The biogeographic reorganization at about 4.4 Ma allowed *G. puncticulata* of the Tasman Sea to disperse into the cool subtropical oceans. The migration might have resulted in the extinction of the local species, *G. pliozea*, due to biotic competition.

(3) The enhancement of the Antarctic Intermediate Water (AAIW) at about 3.5 Ma created new ecological niches in upper water column and triggered the cladogenesis of *G. inflata* from *G. puncticulata*.

(4) The onset of significant glaciations in the northern Hemisphere at 2.4 Ma might be responsible for the extinction of *G. puncticulata* and the pre-displacement mode of heterochrony in *G. inflata*.

(5) The intensification of glacial/interglacial fluctuations at 0.9 Ma caused the heterochrony of *G. inflata* to change from the pre-displacement trend to a stasis.

Acknowledgements

Much of this report resulted from my Ph.D research project at the Graduate School of Oceanography, University of Rhode Island, U.S.A. Additional studies were conducted at the University of California at Santa Barbara and at Yale University. My deepest gratitude is dedicated to my major professor, Dr. James P. Kennett, for his continuous guidance and support during the various stages of this study. The assistance of Linda Stathoplos in isotopic analysis is gratefully appreciated. James P. Kennett, Jon Moore, and Elisabeth Vrba made many helpful suggestions. This research was supported by NSF grants OCE84-20671 and OCE87-13505 (Marine Geology and Geophysics) to James P. Kennett and a starting fund provided by the Yale University.

References

Abelmann, A., Gerosonde, R., and Spiess, V., 1990. Pliocene-Pleistocene paleoceanography in the Weddel Sea — siliceous microfossil evidence. *In:* U. Bleil and J. Thiede (Eds.), Geological History of the Polar Oceans: Arctic Versus Antarctic, pp. 729–759. Kluwer, Amsterdam.

Andrews, J.C., Lawrence, M.W., and Nilsson, C.S., 1980. Observations of the Tasman Front. *Jour. of Physical Oceanography*, 10: 1854–1869.

Barton, C.E. and Bloemendal, J., 1986. Paleomagnetism of sediments collected during Leg 90, southwest Pacific. *In:* J.P. Kennett, and C.C. von der Borch et al. (Eds.), Init. Repts. DSDP, 90. U.S. Govt. Print. Office; Washington, D.C. pp. 1273–1316.

Bé, A.W.H., 1977. An ecological, zoogeographic and taxonomic review of recent planktonic foraminifera. *In:* T.S. Ramsay, (Ed.), Oceanic Micropaleontology, Vol. 1. Academic Press, London. pp. 1–100.

Bé, A.W.H. and Toderlund, D.S., 1971. Distribution and ecology of living planktonic foraminifera in surface waters of the Atlantic and Indian Oceans. *In:* B.M. Funnell, and W.R. Riedel, (Eds.), The Micropaleontology of Oceans. Cambridge University Press. pp. 105–149.

Berggren, W.A., Kent, D.V., and Van Couvering, J.A., 1985. Neogene geochronology and chronostratigraphy. *In:* N.J. Snelling (Ed.), The Chronology of the Geological Record. Geological society of London, London, pp. 211–260.

Cifelli, R. and Scott, G., 1986. Stratigraphic Record of the Neogene Globorotalid Radiation (Planktonic Foraminiferida). Smithsonian Institution Press, Washington DC. 101 pp.

Corfield, R.M. and Shackleton, N.J., 1988. Productivity change as a control on planktonic foraminiferal evolution after the Cretaceous/Tertiary boundary. *Hist. Biology*, 1: 323–343.

Eade, J.V., 1973. Geographical distribution of living planktonic foraminifera in the Southwest Pacific. *In:* R. Fraser (Ed.), Oceanography of the South Pacific. New Zealand National Commission for UNESCO, Willington, pp. 249–256.

Elmstrom, K.M., 1985. Late Neogene paleoceanography of the Southwest Pacific: from oxygen and carbon and planktonic foraminiferal faunal evidence of DSDP Sites 588, 590A, and 284. MS Thesis, University of Rhode Island, 192 pp.

Elmstrom, K.M. and Kennett, J.P., 1986. Late Neogene paleoceanographic evolution of Site 590: Southwest Pacific. pp. 1361-1381. *In:* J.P. Kennett and C.C. von der Borch et al., (Eds.), Init. Repts. DSDP, 90. U.S. Govt. Print. Office, Washington, D.C.

Heath, R.A., 1985. A review of the physical oceanography of the seas around New Zealand. *New Zeal. J. Mar. Fresh. Res.*, 19: 79-124.

Hodell, D.A. and Kennett, J.P., 1986. Late Miocene-early Pliocene stratigraphy and paleoceanography of the south Atlantic and southwest Pacific oceans: a synthesis. *Paleoceanog.*, 1: 285–311.

Hodell, D.A. and Warnke, D.A., 1991. Climatic evolution of the Southern Ocean during the Pliocene Epoch from 4.8 to 2.6 million years ago. *Quaternary Science Reviews*, 10: 205–214.

Hoffman, A. and Kitchell, J.A., 1984. Evolution in a pelagic planktic system: a paleobiologic test of models of multispecies evolution. *Paleobio.*, 10: 9–33.

Hoskins, R.H., 1991. Planktic Foraminiferal Correlation of the Late Pliocene to Early Pleistocene of DSDP Sites 284 and 593 (Challenger Plateau) with New Zealand Stages. New Zealand Geological Survey Report PAL 149, 33 pp.

Hornibrook, N. de B., 1981. *Globorotalia* (planktonic foraminifera) in the late Pliocene and early Pleistocene of New Zealand. *New Zeal. J. Geol. Geophy.*, 24: 263–292.

Hornibrook, N. de B., 1982. Late Miocene to Pleistocene *Globorotalia* (foraminifera) from Deep Sea Drilling Project Leg 29, Site 284, Southwest Pacific. *New Zeal. J. Geol. Geophy.*, 25: 83–99.

Hornibrook, N. de B., Brazier, R.C., and Strong, C.P., 1989. Manual of New Zealand Permian to Pleistocene Foraminiferal Biostratigraphy. New Zealand Geological Survey, Lower Hutt, 175 pp.

Huxley, J., 1932. Problems of Relative Growth. 276 pp. Cambridge Univ. Press, Cambridge.

Jansen, E. and Seirup, H.P., 1986. Stable isotope stratigraphy and amino-acid epimerization for the last 2.4 M.Y. at site 610, holes 610 and 610A. pp. 879-888. *In:* W.F. Ruddiman, R.B. Kidd, and E. Thomas (Eds.), Init. Repts. DSDP, 94. U.S. Govt. Print. Office, Washington, D.C.

Keany, J., 1978. Paleoclimatic trends in early and middle Pliocene dee-sea sediments in the Antarctic. *Marine Micropaleontology*, 3: 35–49.

Kennett, J.P., 1966. The *Globorotalia crassaformis* bioseries in north Westland and Marlborough. *Micropaleontology*, 12: 235–245.

Kennett, J.P., 1973. Middle and late Cenozoic planktonic foraminiferal biostratigraphy of the southwest Pacific — DSDP Leg 21. pp. 575–640. *In:* R.E. Burns, et al. (Eds.), Init. Repts. DSDP, 21. U.S. Govt. Print. Office, Washington, D.C.

Kennett, J.P., 1986. Miocene to early Pliocene oxygen and carbon isotope stratigraphy in the southwest Pacific, Deep Sea Drilling Project Leg 90. pp. 1384–1411. *In:* J.P. Kennett and C.C. von der Borch et al. (Eds.), Init. Repts. DSDP, 90, U.S. Govt. Print. Office, Washington, D.C.

Kennett, J.P. and von der Borch, C.C., 1986. Southwest Pacific Cenozoic paleoceanography. pp. 1493–1517. *In:* J.P. Kennett and C.C. von der Borch et al. (Eds.), Init. Repts. DSDP, 90. U.S. Govt. Print. Office, Washington, D.C.

Kennett, J.P. and Srinivasan, M.S., 1983. An Atlas of Neogene Planktonic Foraminifera: Phylogenetic Approach. Hutchinson and Ross, Stroudsburg, 253 pp.

Kennett, J.P. and Vella, P., 1975. Late Cenozoic planktonic foraminifera and paleoceanography at DSDP Site 284 in the cool subtropical south Pacific. pp. 769–799. *In:* J.P. Kennett and R.E. Houtz (Eds.), Init. Repts. DSDP, 29. U.S. Govt. Print. Office, Washington, D.C.

Leckie, R.M., 1989. A paleoceanographic model for the early evolutionary history of planktonic foraminifera. *Palaeogeogr. Palaeoclimatol. Palaeoecol.*, 73: 107–138.

Ludwig, J.A. and Reynolds, J.F., 1988. Statistical Ecology. John Wiley & Sons, New York, 337 pp.

Maasch, K.A., 1988. Statistical detection of the mid-Pleistocene transition. *Climate Dynamics*, 2: 133–143.

Maiya, S.T., Saito, T., and Sato, T., 1976. Late Cenozoic planktonic foraminiferal biostratigraphy of northwest Pacific sedimentary sequences. pp.395-422. *In:* Y. Takayanagi and T. Saito (Eds.), Progress in Micropaleontology. Amer. Muse. Nat. Hist., New York.

Malmgren, B.A. and Berggren, W.A., 1987. Evolutionary changes in some late Neogene planktonic foraminiferal lineages and their relationships to paleoceanographic changes. *Paleoceanog.*, 2: 445–456.

Malmgren, B.A. and Kennett, J.P., 1981. Phyletic gradualism in a late Cenozoic planktonic foraminiferal lineage: DSDP Site 284, southwest Pacific. *Paleobio.*, 7: 230–240.

Malmgren, B.A. and Kennett, J.P., 1982. The potential morphometrically based phylozonation: application of a late Cenozoic foraminiferal lineage. *Mar. Micropaleont.*, 7: 285–296.

McKinney, M.L., 1986. Ecological causation of heterochrony: test and implications for evolutionary theory. *Paleobio.*, 12: 282–289.

McNamara, K.J., 1986. A guide to the nomenclature of heterochrony. *J. Paleont.*, 60: 4–13.

Mulhearn, P.J., 1987. The Tasman Front: a study using satellite infrared imagery. *J. Phy. Ocean.*, 17: 1148–1155.

Nelson, C.S., Hendy, C.H., and Dudley, W.C., 1986. Quaternary isotope stratigraphy of Hole 593, Challenger Plateau, south Tasman Sea: preliminary observations based on foraminifers and calcareous nannofossils. pp. 1413–1426. *In:* J.P. Kennett and C.C. von der Borch et al. (Eds.), Init. Repts. DSDP, 90. U.S. Govt. Print. Office, Washington, D.C.

Pielou, E.C., 1984. The Interpretation of Ecological Data. Wiley, New York.

Raymo, M.E., Ruddiman, W.F. Backman, J., Clement, B.M., and Martinson, D.G., 1989. Late Pliocene variation in North Hemisphere ice sheets and North Atlantic Deep Water Circulation. *Paleoceanog.*, 4: 413–446.

Rea, D.K. and Schrader, H., 1985. Late Pliocene onset of glaciation: ice-rafting and diatom stratigraphy of North Pacific DSDP cores. *Palaeogeogr., Palaeoclimatol., Palaeoecol.*, 49: 313–325.

Ruddiman, W.F. and Raymo, M.E., 1988. Northern hemisphere climate regimes during the last 3 Ma: possible tectonic connections. *Phil. Trans. R. Soc. London*, B 318: 411–430.

Scott, G.H., 1979. The late Miocene to early Pliocene history of *Globorotalia miozea* plexus from Blind River. *Mar. Micropaleont.*, 4: 341–361.

Scott, G.H., 1982. Tempo and stratigraphic record of speciation in *Globorotalia puncticulata, Jour. Foram. Res.*, 12: 1–12.

Shackleton, N.J., Backman, J., Zimmerman, H., and others, 1984. Oxygen isotope calibration of the onset of ice-rafting and history of glaciation in the North Atlantic region. *Nature*, 307: 620–623.

Stanton, B.R., 1969. Hydrological observations across the Tropical Convergence north of New Zealand. New Zeal. *J. Mar. Fresh. Res.*, 3: 124–146.

Stanton, B.R., 1979. The Tasman Front. New Zeal. *J. Mar. Fresh. Res.*, 13: 201–214.

Stein, R., 1986. Late Neogene evolution of paleoclimate and paleoceanographic circulation in the Northern and Southern Hemisphere- A comparison. *Geologische Rundschau*, 75: 125–135.

Stein, R. and Robert, C., 1986. Siliciclastic sediments at sites 588, 590, and 591: Neogene and Paleogene evolution in the southwest Pacific and Australia climte. pp. 1437–1455. *In:* J.P. Kennett and C.C. von der Borch et al. (Eds.), Init. Repts. DSDP, 90. U.S. Govt. Print. Office, Washington, D.C.

Wei, K.-Y., 1987a. Multivariate morphometric differentiation of chronospecies in the late Neogene planktonic foraminiferal lineage *Globoconella. Mar. Micropaleont.*, 12: 183–202.

Wei, K.-Y., 1987b. Tempo and mode of evolution in Neogene planktonic foraminifera: Taxonomic and morphometric evidence. Ph.D. Dissertation, University of Rhode Island, 397 pp.

Wei, K.-Y., in press. Statistical pattern recognition in paleontology using SIMAC-MACUP. *Jour. Paleontology*, Vol. 67.

Wei, K.-Y. and Kennett, J.P., 1986. Taxonomic evolution of Neogene planktonic foraminifera and paleoceanographic relations. *Paleoceanog.*, 1: 67–84.

Wei, K.-Y. and Kennett, J.P., 1988. Phyletic gradualism and punctuated equilibrium in the late Neogene planktonic foraminifera clade *Globoconella. Paleobiology*, 14: 345–363.

Late Neogene Planktonic Foraminiferal Events of the Southwest Pacific and Indian Ocean: A Comparison

M.S. Srinivasan and D.K. Sinha

Department of Geology, Banaras Hindu University Varanasi – 221 005, India

Abstract

Late Neogene planktonic foraminiferal events and biostratigraphy have been established at four DSDP sites in the southwest Pacific (sites 586B, 587, 588, and 590) and four in the Indian Ocean (sites 214, 219, 237, and 238). A comparison of the chronological succession of the planktonic foraminiferal events of the southwest Pacific with Indian Ocean DSDP sites indicates a remarkable similarity. The main difference, however, is the complete absence of *Pulleniatina spectabilis* from the Indian Ocean sites, which appears to reflect severing of the surface water connection between the tropical Pacific and the Indian Ocean. In order to identify the synchronous and diachronous planktonic foraminiferal events between the two oceans, Shaw's Graphic Correlation method has been employed. Using this method, Composite Standard Reference Sections (CSRS) for the southwest Pacific and Indian Ocean were developed, which contain the total stratigraphic ranges of the planktonic foraminiferal taxa in the respective oceans. The two Composite Standards were then compared using a Shaw plot to identify the synchronous and diachronous events between the two oceans. The Composite Standard for the southwest Pacific was integrated with the magnetostratigraphic data of Barton and Bloemendal (1986). The paleomagnetic events were calibrated to time using the Neogene geochronology of Berggren et al. (1985), which yielded age estimates for the Late Neogene planktonic foraminiferal events in the southwest Pacific and the Indian Ocean. The result reveals the following events to be synchronous between the southwest Pacific and the Indian Ocean:

Globorotalia tosaensis LA (0.6 Ma), *Globigerinoides fistulosus* LA (1.6 Ma), *Globorotalia tosaensis* FA (3.26 Ma), *Globigerinoides fistulosus* FA (3.42 Ma), *Sphaeroidinellopsis* spp. LA (3.1 Ma), *Globorotalia margaritae* LA (3.46 Ma), *Globigerina nepenthes* LA (3.78 Ma), *Globorotalia crassaformis* FA (4.9 Ma), *Globorotalia tumida tumida* FA (5.01 Ma), *Globoquadrina dehiscens* LA (5.3 Ma), and *Pulleniatina primalis* FA (5.78 Ma).

The data also suggest that several important biostratigraphic markers are highly diachronous between these two oceans. For example, the study shows that *Globorotalia truncatulinoides* and *Globorotalia margaritae* first appear in the southwest Pacific (2.47 and 5.8 Ma, respectively) approximately 0.48 and 0.7 Ma earlier than in the Indian Ocean. On the other hand, *Sphaeroidinella dehiscens* and *Globorotalia tumida flexuosa* appear at least 0.8 and 0.54 Ma earlier in the Indian Ocean than in the southwest Pacific. *Dentoglobigerina altispira, Globigerinoides obliquus*, and *Globorotalia plesiotumida* became extinct earlier in the southwest Pacific than in the Indian Ocean.

Thus, the study enables us to gain a better understanding of the paleobiogeographic distribution pattern of the Late Neogene planktonic foraminiferal species in the southwest Pacific and Indian Ocean, which in turn is useful in paleoceanographic interpretations.

Introduction

The primary objective of biostratigraphic studies has been the attainment of more and more precise correlations. To a great extent this goal has been achieved with the help of stratigraphically important microfossil taxa. Planktonic foraminifera are one of the most significant groups of microfossils extensively used for age determination and correlation. The first and last appearances (FA and LA) of the planktonic foraminiferal species have been widely utilized as "datums" for correlating the Neogene marine sequences on regional as well as on global scale. However, such correlations have been generally attempted with a prior assumption that the first and last appearances of the planktonic foraminiferal species are synchronous in the stratigraphic sections being correlated.

Biostratigraphic and paleoceanographic studies based on deep sea drilling data have revealed the establishment of latitudinal provincialism in planktonic foraminifera in the Neogene period. Recent studies by Kennett and Srinivasan (1983) on the evolution of Neogene planktonic foraminifera have revealed that planktonic foraminifera evolved in different water masses and many of them migrated later to other parts of the world oceans. In view of this, the first and last appearacnes of planktonic foraminiferal species to be used as datums for correlating the sequences over a wide latitudinal range deserve re-evaluation.

To evaluate the effectiveness of faunal events for correlation (Synchroneity and Diachroneity), Shaw (1964) developed a "graphic correlation method" based on the principle of finding total stratigraphic ranges of the fossil taxa. In the present study Late Neogene planktonic foraminiferal events of the southwest Pacific and Indian Ocean have been compared using Shaw's method in order to identify the synchronous and diachronous events. Also the data have been integrated with the paleomagnetic record for approximating the timing of such events.

Though a number of studies have been carried out to correlate Neogene planktonic foraminiferal events of the south Pacific with those of the Atlantic Ocean, little work has been done pertaining to southwest Pacific and Indian Ocean planktonic foraminiferal events. The earliest attempt to correlate the Pacific and Indian Ocean sequences was made by Srinivasan and Azmi (1979), who attempted a correlation based on the comparison of stratigraphic ranges of Neogene planktonic foraminifera from the tropical Pacific and uplifted marine sequences of the Andaman Nicobar islands. Later Berggren (1984) gave a comparative account of the Neogene planktonic foraminiferal events of the Atlantic, Pacific, and Indian Ocean regions. In the present study the Late Neogene planktonic foraminiferal events in the southwest Pacific and Indian Ocean have been compared using the Graphic Correlation method in order to identify the synchronous and diachronous nature of the events in these two oceans. This would enable us to gain a better understanding of the paleobiogeographic distribution of planktonic foraminifera in the Indo-Pacific region during the Late Neogene.

Four DSDP sites from the southwest Pacific (sites 586B, 587, 588, and 590) and four from the Indian Ocean (sites 214, 219, 237, and 238) were selected for the present study (Fig. 1). Sites 586B and 587 lie in tropical waters, while site 588 and 590 are situated in the warm subtropical waters of the southwest Pacific. All the four Indian Ocean sites — 214, 219, 237, and 238 — lie well within the tropical waters (Table 1).

Methodology

Detailed Late Neogene planktonic foraminiferal biochronology for the southwest Pacific DSDP sites (Srinivasan and Sinha, 1991) and sites from the Indian Ocean (Srinivasan and Chaturvedi, 1992; Srinivasan and Singh, 1992) enabled us to develop a data set showing sequential order of planktonic foraminiferal events with depth below the ocean floor for all the examined sites. This data set was used as input for the Graphic Correlation method employed here.

DSDP site 588 in the southwest Pacific and site 214 in the Indian Ocean were selected as the Standard Reference Sections (SRS) for the respective oceans because

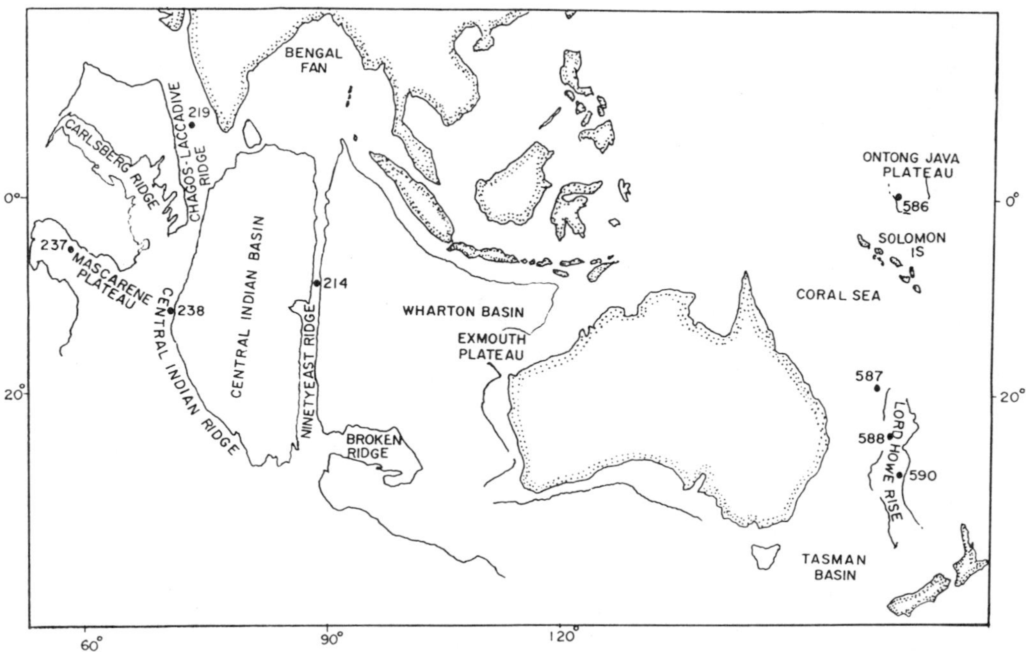

Fig. 1. Locations of the southwest Pacific and Indian Ocean DSDP sites examined.

Table 1. Details of the DSDP sites examined in the present study.

Site	Latitude	Longitude	Water depth (m)	Location
586B	00 29.84'S	158 29.89'E	2208	Ontong Java Plateau Southwest Pacific
587	21 11.87'S	160 19.99'E	1101	Lansdowne Bank Southwest Pacific
588	26 06.07'S	161 13.06'E	1533	Lord Howe Rise Southwest Pacific
590	31 10.02'S	163 21.51'E	1299	Lord Howe Rise Southwest Pacific
214	11 20.21'S	88 43.08'E	1665	Ninetyeast Ridge Indian Ocean
219	09 01.75'N	75 52.67'E	1764	Chagos Laccadive Ridge Indian Ocean
237	07 04.99'S	58 07.48'E	1640	Mascarene Plateau Indian Ocean
238	11 09.21'S	70 31.56'E	2844	Central Indian Ridge Indian Ocean

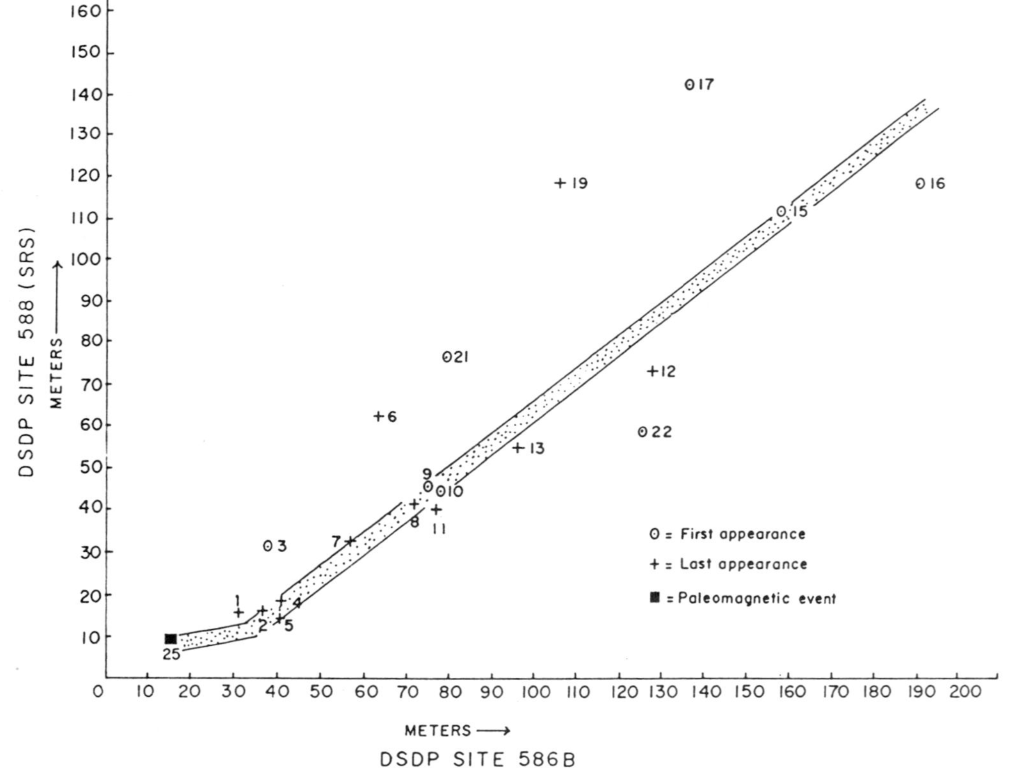

Fig. 2. Shaw plot of important events in sites 588 (SRS) and 586B (See Table 4 for explanation of numbers).

of their relative superiority to other sites with regard to continuous coring and avail-ability of other relevant information. Graphic correlation was first individually applied to the southwest Pacific and Indian Ocean sites in order to develop Composite Standard Reference Sections (CSRS) for both the oceans (Figs. 2–7). The next step was to compare and correlate CSRS for the southwest Pacific with CSRS for the Indian Ocean using Shaw plots (Fig. 8), which in turn led to identifying the synchro-nous and diachronous planktonic foraminiferal events. Detailed magnetostratigraphic data (Fig. 9) for CSRS for the southwest Pacific (Barton and Bloemendal, 1986) enabled us to estimate the ages of the Late Neogene planktonic foraminiferal events in the southwest Pacific (Tables 2–3). Using the LOC (line of correlation) between CSRS (Southwest Pacific) and CSRS (Indian Ocean), age estimates were also made for the Indian Ocean (Tables 2–3).

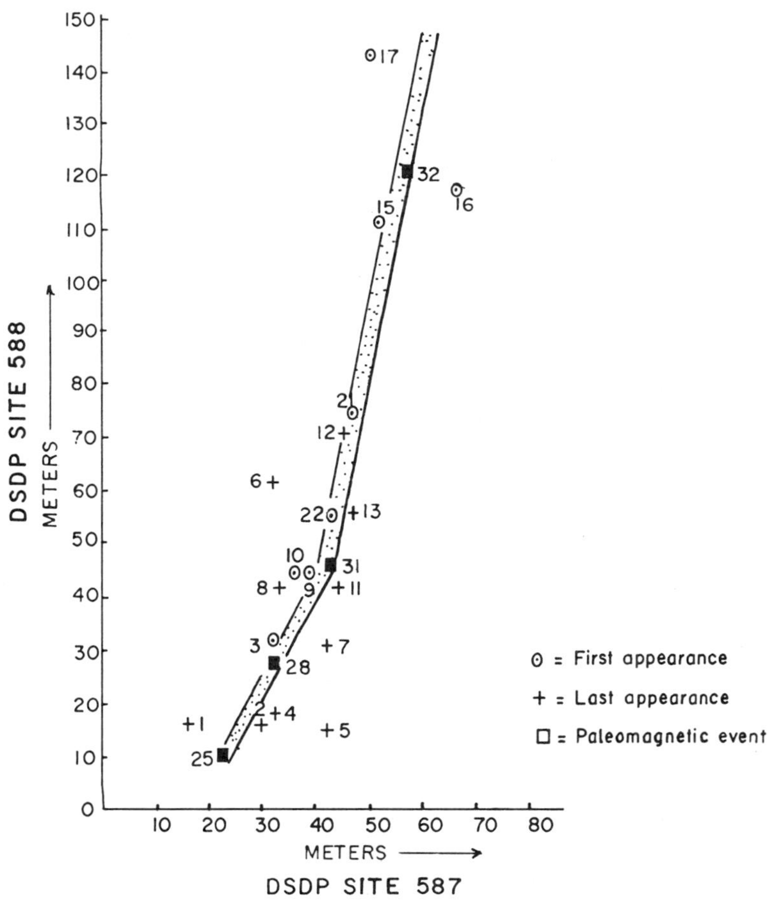

Fig. 3. Shaw plot of important events in Sites 588 (SRS) and 587 (See Table 4 for explanation of numbers).

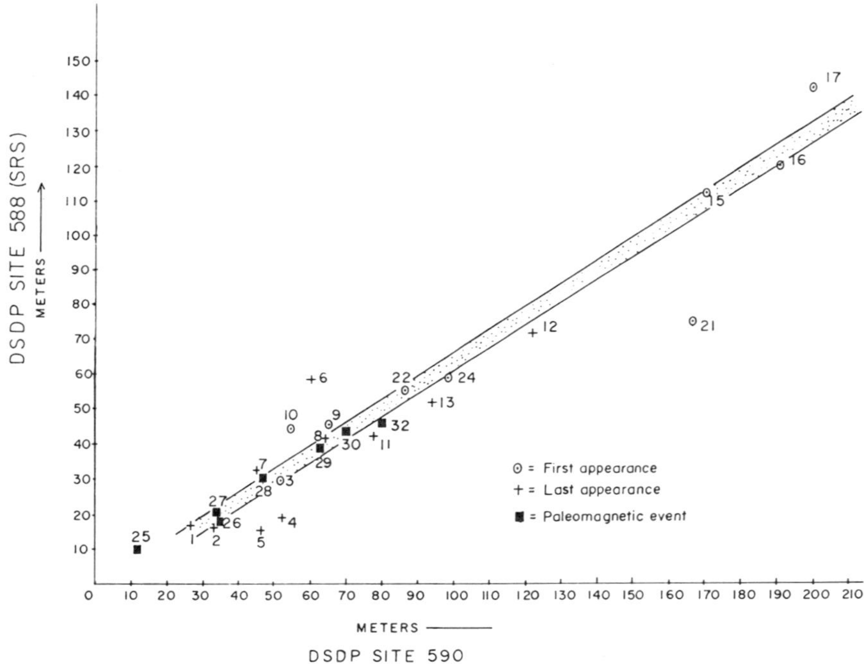

Fig. 4. Shaw plot of important events in Sites 588 (SRS) and 590 (See Table 4 for explanation of numbers).

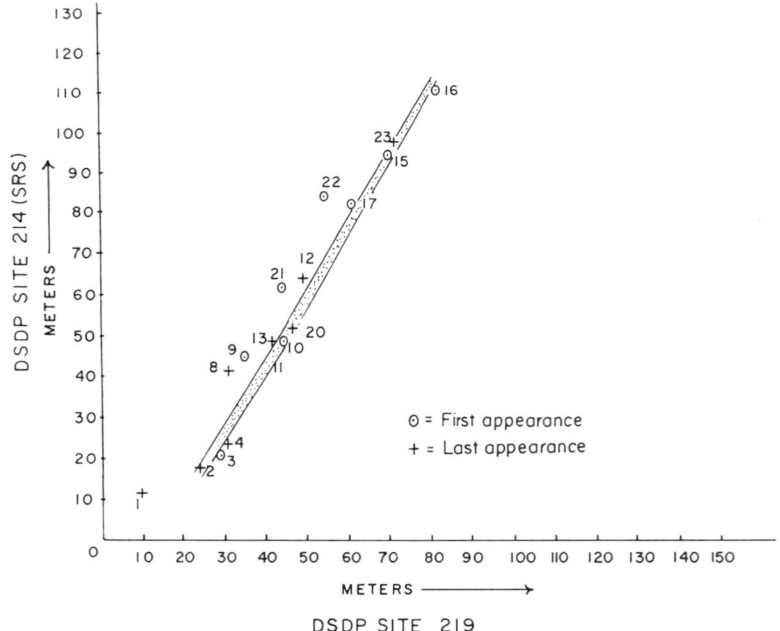

Fig. 5. Shaw plot of important events in Sites 214 (SRS) and 219 (See Table 4 for explanation of numbers).

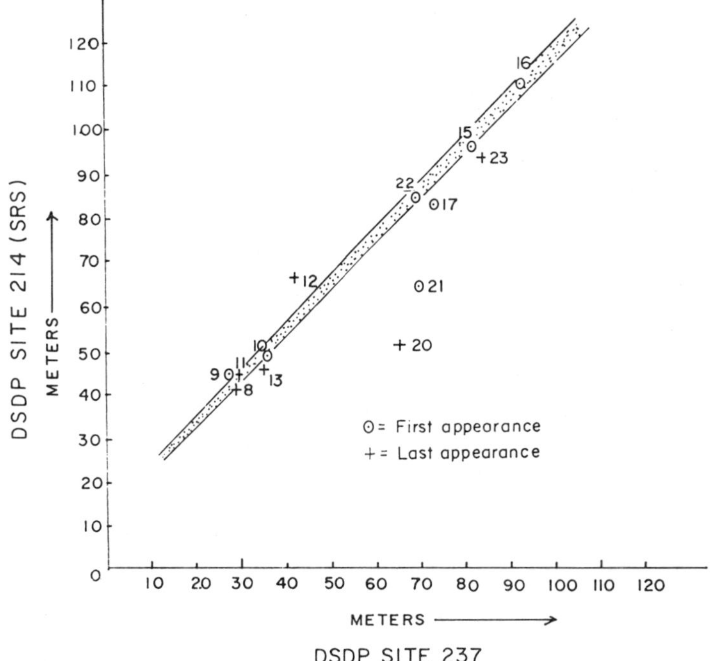

Fig. 6. Shaw plot of important events in Sites 214 (SRS) and 237 (See Table 4 for explanation of numbers).

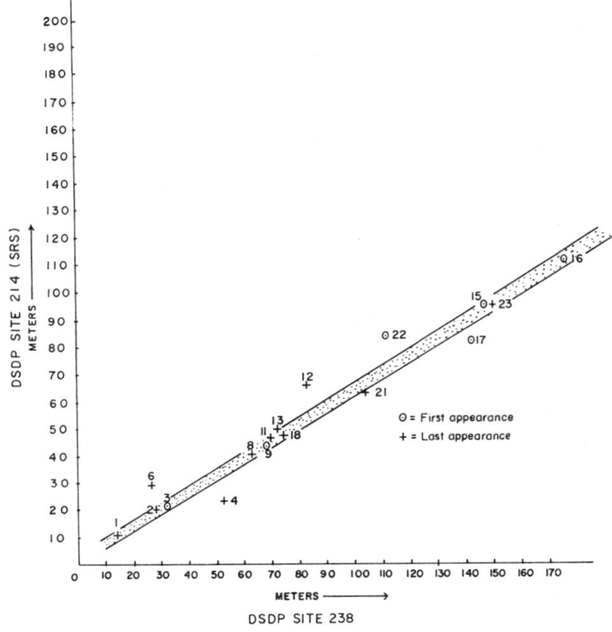

Fig. 7. Shaw plot of important events in Sites 214 (SRS) and 238 (See Table 4 for explanation of numbers).

      M.S. SRINIVASAN and D.K. SINHA

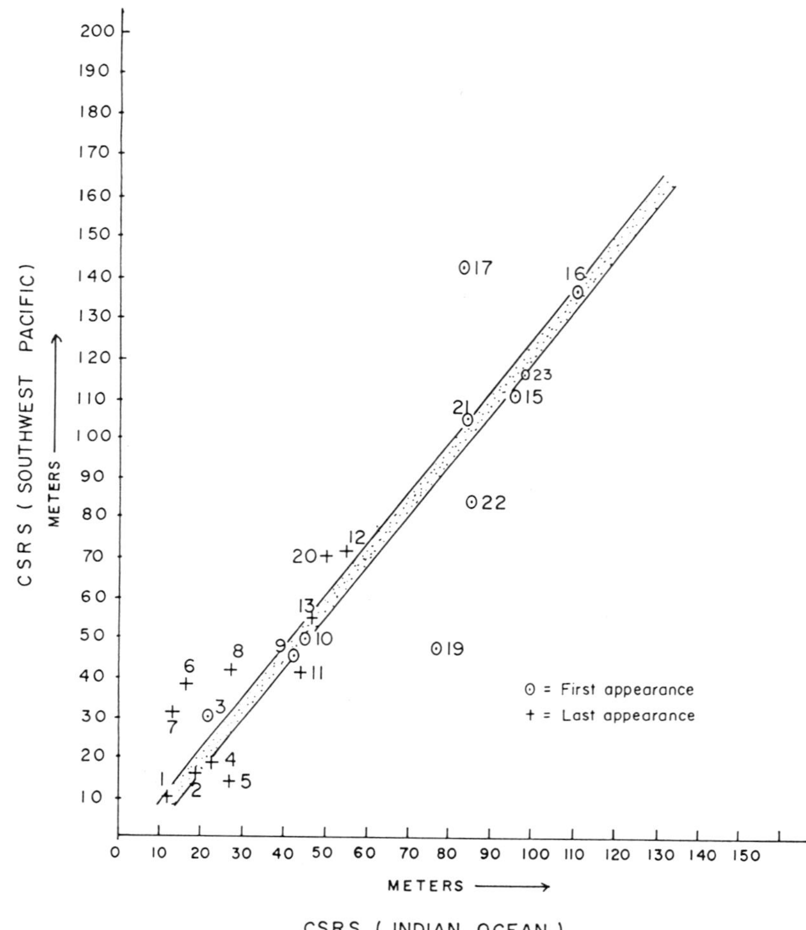

Fig. 8. Shaw plot of important events in CSRS (Southwest Pacific) and CSRS (Indian Ocean) (See Table 4 for explanation of numbers).

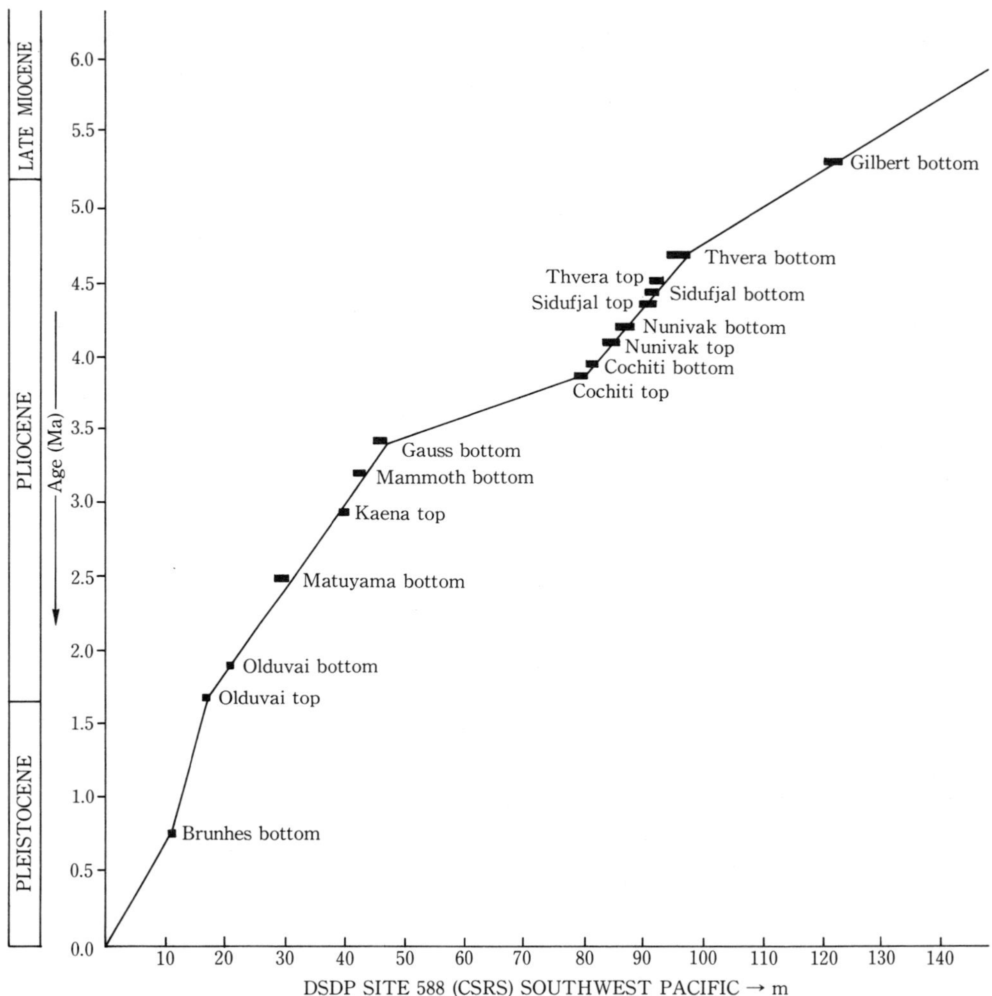

Fig. 9. Plot comparing the ages calculated for Late Neogene magnetic anomalies with the position of these anomalies in DSDP Site 588 (CSRS) Southwest Pacific (Ages after Berggren et al., 1985 and Paleomagnetic data after Barton and Bloemendal, 1986).

Table 2. Estimated ages of the planktonic foraminiferal events which are synchronous in the southwest Pacific and Indian Ocean

Planktonic foraminiferal Events	Estimated Ages	
	Present Work	Berggren et al. (1985)
Globorotalia tosaensis LA	0.6 Ma	0.60 Ma
Globigerinoides fistulosus LA	1.6 Ma	1.60 Ma
Globorotalia tosaensis FA	3.26 Ma	3.10 Ma
Globigerinoides fistulosus FA	3.42 Ma	2.90 Ma
Sphaeroidinellopsis spp. LA	3.10 Ma	3.00 Ma
Globigerina nepenthes LA	3.78 Ma	3.90 Ma
Globorotalia margaritae LA	3.46 Ma	3.40 Ma
Globorotalia crassaformis FA	4.90 Ma	4.30 Ma
Globorotalia tumida tumida FA	5.01 Ma	5.20 Ma
Globoquadrina dehiscens LA	5.30 Ma	5.30 Ma
Pulleniatina primalis FA	5.78 Ma	5.80 Ma

Table 3. Estimated ages and age differences of the diachronous planktonic foraminiferal events in the southwest Pacific and Indian Oceans

Planktonic foraminiferal events	Estimated ages in southwest Pacific	Estimated ages in Indian Ocean	Diff.
Globigerinoides obliquus LA	2.88 Ma	1.80 Ma	1.08 Ma
Globorotalia truncatulinoides FA	2.47 Ma	1.99 Ma	0.48 Ma
Globorotalia tumida flexuosa LA	2.60 Ma	3.50 Ma	0.90 Ma
Dentoglobigerina altispira LA	3.04 Ma	2.60 Ma	0.44 Ma
Globorotalia plesiotumida LA	3.78 Ma	3.60 Ma	0.18 Ma
Globorotalia tumida flexuosa FA	3.46 Ma	4.00 Ma	0.54 Ma
Sphaeroidinella dehiscene FA	4.10 Ma	4.90 Ma	0.80 Ma
Globorotalia margaritae FA	5.80 Ma	5.10 Ma	0.70 Ma

Late Neogene Planktonic Foraminiferal Events in the Southwest Pacific and Indian Ocean

Recently Srinivasan and Sinha (1991) critically evaluated the Late Neogene planktonic foraminiferal events in the southwest Pacific employing graphic correlation and grouped the events into two orders representing synchronous and diachronous events within the southwest Pacific. Srinivasan and Chaturvedi (1992) and Srinivasan and Singh (1992) discussed in detail the Neogene planktonic foraminiferal events in the Indian Ocean based solely on the biochronologic data (sequential order of events). In the present paper the Late Neogene planktonic foraminiferal events of the southwest Pacific and the Indian Ocean have been compared using graphic correlation and magnetostratigraphic data in order to achieve precise correlation between the two oceans. The study reveals the following Late Neogene planktonic foraminiferal events to be synchronous between the southwest Pacific and Indian Ocean:

Globorotalia tosaensis LA
Globigerinoides fistulosus LA
Globorotalia tosaensis FA
Globigerinoides fistulosus FA
Sphaeroidinellopsis spp. LA
Globorotalia margaritae LA
Globigerina nepenthes LA
Globorotalia crassaformis FA
Globorotalia tumida tumida FA
Globoquadrina dehiscens LA
Pulleniatina primalis FA

On the other hand, a number of planktonic foraminiferal events appear to be diachronous (Fig. 10) between the southwest Pacific and Indian Ocean. These include:

Globorotalia truncatulinoides FA
Globigerinoides obliquus LA
Dentoglobigerina altispira LA
Globorotalia tumida flexuosa LA
Globorotalia tumida flexuosa FA
Globorotalia plesiotumida LA
Sphaeroidinella dehiscens FA
Globorotalia margaritae FA

In the following section the above Late Neogene planktonic foraminiferal events are discussed based on the results of graphic correlation plots and magnetostratigraphic data.

Synchronous Planktonic Foraminiferal Events

Globorotalia tosaensis LA

This event lies on the Line of Correlation (LOC) between the Composite Standard Reference Sections (CSRS) for the southwest Pacific sites and the Indian Ocean sites (Fig. 8) and thus appears to be a synchronous event between the two oceans. Comparing the depth of this event in the CSRS (southwest Pacific) with magnetostratigraphic data, an age of 0.6 Ma for this event seems to be appropriate. Thus *Globorotalia tosaensis* LA appears to be a synchronous event between the southwest Pacific and Indian Ocean occurring at 0.6 Ma. Berggren et al. (1985) gave the same age for the last appearance of *Globorotalia tosaensis*.

Globigerinoides fistulosus LA

Recently Srinivasan and Sinha (1991), based on integrated biochronologic and paleomagnetic data, documented the last appearance of *Globigerinoides fistulosus* to be virtually coincident with the top of the Olduvai Normal Event, estimated at 1.6 Ma (Pliocene / Pleistocene boundary) in the southwest Pacific DSDP sites. The Shaw plot (Fig. 8) between the CSRS (southwest Pacific) and CSRS (Indian Ocean) shows

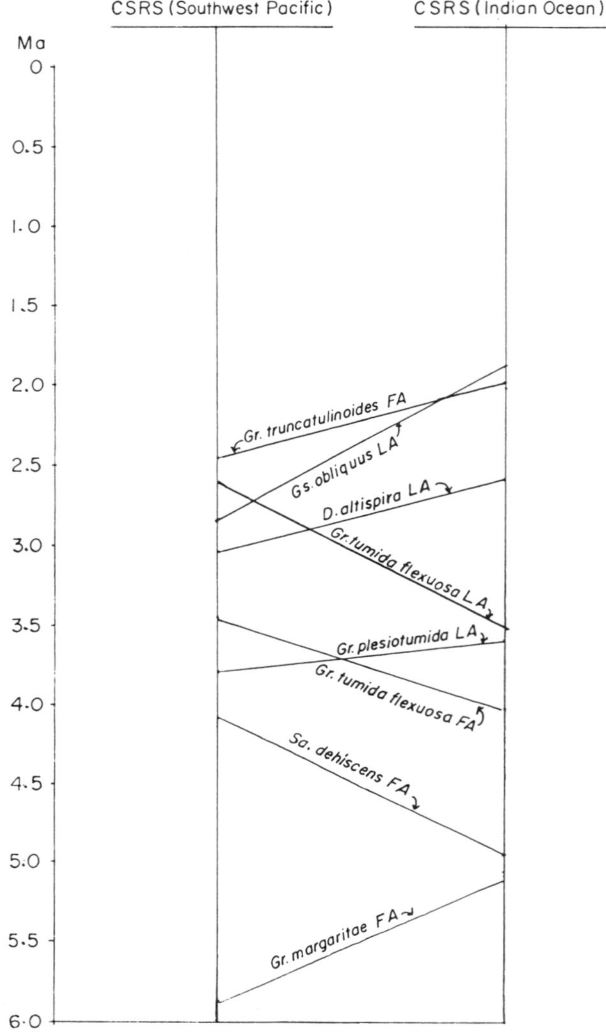

Fig. 10. Diachrony of selected late Neogene planktonic foraminiferal events between the southwest Pacific and Indian Ocean.

this event to lie on the line of correlation (Fig. 9) and hence the last appearance of *Globigerinoides fistulosus* is time-equivalent in the two oceans.

Globorotalia tosaensis FA

Globorotalia tosaensis FA occurs almost on the line of correlation in the Shaw plot between the Composite Standards for the southwest Pacific and Indian Ocean (Fig. 8), and thus appears to be a synchronous event. The magnetostratigraphic data for the CSRS (southwest Pacific) indicate an age of 3.26 Ma for this event; considering its synchronous nature, the same age can be assigned for the Indian Ocean. This age estimate is slightly higher than the age of 3.1 Ma given by Berggren et al. (1985).

Globigerinoides fistulosus FA

Globigerinoides fistulosus FA occurs slightly below *Globorotalia tosaensis* FA in the Composite Standard Reference Sections for the southwest Pacific and the Indian Ocean. Also *Gs. fistulosus* FA falls virtually on the line of correlation between the two composite standards (Fig. 8), thereby indicating its synchronous nature. An age of 3.42 Ma has been estimated for this event based on magnetostratigraphic data; this is slightly higher than the value of 3.2 Ma estimated by Saito (1977) based solely on site 289 (Site 586) data.

Sphaeroidinellopsis spp. LA

This event falls very close to the line of correlation between CSRS (southwest Pacific) and CSRS (Indian Ocean) (Fig. 8), and thus can be considered a nearly synchronous event. An age difference of only 0.1 m.y. has been estimated for this event between the southwest Pacific and Indian Ocean based on magnetostratigraphic data.

Earlier, Saito (1977) and Berggren et al. (1985) estimated an age of 3.0 Ma for the extinction level of *Sphaeroidinellopsis*. Later Saito (1984) gave a revised estimate of age for this event as 3.1 Ma. Both these age estimates appear to be correct in view of the present observation.

Globorotalia margaritae LA

The last appearance of *Globorotalia margaritae* lies on the line of correlation in the Shaw plot between composite standards of the southwest Pacific and the Indian Ocean (Fig. 8), indicating this event to be synchronous in the two oceans. An age of 3.46 Ma has been estimated for the last appearance of *Globorotalia margaritae* based on paleomagnetic data; this is very close to the age estimate given by Berggren et al. (1985).

Globigerina nepenthes LA

The last appearance of *Globigerina nepenthes* occurs earlier than the last appearance of *Globorotalia margaritae* in both the Composite Standard Reference sections for the southwest Pacific and Indian Ocean. The Shaw plot (Fig. 8) between the two composite standards reveals the last appearance of *Globigerina nepenthes* falling on the line of correlation, and thus this event appears to be synchronous between the southwest Pacific and Indian Ocean. The age estimate based on magnetostratigraphic data suggests a value of 3.78 Ma for the last appearance of *Globigerina nepenthes*, which is slightly lower than the age of 3.9 Ma estimated by Berggren et al. (1985).

Globorotalia crassaformis FA

The Shaw plot (Fig. 8) between the CSRS (southwest Pacific) and CSRS (Indian

Ocean) reveals the first appearance of *Gr. crassaformis* falling on the line of correlation, suggesting that this event is synchronous between the southwest Pacific and the Indian Ocean. The magnetostratigraphic data suggest an age of 4.9 Ma for the first appearance of *Gr. crassaformis*.

Berggren et al. (1985) estimated an age of 4.3 Ma for the first appearance of *Gr. crassaformis*, which is slightly younger than the age estimated here.

Globorotalia tumida tumida FA

The first appearance of *Globorotalia tumida tumida* corresponds very closely to the Miocene/Pliocene boundary (Berggren et al., 1985; Jenkins and Srinivasan, 1986; Srinivasan and Sinha, 1991), and thus is an important bioevent. The first appearance of *Globorotalia tumida tumida* lies on the line of correlation between the CSRS (southwest Pacific) and CSRS (Indian Ocean), indicating its synchronous nature between the two composite standards (Fig. 8).

An age of 5.01 Ma has been estimated for the first appearance of *Globorotalia tumida tumida* based on magnetostratigraphic data. This value is very close to the age of 5.2 Ma given by Berggren et al. (1985).

Globoquadrina dehiscens LA

The last appearance of *Globoquadrina dehiscens* falls on the line of correlation between CSRS (southwest Pacific) and CSRS (Indian Ocean), indicating its synchronous nature. The magnetostratigraphic data suggest an age of 5.3 Ma for the last appearance of *Globoquadrina dehiscens*.

Berggren et al. (1985) estimated the same age (5.3 Ma) for *Globoquadrina dehiscens* LA.

Pulleniatina primalis FA

Pulleniatina primalis FA lies very close to the line of correlation in the Shaw plot (Fig. 8) between CSRS (southwest Pacific) and CSRS (Indian Ocean), indicating its synchronous nature between the two composite standards. This event occurs earlier than *Globorotalia tumida tumida* FA in both the CSRS. The magnetostratigraphic data suggests an age of 5.78 Ma for the first appearance of *Pulleniatina primalis*. Berggren et al. (1985) gave an age of 5.8 Ma for this event.

The complete absence of Early Pliocene species *Pulleniatina spectabilis* from the examined Indian Ocean sites, in contrast to its dominant presence in the southwest Pacific, appears to reflect the severing of the surface water connection between the tropical Pacific and the Indian Ocean by the end of the Miocene.

Diachronous Planktonic Foraminiferal Events

Globorotalia truncatulinoides FA

Progressive development of *Globorotalia truncatulinoides* from its ancestor *Globorotalia tosaensis* has been clearly traced in all the DSDP sites studied. *Globorotalia truncatulinoides* FA falls away from the line of correlation (toward the left) in the Shaw plot (Fig. 8) between CSRS (southwest Pacific) and CSRS (Indian Ocean), showing a late or delayed appearance in the Indian Ocean. The magnetostratigraphic data reveals this event to be virtually coincident with the Matuyama/Gauss boundary (~2.47 Ma) in the southwest Pacific. An age extrapolation based on the Shaw plot suggests *Gr. truncatulinoides* FA to occur at least 0.48 Ma later in the Indian Ocean than in the southwest Pacific. Therefore the first appearance of *Globorotalia truncatulinoides* is not a synchronous event and can no longer be used for interoceanic correaltion between the southwest Pacific and the Indian Ocean.

Globigerinoides obliquus LA

This event lies far away from the line of correlation (toward the left) in the Shaw plot (Fig. 8) between CSRS (southwest Pacific) and CSRS (Indian Ocean) and thus appears to be a highly diachronous event. The extent of its diachroneity, estimated on the basis of magnetostratigraphic data, appears to be of the order of 1.08 m.y. Thus the present study suggests that the *Globigerinoides obliquus*, last appearance was much earlier in the southwest Pacific (2.88 Ma) than in the Indian Ocean (1.8 Ma).

Dentoglobigerina altispira LA

Like *Globigerinoides obliquus* LA, the last appearance of *D. altispira* seems to be much earlier in the southwest Pacific than in the Indian Ocean, as indicated by the Shaw plots (Fig. 8) of CSRS for the two oceans. From the paleomagnetic data an age of 3.04 Ma has been estimated for this event in the southwest Pacific, and the age extrapolation based on the line of correlation between the two CSRS suggests a value of 2.6 Ma for this event in the Indian Ocean.

Thus, from the above it is suggested that *D. altispira* survived for a further period of 0.44 Ma in the Indian Ocean after becoming extinct in the southwest Pacific.

Saito (1984) estimated an age of 2.82 Ma for this event, while Berggren et al. (1985) gave an age of 2.9 Ma. The former age estimate is close to what we have estimated for the Indian Ocean while the latter seems to be close to the value calculated for the southwest Pacific.

Globorotalia tumida flexuosa LA

The last appearance of *Globorotalia tumida flexuosa* occurs at the same level as the *Globorotalia margaritae* LA (3.46 Ma) in the CSRS (Indian Ocean). On the other hand, the last appearance of *Globorotalia tumida flexuosa* occurs slightly above the *Sphaeroidinellopsis* LA (3.0 Ma) in the CSRS (southwest Pacific). The Shaw plot between the above two CSRSs reveals that the last appearance of *Gr. tumida flexuosa* lies away from the line of correlation (toward the right), indicating the event to be earlier in the Indian Ocean than in the southwest Pacific. The above observation suggests that *Gr. tumida flexuosa* became extinct in the Indian Ocean at least 0.46 m.y. earlier than in the southwest Pacific.

Globorotalia tumida flexuosa FA

The first appearance of *Globorotalia tumida flexuosa* occurs slightly below the level of *Globigerinoides fistulosus* FA (3.42 Ma) in the CSRS (southwest Pacific). On the other hand *Gr. tumida flexuosa* FA occurs much earlier than *Globigerina nepenthes* LA (3.78 Ma) in the CSRS (Indian Ocean). The graphic correlation between the two composite standards (Fig. 8) shows the first appearance of *Gr. tumida flexuosa* falling far away (toward the right) from the line of correlation. This suggests delayed *Gr. tumida flexuosa* FA in the southwest Pacific as compared to the Indian Ocean.

Magnetostratigraphic data indicate that *Gr. tumida flexuosa* appeared in the Indian Ocean at about 4.0 Ma as compared to the southwest Pacific, where this species appeared much later, at 3.46 Ma.

Globorotalia plesiotumida LA

The last appearance of *Globorotalia plesiotumida* occurs later than that of *Globigerina nepenthes* LA in the composite standards for both the southwest Pacific and Indian Ocean. The Shaw plot (Fig. 8), however, reveals that the last appearance of *Gr. plesiotumida* lying slightly away from the line of correlation between CSRS (southwest Pacific) and CSRS (Indian Ocean) suggests the event to be slightly earlier in the southwest Pacific.

Age estimates based on the magnetostratigraphic data reveals that *Gr. plesiotumida* became extinct 0.18 m.y. earlier in the southwest Pacific (3.78 Ma) than in the Indian Ocean, where it made its last appearance at about 3.6 Ma.

Sphaeroidinella dehiscens FA

The Shaw plot (Fig. 8) between Composite Standard Reference Sections of the southwest Pacific and Indian Ocean reveals the first appearance of *Sphaeroidinella dehiscens* falling away from the line of correlation (toward the right), indicating later first appearance of this taxon in the southwest Pacific than in the Indian Ocean. The magnetostratigraphic data suggest an age of 4.1 Ma for the first appearance of *Sa. dehiscens* in the southwest Pacific and 4.9 Ma in the Indian Ocean.

Berggren et al. (1985) gave an age of 5.1 Ma for the first appearance of *Sa. dehiscens*, which is close to the age estimate of 4.9 Ma of the present study for the Indian Ocean.

Globorotalia margaritae FA

The first appearance of *Globorotalia margaritae* has been recorded much earlier than that of *Gr. tumida tumida* FA (5.01 Ma) in the CSRS (southwest Pacific). On the other hand, *Gr. margaritae* FA occurs slightly above *Gr. tumida tumida* FA (5.01 Ma) in the CSRS (Indian Ocean). The Shaw plot (Fig. 8) between the above two composite standards shows *Gr. margaritae* FA falling far away from the line of correlation (toward the left), suggesting this event to be much later in the Indian Ocean than in the southwest Pacific.

The age estimate based on magnetostratigraphic data suggests that *Gr. margaritae* appeared in the southwest Pacific at 5.8 Ma and in the Indian Ocean at about 5.1 Ma. Hoddell and Kennett (1986) estimated an age of 5.97 Ma for *Gr. margaritae* FA at site 588 (southwest Pacific), while Berggren et al. (1985) estimated an age of 5.6 Ma for this event.

From the available data it appears that *Gr. margaritae* first evolved in the southwest Pacific and later migrated to the Indian and Atlantic Oceans.

Acknowledgements

Thanks are offered to the Deep Sea Drilling Project (DSDP) for providing the core samples. Authors are grateful to Dr. Arun Deo Singh and Mr. Ajai Kumar Rai for diverse technical assistance. Financial support from the Council of Scientific and Industrial Research (CSIR) New Delhi is thankfully acknowledged.

Table 4. Event codes used in the Shaw plots.

Planktonic foraminiferal events	Code
Gr. tosaensis LA	1
Gs. fistulosus LA	2
Gr. truncatulinoides FA	3
Gr. multicamerata LA	4
Gg. decoraperta LA	5
Gs. obliquus LA	6
Gs. extremus LA	7
D. altispira LA	8
Gr. tosaensis FA	9
Gs. fistulosus FA	10
Sphaeroidinellopsis spp. LA	11
Gg. nepenthes LA	12
Gr. margaritae LA	13
Pu. spectabilis FA	14
Gr. tumida tumida FA	15
Pu. primalis FA	16
Gr. margaritae FA	17
Gr. tumida flexuosa LA	18
Gr. tumida flexuosa FA	19
Gr. plesiotumida LA	20
Gr. crassaformis FA	21
Sa. dehiscens FA	22
Gq. dehiscens LA	23
Paleomagnetic Events	
Brunhes/Matuyama	25
Olduvai Top	26
Olduvai Bottom	27
Matuyama/Gauss	28
Kaena Top	29
Mammoth Bottom	30
Gauss/Gilbert	31
Gilbert/Chron–5	32

References

Barton, C.E. and Bloemendal, J., 1986. Paleomagnetism of sediments collected during Leg 90, southwest Pacific. *In: Initial Repts. DSDP*, 90: 1273–1816.

Berggren, W.A., 1984. Neogene planktonic foraminiferal biostratigraphy and biogeography: Atlantic, Mediterranean, and Indo-Pacific regions. *In:* N. Ikebe, and R. Tsuchi, (Eds.), Pacific Neogene Datum Planes: Contributions to biostratigraphy and Chronology. University of Tokyo Press, pp. 111–161.

Berggren, W.A., Kent, D.V., and Van Couvering, J.A., 1985. Neogene geochronology and chronostratigraphy: The Neogene Part II. *In:* N.J. Snelling (Ed.), The Chronology of the Geological Record. Geol. Soc. Mem. no. Blacwell Scientific Publ., Oxford London, pp. 211–260.

Hodell, D.A. and Kennett, J.P., 1986. Late Miocene- Early Plocene Stratigraphy and Paleoceanography of the south Atlantic and southwest Pacific Oceans: A synthesis. *Paleoceanography*, 1(3): 285–311.

Jenkins, D.G. and Srinivasan, M.S., 1986. Cenozoic planktonic foraminifers from the equator to the subantarctic of the southwest Pacific. *Initial Repts. DSDP*, 90: 795–834.

Kennett, J.P. and Srinivasan, M.S., 1983. Neogene planktonic foraminifera: A phylogenetic Atlas., Hutchinson Ross Publ. Co., U.S.A., 265 p.

Saito, T., 1977. Late Cenozoic planktonic foraminiferal datum levels: the present state of knowledge toward accomplishing Pan Pacific correlation *In:* International Pacific Neogene Stratigraphy, First, Tokyo, 1796, Proc., pp. 61–80.

Saito, T., 1984. Planktonic foraminiferal Datum planes for biostratigraphic correlation of Pacific Neogene sequences — 1982 Status Report. *In:* N. Ikebe, and R. Tsuchi, (Eds.), Pacific Neogene Datum Planes, University of Tokyo Press., pp. 3–10.

Shaw, A.B., 1964. Time in stratigraphy: New York. McGraw Hill, 365 p.

Srinivasan, M.S. and Azmi, R.J., 1979. Correlation of Late Cenozoic marine sections in Andaman- Nicobar, northern Indian Ocean, and the equatorial Pacific. *Jour. Paleontol.*, 53(6): 1401–1415.

Srinivasan, M.S. and Chaturvedi, S.N. (1992), Neogene planktonic foraminiferal biochronology of the DSDP sites along ninetyeast Ridge, *In:* K. Ishizaki, and T. Saito, (Eds.), Northern Indian Ocean. Centenary of Japanese Micropaleontology, pp. 175–188.

Srinivasan, M.S. and Shingh, A.D. (1992), Neogene planktonic foraminiferal biochronology of DSDP site 210 (Chagos Laccadive Ridge), Arabian Sea. Proceedings. Indian National Science Academy, New Delhi, 4(58): 335–354.

Srinivasan, M.S. and Shinha, D.K., 1991. Improved correlation of the Late Neogene planktonic foraminiferal datums in the equatorial to cool subtropical DSDP sites, southwest Pacific: Application of the Graphic Correlation Method. *Geol. Soc. India Mem.*, 20: 55–93.

Larger Foraminifera and the Dating of Neogene Events

C.G. Adams

Department of Palaeontology, British Museum (Natural History), London SW7 5BD, U.K.

Abstract

The nature of geological events is briefly considered and Cenozoic examples susceptible to evaluation by means of larger foraminifera are discussed. Attention is drawn to deficiencies in the quantity and quality of biostratigraphical and distributional information employed in dating and evaluating the Mid-Neogene Climatic Optimum, and it is concluded that the uneven spread of critical data affects our perception of this event. Reviews of three currently accepted extinction datums (those of *Austrotrillina*, *Lepidocyclina*, and *Miogysina*) show that they are not reliable when applied across the entire Indo-Pacific region, and suggest that late occurrences of genera, such as *Lepidocyclina* from the Izu Peninsula, Japan, cannot necessarily be regarded as satisfactory indicators of surface-water palaeotemperatures.

Introduction

IGCP project 246 has been concerned with the recognition, description and dating of Neogene events, evaluation and assessment of which depend upon accurate correlation of sediments over wide areas.

Events, in everyday language, are happenings, usually of short duration, each of which has a clearly marked beginning and an end (birthday parties provide familiar sociological examples). Some naturally occurring events, such as earthquakes, are similarly defined, since they are of short duration and can be seen, heard and felt by on-the-spot observers, inferred soon afterwards from their destructive surface effects, or deduced millions of years later from the structures (e.g., faults) which they produce. Most geological events are, however, less easily defined because they are of long duration and neither commence nor terminate at clearly recognizable points in time.

Larger foraminifera are invaluable in assessing the scale of events such mass extinctions, eustatic, climatic, and sea-surface temperature changes, and tectonically induced palaeogeographic changes previously determined or hypothesized from other palaeontological or geological evidence.

Cenozoic biotic events (first appearances and extinctions) began in the immediate aftermath of the terminal Cretaceous extinction, a catastrophe which left numerous ecological niches to be filled in the World Ocean. The scale of this extinction provides a quantifiable standard against which subsequent events of this kind can be measured (for a review see Adams, 1989). Unfortunately, the most important and easily recog-

221

nizable Tertiary biotic events (so-called mass extinctions) occurred in Palaeogene times and cannot be considered here. Similarly, the most dramatic of sea-level changes postulated by Vail et al. (1977) also occurred during the Palaeogene (Chattian). The closure (tectonic) of the Tethys Seaway and the terminal Miocene event (Messinian Salinity Crisis) were reviewed relatively recently (Adams et al., 1983; Cita and McKenzie, 1986), and are not reconsidered since neither could have affected the Pacific region other than marginally.

Neogene events relevant to the Pacific region, and susceptible to investigation by palaeontological methods, include sea-surface temperature fluctuations and palae-oceanographic and climatic changes, all of which tend to affect the geographical distribution of marine and terrestrial organisms in general, and larger foraminifera in particular. The evidence for them is not wholly preserved in one locality or stratigraphic section, but is scattered across the entire region, or even over the whole world. It is, therefore, necessary to date and correlate many individual sequences in order to determine the magnitude of the events themselves, and final understanding of any one of them depends on how well this can be done.

Correlation is undertaken using a variety of methods and techniques (biostrati-graphical, lithostratigraphical, radiometric, geochemical, palaeomagnetic, isotopic and geophysical), all of which have their advantages and limitations. Unfortunately, the most widely used and longest established correlative technique (biostratigraphy) tends also to be applied with the least rigour. This paper therefore examines the way in which some biostratigraphical information has been employed in the interpretation of events during recent years. It questions whether current levels of distributional data always provide an adequate basis for the conclusions drawn from them, and discusses the general reliability of extinction datums by reference to three particularly important Miocene genera. Its aim is to draw attention to deficiencies in the quantity and quality of data employed in the relative dating and evaluation of Neogene events, and by so doing to encourage the publication of additional information.

The Mid-Neogene Climatic Optimum

This term was introduced by Tsuchi (1987) when age determinations based on planktonic foraminifera showed that the seas around Japan were probably at their warmest during a one-million-year interval straddling the Lower/Middle Miocene boundary.

Larger foraminifera, which are essentially circumtropical in distribution (Fig. 1), have long played an important role in discussions on Cenozoic climatic optima. Changes in latitudinal range are rather easily recognized. Generic diversity is greatest within the tropics and decreases rapidly when the mean sea-surface temperature for the warmest month of the year falls below 24°–25°C. Larger foraminifera disappear altogether when temperatures drop below 18°–20°C (for a detailed discussion see Adams et al., 1990).

If shallow-water benthic foraminifera have not changed physiologically during

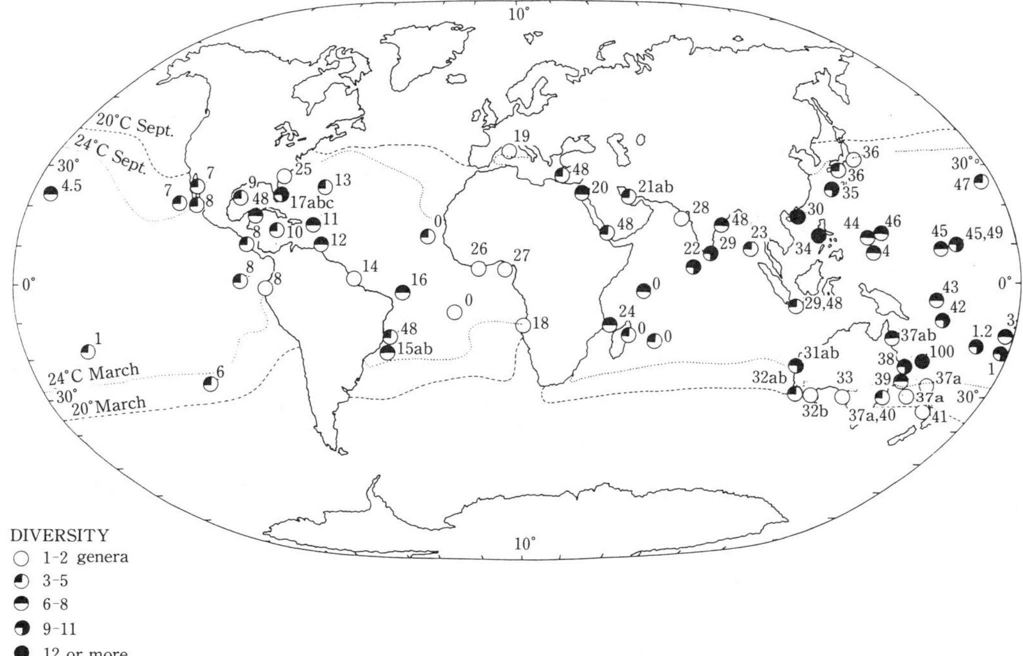

Fig. 1. Diversity distribution of Recent larger foraminiferal genera. Note that very low tropical/subtropical diversities, e.g. at 13 (Bermuda), usually reflect inadequate sampling through the euphotic zone. For full explanation see Adams et al. (1990).

Cenozoic times (and there is no reason to suppose that they have), it follows that similar diversity distribution patterns should be recognizable in all epochs. It might be expected that the latitudinal limits of larger foraminifera would expand polewards when sea-surface temperatures and climates were warmer than today, and would contract toward the equator when global conditions were cooler. It would, of course, be unsafe to draw general conclusions about climatic or sea-surface temperature changes from foraminiferal distributions within a single region, since these are locally dependent on current patterns (Fig. 2) which may themselves be affected by major tectonic movements.

Maps showing the diversity distributions of larger foraminifera of Middle and Late Eocene, Early and Middle Oligocene, and Middle Miocene age have already been published (Adams et al., 1990) and are not repeated here. Attention is, however, drawn to the great difference in the quantity of distributional data available for present-day, Middle and Late Eocene taxa, compared with that for the Middle Miocene. The numerous Eocene records (Adams et al., 1990, Figs. 3 and 4) reflect both the world-wide distribution of shallow-water carbonate sediments and the amount of research done on them, whereas the relatively few M. Miocene records (*op. cit.*, Fig. 7) expose the paucity of published Middle Miocene data. Although

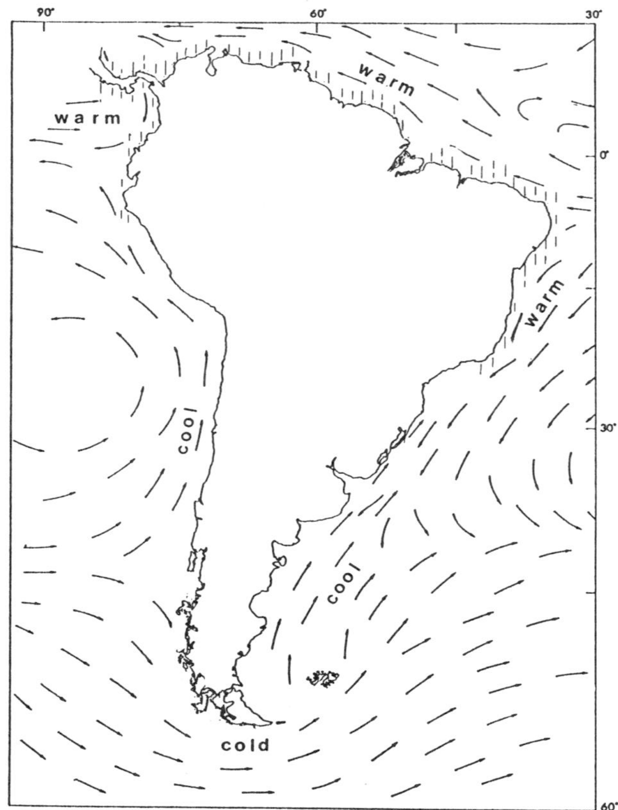

Fig. 2. Map showing the relationship between current patterns and the distribution of larger foraminifera around South America. Vertical lines mark the presence of larger foraminifera. Base map modified after Boltovskoy (1976).

carbonates of Mid to Late Miocene age are well-developed and widely distributed, their larger foraminiferal faunas are poorly described. This is almost certainly because assemblages of this age rarely include new taxa, and have not therefore been thought to be of special taxonomic or stratigraphical significance. Correlation is, however, not the only purpose served by larger foraminifera, and one consequence of recent failures to list or illustrate additional occurrences is that we now lack sufficient biostratigraphical and palaeobiogeographical data to adequately date and evaluate events such as climatic optima on the basis of larger foraminifera alone. Most existing records of mid-Miocene larger foraminifera are concentrated on the western side of the Pacific Basin (Adams et al., 1990, Fig. 7), and although they certainly suggest a poleward expansion of latitudinal limits at this time, they are too few for this to be certain. Furthermore, it is unsafe to assume that the critical northerly and southerly records are isochronous to within 1 million years. Coeval American records are sparse, and those from the Mediterranean and Europe are of limited value because the closure of the Tethys Seaway greatly reduced generic diversity in the Mediterranean region (Adams et al., 1983; Adams, et al., 1990). Although two of the genera (*Borelis* and *Heterostegina*) which survived into the Middle Miocene extended northwards in Europe to the latitude of northern Japan, so did others earlier in the Miocene, and even during the Early to Mid-Oligocene when climatic conditions are thought to have been cooler. The absolute latitudinal limits of warm-water taxa are not, therefore, satisfactory guides to palaeotemperature, and, insofar as the larger foraminifera are concerned, only generic diversity is currently regarded as acceptable.

Our perception of the Mid-Neogene Climatic Optimum is therefore based on very limited data. Although it appears that warm-water genera occupied a somewhat broader latitudinal belt than today, this may not have been significantly wider than during the Late Oligocene and Early Miocene, for which subepochs relevant distributional data have yet to be collated. Sato (1987) has pointed out that information on the faunas and floras of Japan is particularly poor for the Early Miocene, and has questioned whether the pollen evidence supports a maximum warming at the N8/N9 boundary. The Mid-Neogene Climatic Optimum may, indeed, have been no more than a minor blip on a warming trend that began in Chattian time.

Various authors, e.g. Ikebe and Chiji (1971), have used changes in the geographical and stratigraphical distributions of *Lepidocyclina* and *Miogypsina* in Japan to infer climatic and/or sea-water temperature changes. This procedure is satisfactory only when the taxa concerned can be shown (by other evidence) to have lived in the area at the critical times, thus reducing the probability that they could be reworked. The long-running debate about the significance of these genera in Middle to Late Miocene sediments of Japan can be very largely attributed to the fact that this condition has not yet been met.

Attention must be drawn to occurrences of mangroves in the Miocene, since their distributional pattern today is very similar to that of the larger foraminifera (Adams et al., 1990). Clearly, we need to know as much as possible about the overall distribution of Miocene mangroves (and, indeed, of other groups of fossil organisms) before attempting to draw firm conclusions about climatic conditions and sea-surface temperatures either from limited planktonic data or from restricted assemblages of larger foraminifera, some of which may very well be reworked.

Extant hermatypic corals are also good sea-surface temperature and climatic indicators (Rosen, 1984; Adams et al., 1990), but their fossil representatives are difficult to use in event studies since most known assemblages have not yet been dated with sufficient precision. Geochemical techniques using strontium isotopes may, however, soon remedy this.

Extinction Datums

As already mentioned, the amount of biostratigraphic information available for dating Cenozoic events differs from epoch to epoch and region to region. This poses a problem which, for the Neogene at least, is exacerbated by the reliance placed on a small number of datum levels defined by first and last occurrences. The writer

(1989) recently plotted the last-known occurrences of *Austrotrillina* and *Lepidocyclina* for the three main faunal provinces, and showed that each genus disappeared at different times in different areas. Their generally accepted extinction datums are thus unreliable circumtropically, and even create problems within the Indo- West Pacific region.

Since no new evidence for the extinction of *Austrotrillina*, a miliolacean commonly found associated with larger foraminifera, seems to have been published recently, the writer's (1968) records are used here (Fig. 3).

Determination of the final extinction level of *Lepidocyclina* is complicated by the following factors:

1. Extinctions in the major provinces were not isochronous (Fig. 3). The lepido-cyclinids of the western Tethys, for example, disappeared at about the time the Mediterranean lost its connection with the Indian Ocean (Adams et al., 1983), whereas those in the Indo- West Pacific survived longer.

2. The last species to become extinct in each region were different. The youngest American species is unknown from the Indo-West Pacific and *vice versa*.

3. The geologically young records (Fiji excepted) are not from well-described limestones of unequivocal shallow-water origin, but from geographically isolated samples, or from thin limestones in deeper-water successions.

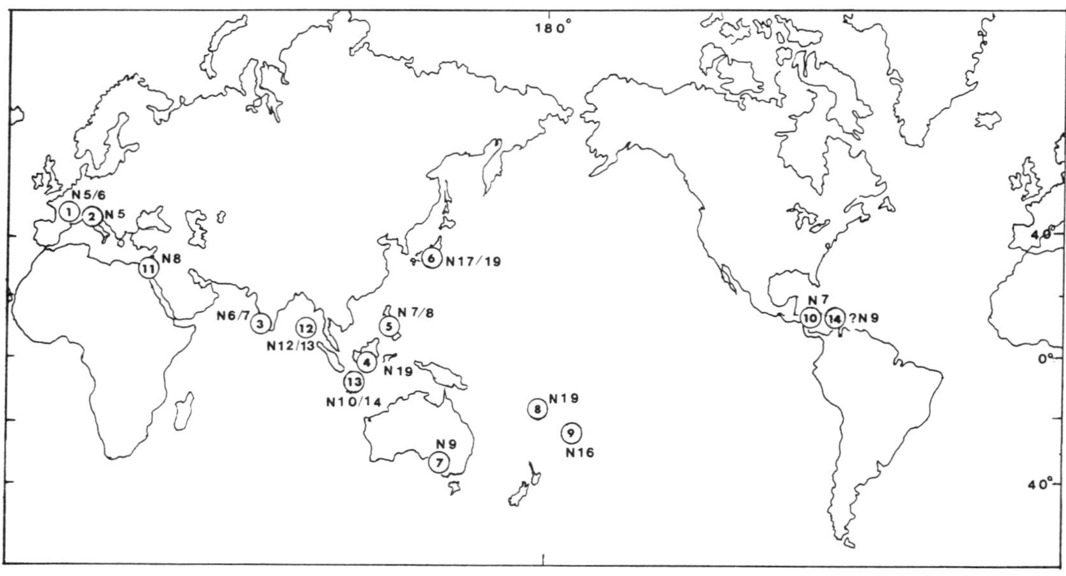

Fig. 3. Some later occurrences of *Austrotrillina*, *Lepidocyclina* and *Miogypsina* in the three major circumtropical faunal provinces. Note that many of the youngest records of *Lepidocyclina* and *Miogypsina* may represent reworked material (see text). *Austrotrillina* (2, 3, 5, 7), *Lepidocyclina* (1, 4, 6, 8, 9, 10), *Miogypsina* (11–14). Loc. 11 (Souaya, 1961), 12 (Raju & Chidambaram, 1989), 13 (Van Vessem, 1978), 14 (Butterlin, 1981 & pers. comm. 1984). For other details see Adams (1989).

A further problem arises because some geologically young records of *Lepidocyclina*, e.g. those from Japan and Borneo (Fig. 3), could be of reworked specimens, although Japanese authorities such as Ibaraki and Tsuchi (1978) and Ibaraki (1988, 1990) believe otherwise. The Fijian occurrences (see Adams et al., 1979) occur in facies favorable for the occurrence of the genus, but should nevertheless be treated with caution until other records of N17 − N19 age have been adequately documented. Asano's (1965) Late Miocene or Early Pliocene record of *Trybliolepidina* from the basal part of the Carcar Limestone, Cebu, Philippines, was unfigured and, according to Hashimoto et al. (1977), should be checked for reworking. It is evident from Hashimoto and Balce (1977) that there is confusion over the age and provenance of faunas said to be from this limestone, and it is by no means certain that the older Barili Limestone (containing only *Trybliolepidina rutteni*) is any younger than Middle Miocene. Positive information from Middle to Late Miocene sediments in the equatorial part of the Indo-Pacific region is clearly needed if the extinction pattern of *Lepidocyclina* is to be determined satisfactorily.

The danger of accepting extinction levels which have not been fully substantiated is demonstrated by Ibaraki's use (1988) of a *Lepidocyclina* extinction datum within N19. She contends that the occurrence of *Lepidocyclina* (*Nephrolepidina*) in beds of M. Miocene to Early Pliocene age in the Izu Peninsula, Japan, supports the claim by Hirooka et al. (1985) that this small area drifted northwards through approximately 20° of latitude and collided with the Japanese mainland some 3–5 m.y. ago. Her thoughtful argument rests on the absence of *Lepidocyclina* from Upper Miocene sediments on the mainland proper, and on acceptance of an extinction datum within N19 for *Lepidocyclina*. But this datum presents a problem. The best-documented N19 occurrence of *Lepidocyclina* is from the Lami Limestone of Fiji (Adams et al., 1979), but the species in Fiji and Japan are different, and there are no satisfactory N15–N18 records from the North Pacific region to support the contention that *L. (Nephrolepidina)* persisted through the Late Miocene. Until the regional extinction level of *Lepidocyclina* can be firmly established, it is unwise to assume that it was everywhere (or, indeed, anywhere) in N19. Reworking provides a more likely explanation for species found in Late Miocene and younger sediments, especially in areas known to have been active tectonically. In this connection, attention is drawn to the association of *Miogypsina* with *Lepidocyclina* in the Nashimoto locality (Ibaraki, 1981, 1988) at a level in Zone N17, i.e., far higher than any previously postulated for the extinction of the former genus (see below).

An even better demonstration of the unreliability of many currently accepted Neogene extinction datums is provided by *Miogypsina*, a genus well known through the evolutionary studies of Tan Sin Hok (1936, 1937), Drooger (1952, 1963) and Raju (1974).

During the last 20 years it has been widely accepted that *Miogypsina* s.s. became extinct during the Middle Miocene. (in planktonic zone N12 according to Clarke and Blow, 1969; in N14 according to Ikebe and Chiji, 1971; in N12–N14 according to van Vessem, 1978; and at approximately equivalent levels in terms of the East Indies

Letter Classification (Adams, 1984; Hashimoto et al., 1977). However, in 1963, Drooger stated that he knew of no valid records of *Miogypsina* after the appearance of *Orbulina*, and in 1979 still regarded this as true, at least for the Mediterranean region. In this connexion, it is worth noting that Souaya (1963) showed that the last-known occurrences of *Miogypsina* in Egypt were below the earliest *Borelis melo curdica*, and therefore probably below the *Orbulina* datum (see Adams et al., 1983). Post-Lower Miocene records from outside the Mediterranean region are, therefore, reviewed below.

The Indo-West Pacific

In this region, it is necessary to distinguish between the extinction of *Miogypsina* (*Miogypsina*) and *M.* (*Miogypsinoides*) since the disappearance of the latter subgenus has been the cause of some confusion. It is known (Drooger, 1963) that *Miogypsinoides* became extinct during the Early Miocene elsewhere in the world, but the indigenous Indo-West Pacific species, *Miogypsinoides cupulaeformis* (Zuffardi-Comerci), poses a problem. Cole (1954, 1958) recorded this species from the upper part of the Tf limestones penetrated in the Bikini and Eniwetak (= Eniwetok) drill holes. He had, however, no evidence to support the Tf$_3$ age which he inferred, and his latest Miocene beds might, in fact, be no younger than Tf$_1$ on the basis of the other foraminiferal data. Cole's (1945) two records of *Miogypsina neodispansa* (Jones and Chapman) from Vanua Mbalavu (Lau Islands) occurred in isolated samples, and were subsequently (Cole, 1954: 602) referred to *Miogypsinoides* by Cole himself; at the same time, he noted their resemblance to *Ms cupulaeformis*. The writer has examined similar material from Lau and is doubtful if it can properly be assigned to the Miogypsinidae. It certainly does not belong to *Miogypsina* (*Miogypsina*), and its occurrence is not, therefore, relevant to the extinction level of this subgenus.

Modern Middle Miocene records of *Miogypsina* s.s. known to the writer are listed in Table 1, problems associated with their uncritical acceptance being noted in the right-hand column.

It should be noted that Van Vessem's valuable paper (1978) does not in itself establish the *in situ* occurrence of *Miogypsina* in post–N9 sediments, and against it must be set the failure of Carozzi et al. (1976) to establish its presence in Middle Miocene carbonates in the Philippines. During the last three years, Raju and Chidambaram (1989) have established the presence of *Miogypsina antillea* in beds as young as N12/N13 in the Andaman Basin, and Raju (1991) has recorded the same species and *M. polymorpha* above the *Orbulina* datum in the Cauvery Basin. These records are, however, from bore-holes through mid to outer shelf sequences, and lack the support of associated larger foraminiferal taxa of known Middle Miocene age. This suggests — but does not, of course, prove — reworking.

In summary, there appear to be no verifiable records of *in situ Miogypsina* s.s. from continuously sedimented shallow-water carbonates of post–N9 (and possibly post–N8) age anywhere in the Indo-West Pacific region. The case for a later extinction rests on circumstantial evidence contained in a series of reports, each of which is flawed in one way or another (usually by incomplete documentation or failure to consider environmental constraints).

The Americas

Knowledge of larger foraminiferal extinction levels in the Miocene of the Americas is very limited. Barker and Blow (1976) cited the disappearance of *Miogypsina* from Mexico as N7, and Cole (1967) as not younger than the *Globigerinatella insueta Zone* (= N8). Clark and Blow (1969) claimed that it became extinct at the N10/N11 boundary in Venezuela, citing (abstract of a paper by W. Schweighauser read at a meeting in 1956) an unpublished record of *M. cushmani* from the Cerro Pelado Formation as denoting N9 and N10, and stating that the overlying Dividive Limestone of the Socorro Formation yielded *M. antillea* (no source given) and that this could not be older than N9 or younger than upper N11. They also referred (authority not given) to the occurrence of *M. antillea* in the Riecito Limestone (N9–N10) of Falcon, Venezuela. Another author has regarded the Riecito and Dividive limestones as coeval (Petzell, 1959, 1978), but without adding to Senn's original statement (1935) that only fragments of *Miogypsina* had been recovered from the up to 10 m thick, lens-like, Dividive Limestone itself. However, Clarke's and Blow's paper has been relied on by subsequent authors (e.g. Ikebe and Chiji, 1971) in discussions of world-wide extinction levels, although there appears to be no verifiable record of *in situ Miogypsina* above N9 in the Americas (as in the Indo-Pacific), and the N9 records themselves require confirmation. Butterlin (1981) gave the extinction of *Miogypsina* as M. Miocene, but in a French translation (personal communication, 1984) queried its occurrence at this level. The writer therefore takes the view that *Miogypsina* probably became extinct in the Americas at or about the level of the *Orbulina* datum, and that proof is required before a later extinction can be accepted.

Summary

The reasons for questioning some N9, and all post–N9, extinction records of *Miogypsina* spp are as follows:

1. Most records of *Miogypsina* from the Middle Miocene are from deepish-water sediments (shales and marls) which are unlikely to have supported species normally restricted to shallow-water carbonates. Although such specimens could have been penecontemporaneously transported down-slope, the possibility of reworking must be seriously considered in the absence of acceptable records from coeval shallow-water carbonates at outcrop.

2. Well-dated, shallow-water carbonates of M. Miocene age (e.g. the type section of the Futuna Limestone, Lau Islands, the Darai Limestone of Papua New Guinea, the sediments of the Capricorn Basin, eastern Australia, the limestones of the Queensland Plateau (Dr. C. Betzler, personal communication) and the Mid- to Late Miocene limestones of the Philippines) have not been shown to contain *Miogypsina*, although it should have persisted in tropical areas if extant in post–N9 times.

3. Evolutionary work, particularly that of Drooger (1963) and Raju (1974), has

Table 1. Records of *Miogypsina* from the Middle Miocene of the Indo-West Pacific region.

Author (s)	Taxon	Location	Supposed age
Coleman & McTavish, 1964	*M. polymorpha*	Charikange Beds, Guadalcanal, Solomon Islands	M. Miocene
Van der Vlerk & Postuma, 1967	*M. cushmani*	Java	M. Miocene (N9)
Clarke & Blow, 1969	*M. bifida*	Central Sumatra	M. Miocene (N11 & N12)
Ikebe & Chiji, 1971	*Miogypsina*	Japan	M. Miocene (N14)
Binnekamp, 1973	*M.* sp.	New Britain	M. Miocene (Top Tf$_2$)
Baumann, 1975	*M.* sp.	Java	M. Miocene (N12)
Haak & Postuma, 1975	*Miogypsina*	Far East	M. Miocene (? top N12)
Carozzi, Reyes & Ocampo, 1976	*M.* sp.	Tanian, Lmst., Panay, Philippines	M. Miocene
Hashimoto et al. (1977, tables)	*M.* sp.	Carcar Fm., Cebu, Philippines	U. Miocene (mid-Tf$_3$)
ditto (p. 113)	ditto	Maingit Fmt., Cebu, Philippines	Basal U. Miocene (= N13/14)
Van Vessem, 1978	*M.* sp.	Borneo and Java	M. Miocene (N10–N14)
Pandey, 1982	*M. cushmani* & *M. antillea*	Muldwarkian Stage, western India	M. Miocene (exit N14)
Belford, 1984	*M.* sp.	Papua New Guinea	M. Miocene
Chaproniere, 1984	*M.* sp.	Trealla Lmst. W. Australia	M. Miocene
Adams, 1984	*Miogypsina*	East Indies	M. Miocene (? N14)
Cosico, Gramann & Porth, 1989	*M.* sp.	Philippines	M. Miocene (early Tf)
Raju & Chidambarum, 1989	*M. cushmani* & *M. antillea*	Andaman Basin	M. Miocene (N9–N12/13)

[1] The only N13/N14 assemblage (Boring Boegal, Java, at 420 m) was from "an unknown location (possibly Tjepu area)" and from a sequence which "probably comprised limestones and marls". Its depositional environment is therefore uncertain.

Comments

In turbidites. *M. polymorpha* usually found in Lower Miocene sediments.

Age based on single specimen of *Orbulina*. Schipper & Drooger (1974) referred all 16 specimens from this sample and from B295 (the one above), to *M. antillea*.

No confirmatory evidence for *M. bifida* provided. Instead, authors refer to occurrences of three L. 'Miocene species (*M. kotoi*, *M. thecideaeformis* & *M. excentrica*) in the Telisa Shale, an unfavourable deep-water palaeoenvironment. See LeRoy (1952).

Relied on unsubstantiated and unverifiable reports of *Miogypsina* in beds above the *G. nepenthes* datum in Taiwan (Chang, 1967) and the Philippines (Hashimoto, 1966; Gonzales, 1969) to support dating in Japan. Reference to Bandy (1963) appears to be an error.

Samples from float blocks. Age based on unsubstantiated assumption that *Flosculinella borneensis* ranges to base of Tf$_3$.

No verifiable evidence provided.

No records cited from the M. Miocene.

Listed fauna could equally well be of E. Miocene (Burdigalian) age.

Drafting error? Fauna shown to be Te$_5$ (= L. Miocene) in text.

No confirmation of age from plankton or other larger foraminifera. Hashimoto & Balce (1977) confirm local extinction of *Miogypsina* with in Maingit Fm. and equate it with N13/N14, but without supporting evidence.

Study based largely on isolated, poorly localized samples from museum collections[1].

Associated taxa (*C. indopacificus* and *P. malabarica*) typical of older horizons (Adams, 1984). Diagnostic planktonic species (*G. nepenthes*) unfigured[2].

None shown to be younger than N9; all could be older.

Praeorbulina can be mistaken for *O. suturalis* in random sections of limestone. *O. universa* not recorded. Associated taxa typical of N8/N9. *Orbulina* not in same samples as *Miogypsina* (Chaproniere, pers. comm. 1990).

Relied heavily on unsatisfactory record (Baumann, 1975).

Much of Tf$_1$ now known to equate with E. Miocene (Adams, 1984).

Reported from relatively deep-water sequences of mid-outer shelf claystones, sandstones, and thin limestones. Occasional *Lepidocyclina* sp.; reworked Palaeogene plankton at some levels.

[2] The Cochin/Quilon limestones, western India, contain *Miogypsina, A. howchini* and *Pseudotaberina malabarica*, an assemblage which suggests an N7–N8 age (Adams, 1984; Banner and Highton, 1989).

shown that known species of *Miogypsina* (*M. globulina* excepted) were short-ranging during the Early Miocene. Yet the named M. Miocene species (Table 1) all appeared in the Early Miocene and must therefore have had rather long ranges. It is, moreover, surprising that no new species evolved during the Middle Miocene.

4. None of the so-called M. Miocene occurrences of *Miogypsina* spp is supported by other larger foraminifera (e.g., identified species of *Alveolinella, Cycloclypeus* or *Lepidocyclina*) characteristic of this subepoch.

In the author's opinion, published evidence shows that *Miogypsina* definitely ranges up into N8, and probably into N9 (van der Vlerk and Postuma, 1967; Belford, 1984; Chaproniere, 1984). Van Vessem's (1978) records of younger occurrences (N10-N14) require confirmation from better-documented carbonate sequences before being accepted, as do all records from off-shore wells in the Indian Ocean. Despite the circumstantial evidence from bore-holes, the *Miogypsina* extinction datum should not, at present, be drawn later than the N8/9 or N9/10 boundary in the Indo-West Pacific, and at the N8/N9 boundary elsewhere. All younger records of the genus require confirmation.

Conclusions

The uneven spread of biostratigraphical, distributional and diversity data for critical Neogene larger foraminifera may be affecting our perception of certain geological events in the Indo-Pacific region. Additional data for larger foraminifera, reef-corals and mangroves are, therefore, required if current determinations of sea-surface palaeotemperatures and climatic changes are to be improved.

Greater rigour is needed in the application of biostratigraphic methods if our understanding of geological events is to be further enhanced through the study of larger foraminifera. Currently accepted extinction levels for *Miogypsina* (N12–N14) and *Lepidocyclina* (N15–N19) are based on too little evidence for them to be safely applicable across the Indo-West Pacific region; all new post–N14 discoveries of *Lepidocyclina*, and post–N9 reports of *Miogypsina*, should, therefore, be fully documented so that the true regional extinction levels of these genera can be determined.

Larger foraminifera in general, and *Miogypsina* and *Lepidocyclina* in particular, should not be used to infer sea-surface palaeotemperatures unless it can be shown that they are from sediments older than their established extinction levels.

Acknowledgements
The author wishes to thank Dr. D.S.N. Raju for his kindness in providing useful information and literature, and Dr. A.H.H. Wonders for helpful discussions.

References

Adams, C.G., 1968. A revision of the foraminiferal genus *Austrotrillina* Parr. *Bull. Brt. Mus. nat. Hist. London (Geology)*, 16: 73–97, pls. 1–6.

Adams, C.G., 1984. Neogene larger foraminifera, evolutionary and geological events in the context of datum planes, pp. 47–68. *In:* I. Ikebe and R. Tsuchi (Eds.), Pacific Neogene Datum Planes, University of Tokyo Press, 288 pp.

Adams, C.G., 1989. Foraminifera as indicators of geological events. *Proc. Geol. Assoc.*, 100(3): 297–311.

Adams, C.G., Gentry, A., and Whybrow, P., 1983. Dating the terminal Tethyan event. *In:* J.E. Meulenkamp, (Ed.), *Utrecht Micropaleont. Bull.*, 30: 273–298.

Adams, C.G., Lee, D., and Rosen, B.R., 1990. Conflicting isotopic and biotic evidence for tropical sea-surface temperatures during the Tertiary. *Palaeogeog., Paleoclim., Palaeoecol.*, 77: 289–313.

Adams, C.G., Rodda, P., and Kiteley, R.J., 1979. The extinction of the foraminiferal genus *Lepidocyclina* and the Miocene/Pliocene boundary problem in Fiji. *Marine Micropalaeontology*, 4: 319–339.

Asano, K., 1965. On the Geology of Cebu, Philippines. 1st Symposium, Assoc. Palaeont. Res. in S.E. Asia, 2 pp., Pre-print [in Japanese].

Bandy, O.L., 1963. Cenozoic planktonic zonation and basinal development in Philippines. *Bull. Amer. Assoc. Petrol. Geol.*, 47: 1733–1745.

Banner, F.T. and Highton J., 1989. On *Pseudotaberina malabarica* (Carter) (Foraminiferida). *J. Micropalaeont.*, 8: 113–129.

Barker, R.W. and Blow, W.H., 1976. Biostratigraphy of some Tertiary formations in the Tampico-Misentla embayment, Mexico. *Jl. Foraminif. Res.*, 6(1): 39–58.

Baumann, P., 1975. The Middle Miocene diastrophism, its influence to the sedimentary and faunal distribution of Java and the Java Sea basin. *Bull. Nat. Instit. Geol. & Mining, Bandung*, 5(1): 13–28.

Belford, D.J., 1984. Tertiary foraminifera and age of sediments, Ok Tedi-Wabag, Papua New Guinea. *Bull. Bur. Min. Resour. Geol., Geophys, Canberra*, 2116: 52 pp. 40 pls.

Binnekamp, J.G., 1973. Tertiary larger foraminifera from New Britain PNG. *Bur. Min. Resour. Geol. Geophys. Aust.*, 140: 1–26.

Boltovskoy, E., 1976. Distribution of recent Foraminifera of the South American Region, 171–236. *In:* R.H. Hedley and C.G. Adams (Eds.), Foraminifera, 2: 265 pp.

Butterlin, J., 1981. Claves para la determinacion de macroforaminiferos de Mexico y del Caribe, del Cretacio superior al Miocene medio. Inst. Mexicano del Petroleo, 219 pp.

Carozi, A.V., Reyes, M.V., and Ocampo, V.P., 1976. Microfacies and microfossils of the Miocene reef carbonates of the Philippines. Spec. Publ. No. 1, Philippine Oil Development Co. Inc., Manila, 80 pp.

Chang, L.S., 1967. Tertiary biostratigraphy of Taiwan and its correlation. *In:* Symposium 25, *Tertiary Correlation & Climatic Changes*, 11th Pacific Science Congress. Tokyo, Sasaki Printing Co., Sendai, pp. 67–65.

Chaproniere, G.C.H., 1984. Oligocene and Miocene larger Foraminiferida from Australia and New Zealand. Bull. Bur. Min. Res., Geol. and Geophys., 188: 98 pp.

Cita, M.B. and McKenzie, J., 1986. The Terminal Miocene Event. *In:* K. Hsü, (Ed.), *Mesozoic and Cenozoic Oceans*. Amer. Gephys. Union, Geodynamics Series, 15: 123–153.

Clarke, W.J. and Blow, W.H., 1969. The interrelationships of some Late Eocene, Oligocene and Miocene larger foraminifera and planktonic biostratigraphic indices. *In:* Proceedings of the First International Conference on Planktonic Microfossils, Geneva, 1967. *E.J. Brill, Leiden*, 11: 82–96.

Cole, W.S., 1945. Larger Foraminifera of Lau, Fiji. *In:* H.S. Ladd, and J.E. Hoffmeister, The Geology of Fiji. *Bull. Bernice P. Bishop Mus.*, 181: 272–297.

Cole, W.S., 1954. Larger Foraminifera and Smaller Diagnostic Foraminifera from Bikini Drill Holes. *Prof. Pap. U.S. geol. Surv.*, 260-O: 569–608.

Cole, W.S. 1958 (dated 1957). Larger Foraminifera from Eniwetok Atoll Drill Holes. *Ibid.*, 260-V: 743–784.

Cole, W.S., 1967. A review of American species of Miogypsinids (Larger Foraminifera). *Contr. Cush. Found. Foraminif. Res.*, 18(3): 99–117.

Coleman, P.J. and McTavish, R.A., 1964. Association of Larger and Planktonic foraminifera in single samples from Middle Miocene sediments, Guadalcanal, Soloman Islands, Southwest Pacific. *Jl. Roy. Soc. W. Australia*, 47(1): 13–24.

Cosico, R., Gramann, F., and Porth, H., 1989. Larger foraminifera from the Visayan Basin and Adjacent Areas of the Philippines (Eocene through Miocene). *Geol. Jb.*, 70: 147–205.

Drooger, C.W., 1952. *Study of American Miogypsinidae*. Thesis, Univ. of Utrecht, 80 pp.

Drooger, C.W., 1963. Evolutionary trends in the Miogypsinidae. *In:* G.H.R. von Koeningswald, et al. (Eds), *Evolutionary Trends in Foraminifera*. Elsevier, London, pp. 315–349.

Gonzales, B.A., 1969. Development and status of micropalaeontological research in the Philippines. *Philip. Geologist*, 23(4): 183–195.

Haak, R. and Postuma, J.A., 1975. The relation between the tropical foraminiferal zonation and the Tertiary Far East Letter Classification. *Geol. en Mijnb.*, 54(3–4): 195–198.

Hashimoto, W., 1966. Miscellaneous notes on geology of Taiwan (3). *Jour. Geogr. Tokyo*, 75: 84–92 (in Japanese).

Hashimoto, W. and Balce, G.R., 1977. A new correlation scheme for the Philippine Cenozoic Formations. Proc. First Internat. Congr. Pacific Neogene Stratigraphy, Tokyo, 119–132.

Hashimoto, W., Matsumaru, K., Kurihara, K., David, P.P., and Balce, G.R., 1977. Larger foraminiferal assemblages useful for correlation of Cenozoic marine sediments in the Mobile Belt of the Philippines. *Geol. & Palaeont. of S. E. Asia*, 18: 103–124.

Hirooka, K., Takahashi, T., Sakai, H., and Nakajima, T., 1985. Paleomagnetic evidence of the northward drift of the Izu Peninsula, central Japan. *In:* N. Nasu et al. (Eds), Formation of active ocean margins, pp. 775–787.

Ibaraki, M., 1981. Geologic ages of "*Lepidocyclina*", *Miogypsina* horizons in Izu Peninsula as determined by planktonic foraminifera. *Jl. geol. Soc. Japan*, 87: 417–420. In Japanese.

Ibaraki, M., 1988. *Lepidocyclina* from the Izu Peninsula and its implications in the northward drift of the area. *In:* M. Ibaraki (Ed.), Guidebook for Excursion "Izu Peninsula and its collision with the mainland of Japan". Oji International Seminar for IGCP–246, Shizuoka, Japan, pp. 7–17.

Ibaraki, M., 1990. Geologic ages of *Lepidocyclina* in Japan with implications in the northward drift of the Izu Peninsula, pp. 137–149. *In:* R. Tsuchi (Ed.), *Pacific Neogene Events*, University of Tokyo Press, 206 pp.

Ibaraki, M. and Tsuchi, R., 1978. Planktonic foraminifera from *Lepidocyclina* horizon at Namegawa in the southern Izu Peninsula, central Japan. *Rep. Fac. Sci. Shizuoka Univ.*, 12: 115–130.

Ikebe, N. and Chiji, M., 1971. Notes on Top-datum of *Lepidocyclina* sensu lato in reference to planktonic foraminiferal datum. *J. Geosc. Osaka City University*, 14(2): 19–52.

LeRoy, L.W., 1952. *Orbulina universa* d'Orbigny in Central Sumatra. *J. Paleont.*, 7: 175–177.

Pandey, J., 1982. Chronostratigraphic correlation of the Neogene sedimentaries of Western Indian Shelf, Himalayas and Upper Assam. *Pal. Soc. India, Spec. Publ.*, 1: 95–129.

Petzell, C., 1959. Estudio de una sección de la Formación Caujaro en el anticlinal de la Vela, Estada Falcón. *Asoc. Venez. Geol. Min. Petról. Bol. Inform.*, 2: 269–320.

Petzell, C., 1978. Lexique Stratigraphique International, vol. 5, Amérique Latine, fasc. 3A, Vénézuéla (2nd ed. revised by Gayrard, Y.), Paris.

Raju, D.S.N., 1974. Study of Indian Miogypsinidae. *Utrecht Micropal. Bull.*, 9: 148 pp.

Raju, D.S.N., 1991. Miogypsina scale and Indian chronostratigraphy. *Geoscience Jl.*, 12: 53–65.

Raju, D.S.N. and Chidambaram, L., 1989. Cretaceous and Cenozoic foraminiferal biostratigraphy of offshore deep well sections, East Coast of Andaman Islands. *In:* P. Kalia (Ed.), Proc. XII Colloq. on Micropalaeontology, pp. 227–240.

Rosen, B.R., 1984. Reef coral biogeography and climate through the late Cainozoic: just islands in the sun or a critical pattern of islands? *In:* P.J. Brenchley (Ed.), *Fossils and Climate*, , John Wiley & Sons, Chichester, pp. 201-262.

Sato, S., 1987. On the so-called Mid-Neogene climatic optimum at 16 Ma in Hokkaido, northern Japan, *In:* R. Tsuchi (Ed.), *Pacific Neogene Event Studies*. Shizuoka pp. 88-93.

Senn, A., 1935. Die stratigraphische Verbreitung der tertiären Orbitoiden, mit spezieller Berücksichtigung ihres Vorkommen in Nord-Venezuela und Nord- Morokko. *Eclog. geol. Helvet.*, 28: 51–113.

Schipper, J. and Drooger, C.W., 1974. Miogypsinidae from East Java and Madura. *Proc. Konikl. Nederl. Akad. van Wetensch*, Amsterdam Ser., B 77(1): 1-14.

Schweighauser, J., 1956. The stratigraphic significance of the morphogenetic evolution of *Miogypsina* Tan Sin Hok. Proc. Geol. Conf. Maracaibo-Venezuela. Sept. 1956, Paper 16, 2 pp. (Abstract only).

Souaya, F.J., 1963. On the Foraminifera of Gebel Ghara (Cairo-Suez Road) and some other Miocene samples. *Jour. Paleont.*, 37: 433–457, pls. 53–58.

Tan Sin Hok, 1936. Zur Kenntnis der Miogypsiniden (1. Fortsetzung). *Ing. in Ned. Indie, Bandung*, 3(5): 84–98.

Tan Sin Hok, 1937. Weitere Untersuchungen über die Miogypsiniden II. *Ibid., Bandung*, 4(6): 87–111.

Tsuchi, R., 1987. Neogene biochronostratigraphy and events in Japan and the Pacific. IVth Internat. Cong. Pac. Neog. Strat. Abstr. Vol., Berkeley, pp. 121–122.

Vail, P.R., Mitchum, R.M. Jr., and Thompson, S., 1977. Seismic stratigraphy and global changes in sea level, pt. 4: Global cycles of relative changes of sea level. *In:* C.E. Payton, (Ed.), Seismic stratigraphy — Applications to hydrocarbon exploration. *Mem. Am. Ass. Petrol. Geol.*, 26: 83–99.

Vessem, E.J. van, 1977. The internal structure of *Miogypsina polymorpha* and *Miogypsina bifida*. *Proc. Konink. Ned. Akad. Wetensch.* Amsterdam, Ser. B, 80: 421–428.

Vessem, E.J. van, 1978. Study of Lepidocyclinidae from Southeast Asia, particularly from Java and Borneo. *Utrecht Micropal. Bull.*, 19: 163 pp.

Vlerk, I.M. van der and Postuma, J.A., 1967. Oligo-Miocene lepidocyclinas and planktonic foraminifera from East Java and Madura, Indonesia. *Proc. Kon. Ned. Akad. Weten.*, Amsterdam, ser. B, 70: 391–398.

Pacific Neogene Climatic Optimum and Accelerated Biotic Evolution in Time and Space

Ryuichi Tsuchi

Prof. Emer. Shizuoka University in Geoscience, Shizuoka 422, Japan

Abstract

Two pronounced Pacific Neogene events which are effectively revealed in middle latitude areas are discussed. The Mid-Neogene climatic optimum is recognized in Japan with the maximum temperature of surface seawater during planktonic foraminiferal subzone N8b of the earliest Middle Miocene around 16 Ma, on the basis of occurrences of marine Mollusca, larger Foraminifera and associated planktonic foraminifera. This marine climatic event is considered to occur synchronously on a Pan-Pacific scale, as examined in Peru and northern Chile, and as known from southern Australia and New Zealand. An accelerated evolution of endemic marine Mollusca is recognized in Japan through successional species during the latest Neogene, from 3 to 1 Ma. Considering changes in the associated planktonic foraminiferal composition, the modification of shell form seems to occur in response to phased decline of surface seawater temperature during the interval. Similar accelerated evolutions are likely to have happened in northern Chile and New Zealand in endemic Mollusca during the latest Neogene since 3 Ma. As marine faunal changes in middle latitudes are effective in accordance with the rise and fall of seawater temperature, the two above-mentioned events may also be notable in middle latitudes.

Introduction

The recent progress in biochronology which combines detailed biostratigraphy by means of planktonic microfossils with radiometric dating and magnetostratigraphy has made it possible to define a more precise distribution of associated benthic faunas and species in time and space. The results provide a more precise correlation of events and a more accurate estimate of the rate of biotic evolution. During the Neogene, various events have been recognized throughout the Pacific region. In middle latitudes, tropical and boreal water faunas are considered to appear and disappear in accordance with the rise and fall of seawater temperature. This article attempts to discuss two pronounced Neogene events which are effectively revealed in middle latitudes.

Neogene Molluscan Faunas of the Japanese Islands

The Japanese Islands, located in middle latitudes of the northwest Pacific, are strongly affected by the warm Kuroshio current from the south and by the cold Oyashio current from the north. During the Neogene, the oceanic surface circulation pattern surrounding the Japanese Islands was essentially similar to the present one based on studies by Kennett et al. (1985).

Neogene marine molluscan faunas of Japan are, therefore, clearly seperated into warm- and cold-water species groups in southern and northern Japan, respectively, in the same way as are living faunas of the warm Kuroshio current and the cold Oyashio current. These faunas are divided chronologically into five phases, from older to younger: 1) Ashiya and Asagai-Poronai, 2) Kadonosawa and Chikubetsu, 3) Sagara and Shiobara-Yama, 4) Kakegawa and Omma-Manganji or Tatsunokuchi, and 5) Recent Kuroshio and Oyashio (Tsuchi, 1986). The chronologic and geographic distribution of marine molluscan faunas of Japan is shown in Fig. 1.

From the chronologic succession of these faunas and changes in their specific composition and geographic distribution, major Neogene events documenting the rise and fall of surface seawater temperature can be recognized. The ratio of warm-water specimens to the total planktonic Foraminifera examined in sections of central Japan (Ibaraki, 1990) are combined with the figure, and presumed marine climatic events are attached. Of these events, the climatic optimum around 16 Ma and an accelerated biotic evolution related to the phased dropping of seawater temperature during 3–1 Ma are discussed below with reference to the Pacific region.

Mid-Neogene Climatic Optimum around 16 Ma

The mid-Neogene maximum of the surface seawater temperature is one of the most pronounced Neogene events in Japan during the earliest Middle Miocene time, around 16 Ma. We call it the "Mid-Neogene climatic optimum," as it is considered to be the climatic maximum of the Neogene (Tsuchi, 1987, 1990a). During the interval of maximum warmth, a tropical-subtropical marine fauna called the Kadonosawa fauna migrated into northern Japan. The fauna includes tropical gastropods, *Vicarya,* derived originally from the Tethyan Sea, and typical mangrove swamp dwellers, *Telescopium* and *Geloina.* The distribution of such mangrove swamp inhabitants is restricted to south of the Okinawa Islands at present. Several kinds of pollen of mangrove trees have been obtained in sequences of the Sea of Japan side in association with these mangrove swamp inhabitants (Yamanoi, 1984). Tropical larger Foraminifera, *Lepidocyclina* and *Miogypsina,* are frequently found associated with these tropical Mollusca. Fossil reef-building corals are also examined in places on the Pacific side of central Japan.

Most of such tropical elements are included in a limited interval of the planktonic foraminiferal zone N8 and the base of Zone N9, spanning 1 m.y. around 16 Ma (Tsuchi et al., 1984). From rich occurrences of the tropical elements and the maximum ratio of warm-water planktonic Foraminifera in the later half of Zone N8 (or Subzone N8b after Tsuchi, 1986), this warm event seems to have culminated in the earliest Middle Miocene.

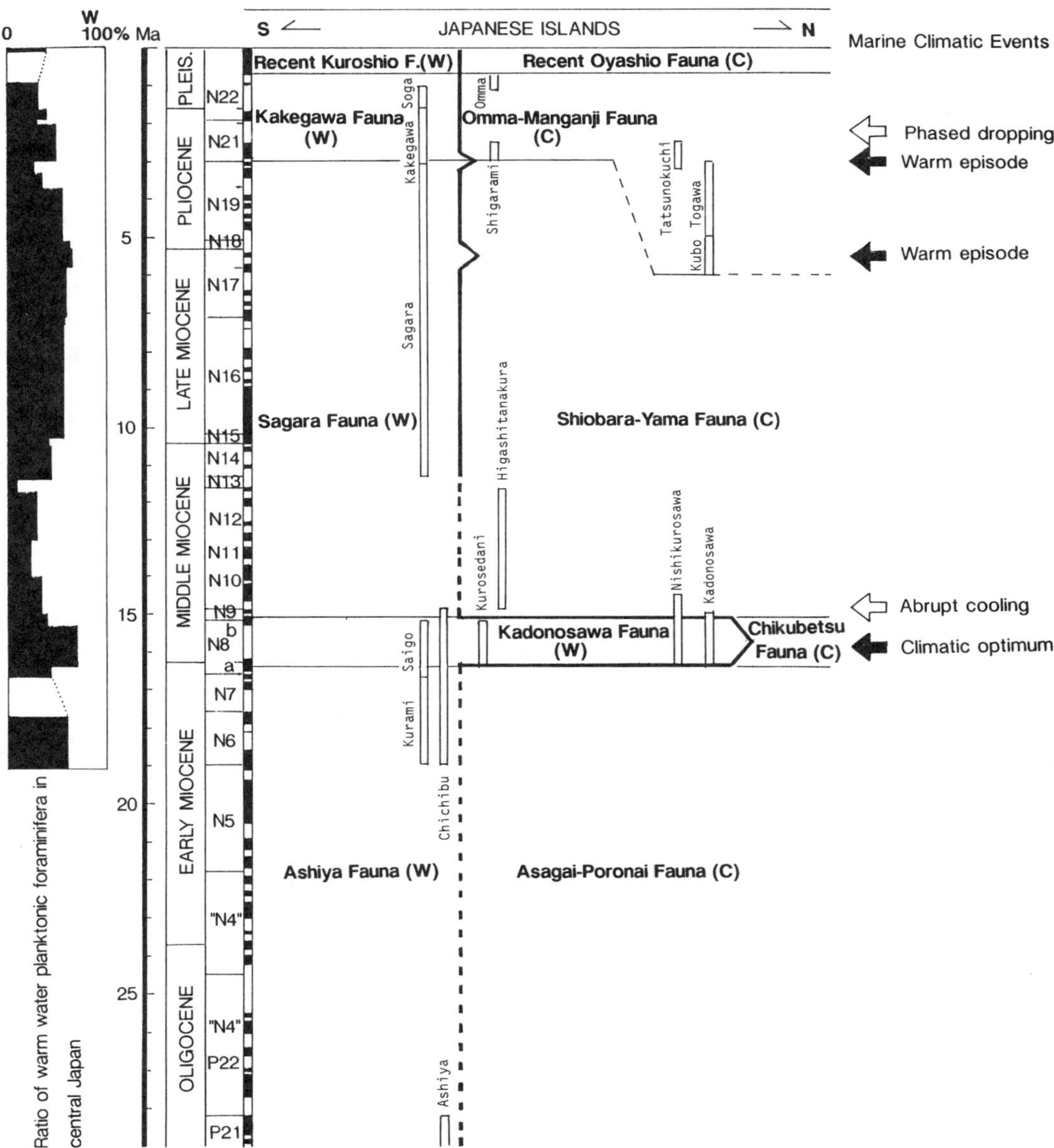

Fig. 1. Chronologic and geographic distribution of marine molluscan faunas of Japan. (W) and (C) indicate warm and cold water faunas, respectively. Chronologic calibrations of planktonic foraminiferal datum levels are those given by Berggren et al. (1985) for ages earlier than 5 Ma, and Tsuchi et al. (1984) for the interval 0–5 Ma. The ratio of warm water planktonic foraminifera to the total fauna examined in the section of central Japan (Ibaraki, 1990) is combined on the left side, and presumed marine climatic events are given on the right side of the figure. Key stratigraphic sections and paleomagnetic time scale are also shown. Slightly modified from Tsuchi (1990a).

Based on the extensive distribution of marine sequences of the lowermost Middle Miocene in Japan, the Japanese Islands at that time were an assembly of small isles due to a relative high stand of sea level, and were located in the periphery of the Asian continent, although the initial phase of the opening of the Sea of Japan had begun around 16 Ma (Chiji et al., 1990; Tsuchi, 1990a) (Fig. 2).

On the Pacific coast of South America, sequences assignable to Zones N8–9 are frequently encountered in such places as the San Vicente and the Jaramijo sections near Manta in Ecuador, the Camana section in southern Peru and the Caleta Herradura de Mejillones section, north of Antofagasta in Chile, where abundant and diverse warm-water planktonic Foraminifera have been obtained (Tsuchi et al., 1990) (Figs. 3 and 4).

In Peru, Lower Miocene sequences bear only intercalations of diatomaceous layers,

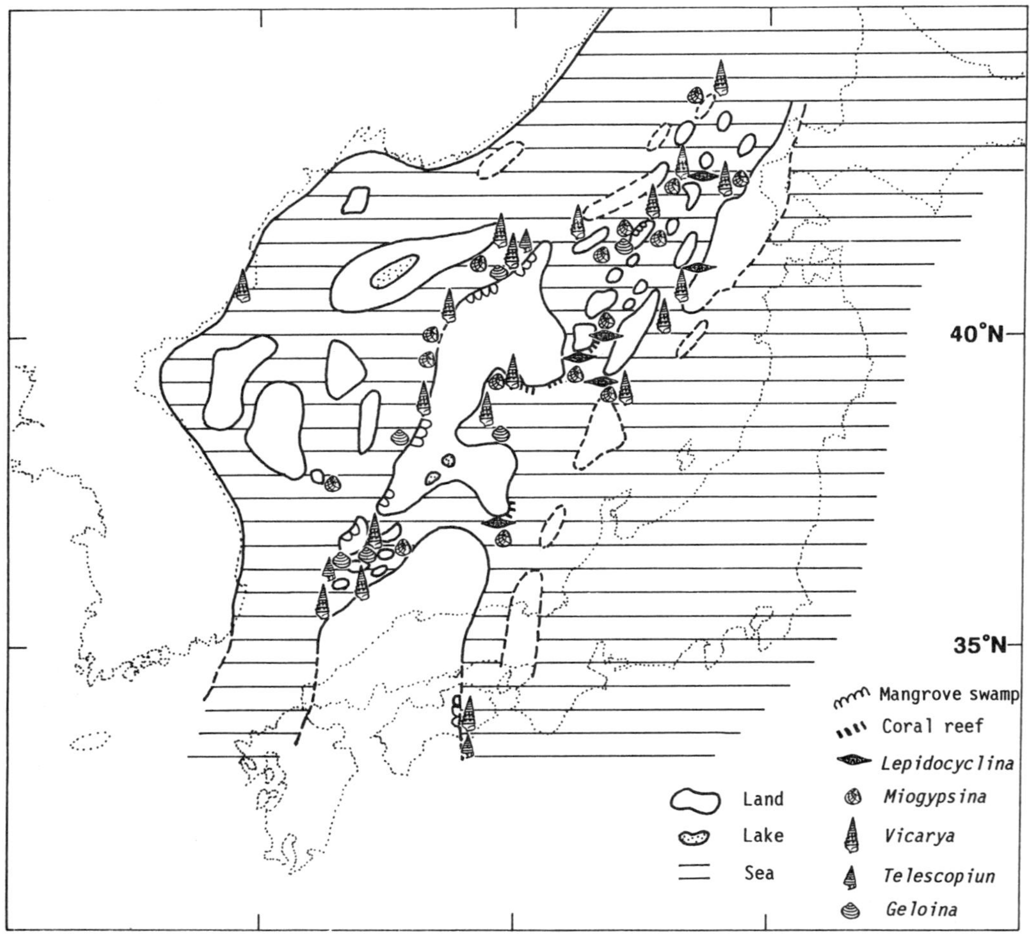

Fig. 2. Paleogeographic map of the Japanese Islands around 16 Ma. The distribution of tropical biota at that time is briefly given. (Slightly modified from Chiji et al., 1990.)

whereas Middle Miocene and later sequences are dominated by biosiliceous sediments associated with coastal upwelling of the cold Peru current. Study of ODP Leg 112 Site 682 on the lower continental slope off Peru disclosed that the biosiliceous facies became predominant ca. 17 Ma in the Early Miocene (Suess, von Huene et al., 1988), although sediments probably assignable to Zones N8–9 were poorly recovered. The Camana Formation at Camana on the coast of southern Peru, which consists of calcareous sandstone and siltstone with rich occurrences of miogypsinid larger Foraminifera, indicating a warm water condition, is assignable to Zones N6–8 from the Early Miocene to the earliest Middle Miocene in age (Ibaraki, 1992). Miogypsinids are included in horizons of Zones N7–8. The present distribution of larger Foraminifera on the Pacific coast of South America is known only from the coast of Ecuador (Adams et al., 1990). Based on the maximum ratio of warm-water planktonic Foraminifera examined in the Camana section (Ibaraki, 1992), this warm-water condition seems to have culminated in the interval of the later part of Subzone N8a and Subzone N8b from the latest Early Miocene to the earliest Middle Miocene, around 16 Ma. The high ratio of warm-water planktonic Foraminifera to the total fauna during the interval of Subzones N8a, N8b, and the basal Zone N9 around 16 Ma has also been examined in the Caleta Herradula de Mejillones section near Antofagasta in northern Chile (Ibaraki, 1990).

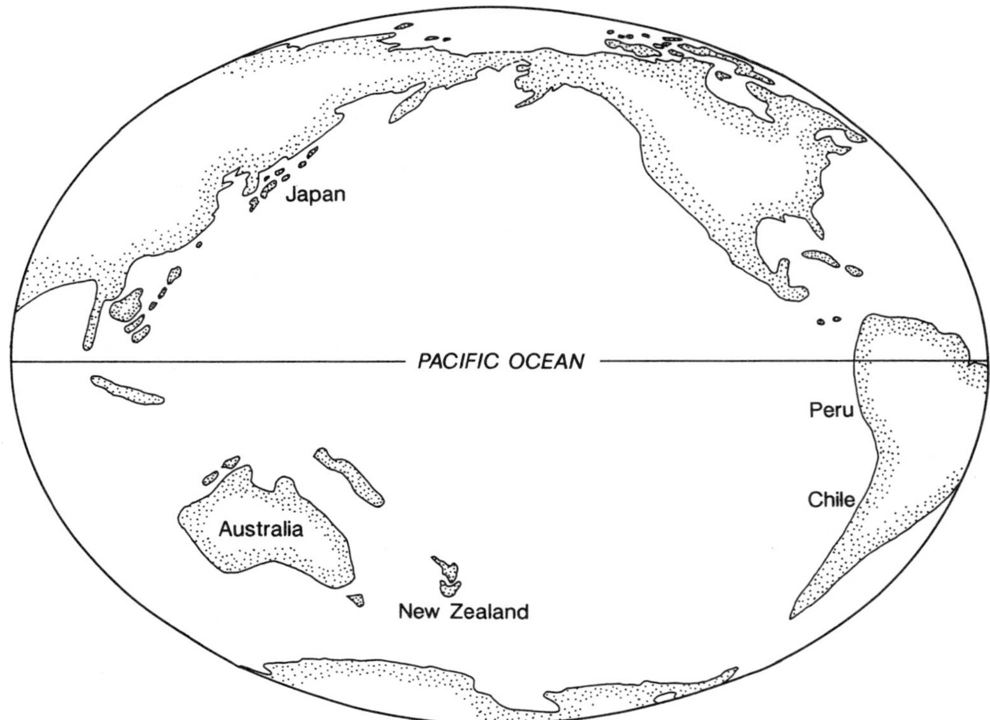

Fig. 3. Studied areas in a sketch map of the Pacific region around 16–17 Ma.

    R. Tsuchi

Fig. 4. Location map of the studied areas on the Pacific coast of South America. The distribution of marine Cenozoic strata (shaded area) is briefly given.

The maximum southward distribution of tropical larger Foraminifera in the Australasia region in the South Pacific during early middle Miocene time has been pointed out by Chaproniere (1980). Benbow and Lindsay (1988) disclosed that ages of some larger Foraminifera-bearing horizons in South Australia are assigned to Subzone N8b of the earliest Middle Miocene based on associated planktonic Foraminifera. In New Zealand, benthic Mollusca and larger Foraminifera reveal that the thermal maximum of the surface seawater temperature occurred in earliest Middle Miocene time (Hornibrook, 1990).

As mentioned above, according to recent studies in the Pacific region, the Mid-Neogene climatic optimum is considered to be a prominent marine climatic event occurring synchronously on the Pan-Pacific scale, and changes of marine faunal aspects are likely to be most evident in middle latitudes.

Latest Neogene Accelerated Biotic Evolution during 3–1 Ma

Another prominent event is an accelerated biotic evolution which occurred in the latest Neogene. As shown in the distribution of marine molluscan faunas of Japan in time and space, some faunas have relatively short ranges, whereas others occupy long ranges. The late Pliocene-Pleistocene faunas on the Pacific coast of southwestern Japan and the Sea of Japan side, which are called the Kakegawa and Omma-Manganji faunas, respectively, both have a short time range, from 3 to 1 Ma.

The Kakegawa fauna consist of two elements: 1) taxa allied to fossil or living forms occurring in the tropical region, such as *Amussiopecten praesignis*, and 2) taxa allied to living forms in adjacent seas. The former taxa disappeared in later stages, and the latter evolved into Recent forms. Among the latter group, an endemic bioseries of *Suchium suchiense-giganteum* demonstrated an accelerated evolution in shell form during the period from 3.0 to 1.6 Ma. *Suchium suchiense* and its related forms occur in abundance in many horizons of shallow sandy facies of the Kakegawa and Soga Groups. The Recent descendant, *S. giganteum*, is commonly found living in the shallow sandy bottom of the open sea in adjacent areas. The surface ornamentation of these ancestral forms is characterized by many spiral cords and tubercles which change gradually and diminish in descendant taxa (Tsuchi, 1990b). *S. suchiense* subsp. A, appearing at 3.0 Ma, successively evolves through *S. suchiense* s.s. at 2.4 Ma, *S. s. subsuchiense* at 1.9 Ma, and *S. giganteum naganumanum* at 1.7 Ma to Recent species, *S. giganteum* s.s. at 1.6 Ma (Fig. 5). Thus, the *S. suchiensesgiganteum* bioseries modifies its shell form every 0.35 m.y. on average, from 3.0 to 1.6 Ma. Such accelerated evolutionary changes in shell form are typically revealed in Japonic endemic taxa. Tropical elements of the Sagara and Kakegawa Groups, such as *Amussiopecten iitomiensis* and *A. praesignis*, have long ranges covering 8 and 1.5 m.y., respectively, without any modification of their shell form or surface ornamentation (Fig. 6). Considering changes in the associated planktonic foraminiferal composition (Ibaraki, 1986), the accelerated evolution recognized here seems to be closely related to phased declines of surface seawater temperature during the interval (Fig. 5).

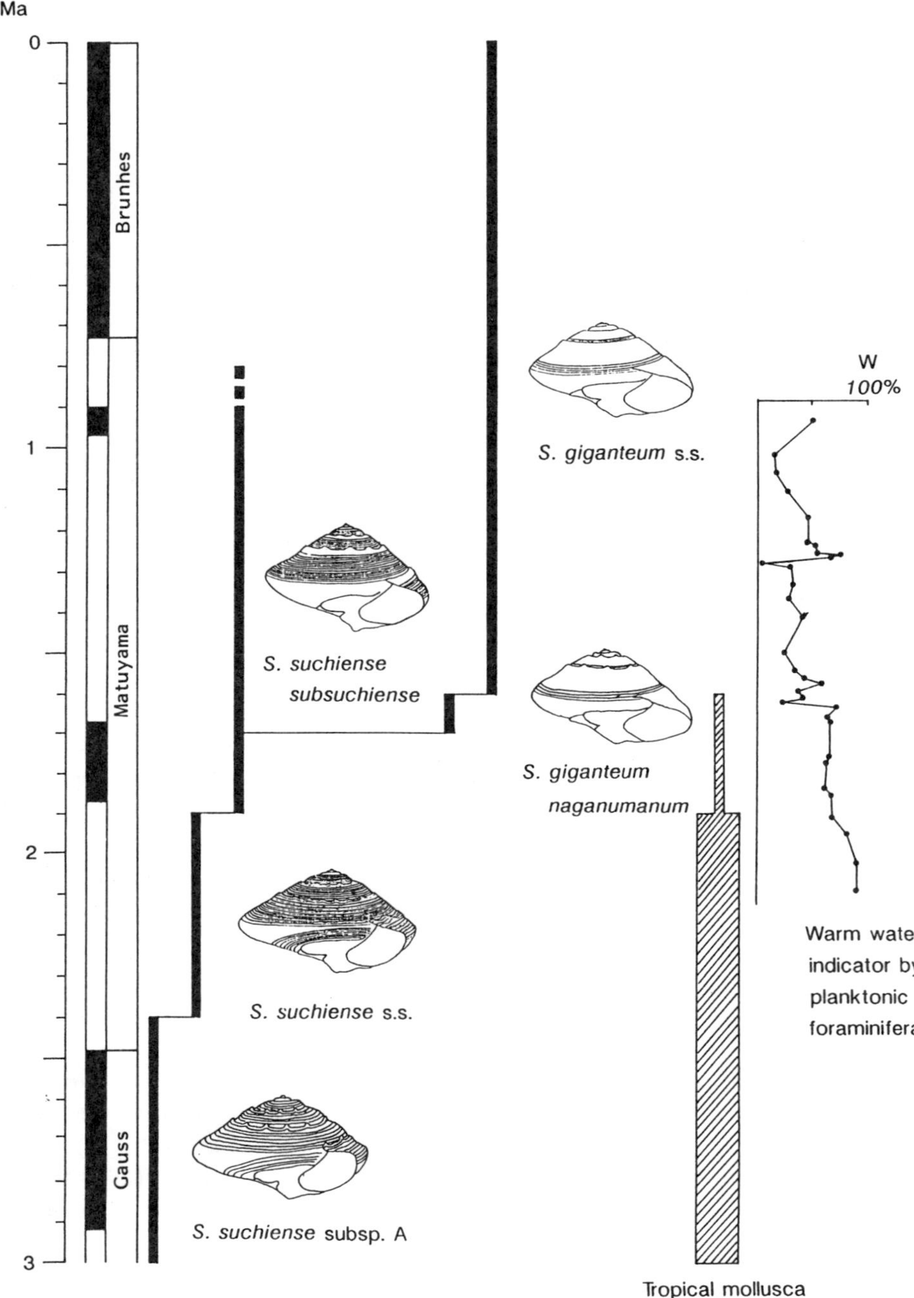

Fig. 5. Evolution of *Suchium suchiense-giganteum* bioseries. The chronologic calibration of evolutionary changes, the reduction of tropical forms within the Kakegawa fauna, and phased declines of seawater temperature estimated from planktonic foraminifera examined in the same section by Ibaraki (1986) are given.

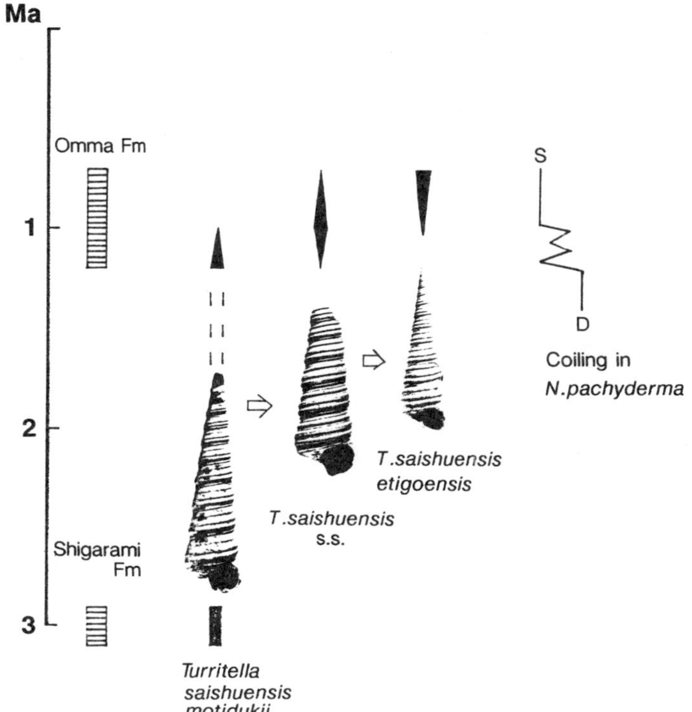

Fig. 6. Evolution of *Turritella saishuensis* bioseries in the Omma-Manganji fauna on the Sea of Japan side. Abrupt coiling change in associated *Neogloboguadrina pachyderma* from dextral to sinistral is also given.

On the Sea of Japan side, the Omma-Manganji fauna includes a typical endemic bioseries of *Turritella saishuensis*, which also undergoes an accelerated evolutionary change in shell form from 3 to 1 Ma. *Turritella saishuensis motidukii*, appearing at 3.0 Ma, evolves through *T. saishuensis* s.s. at ca. 1.2 Ma to *T. saishuensis etigoensis* at ca. 1.0 Ma. Their spiral cords are modified gradually from those of *T. s. motidukii* to those of the descendants. Considering the abrupt coiling change from dextral to sinistral in the planktonic foraminiferal species, *Neogloboquadrina pachyderma*, during the interval, this accelerated evolution seems to occur in response to a sudden drop in seawater temperature at that time (Tsuchi and Ibaraki, 1988; Tsuchi, 1990b) (Figs. 6 and 7).

On the Pacific coast of northern Chile, an endemic gastropod, *Turritella cingulata*, is now living on shallow, sandy substrate. In the Quebrada Blanca section near Caldera, northern Chile, specimens of the *Turritella cingulatiformis-cingulata* bioseries have been successively and stratigraphically examined upwards (Herm, 1969; Tsuchi et al., 1988). Their shell forms and spiral cords gradually change from those of *T. cingulatiformis* in the lowest horizon at 3 Ma through those of *T. cf. cingulatiformis* in upper horizons, those of *T. cf. cingulata* occurring in the high-level

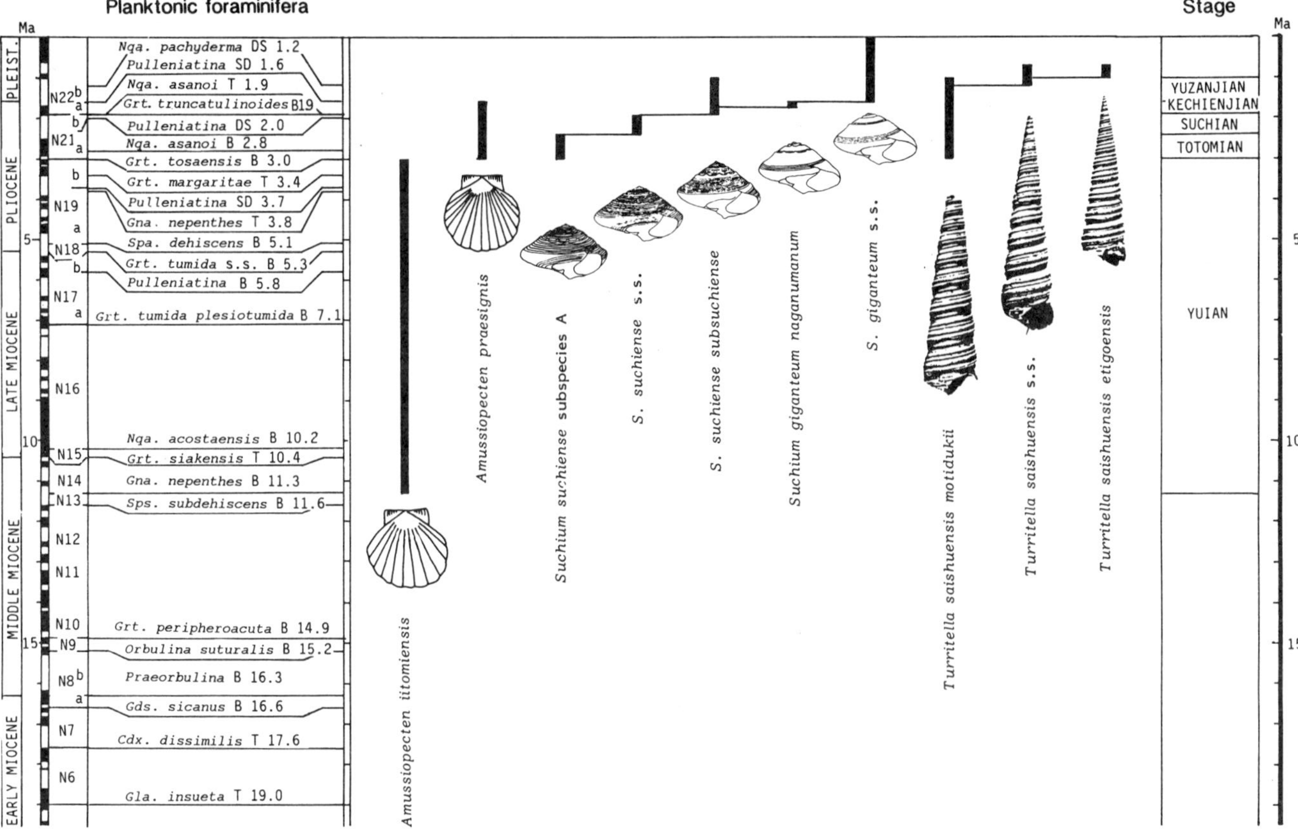

Fig. 7. Chronologic distribution of some representative mollusca of Japan in the late Neogene. The figure indicates accelerated evolutionary changes in endemic bioseries of *Suchium suchiense-giganteum* and *Turritella saishuensis* during 3–1 Ma. Ages of bases or tops of planktonic foraminiferal zones are shown. Stages in Japan after Tsuchi (1986) are also given.

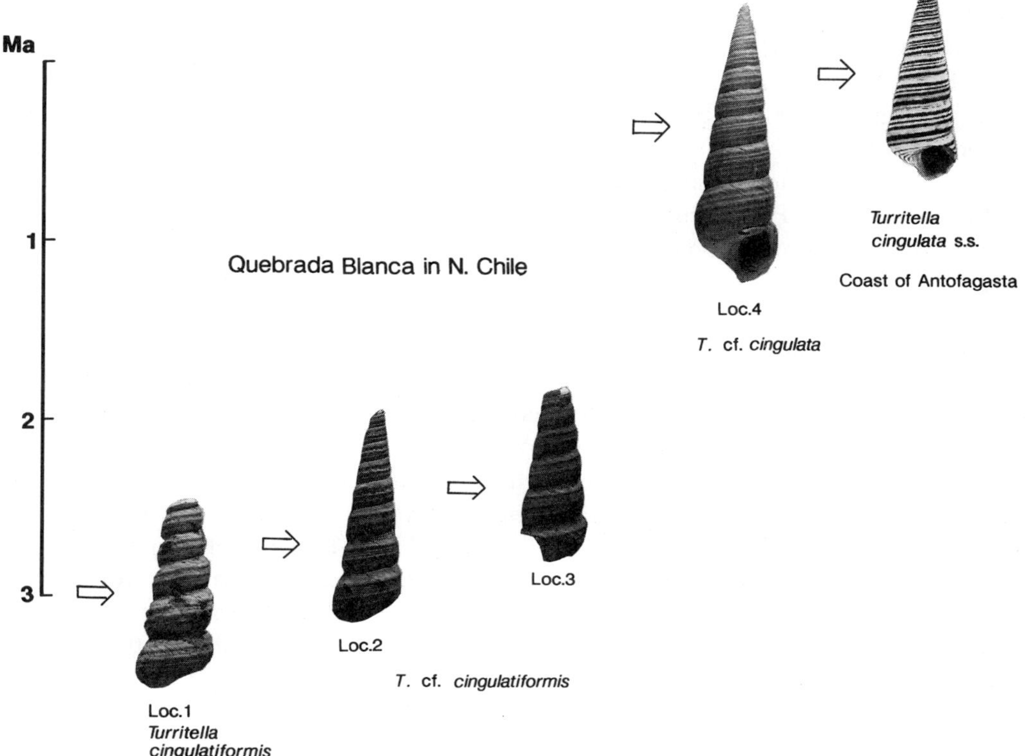

Fig. 8. Evolution of *Turritella cingulatiformis-cingulata* bioseries in northern Chile. Specimens were examined in the section of Quebrada Blanca near Caldera   Loc. 1: The base of the section. Loc. 2 and 3: Upper horizons of the section.   Loc. 4: Overlying high-level terrace deposit.

terrace deposits at ca. 0.3 Ma to those of living *T. cingulata* in the adjacent sea. Thus, the bioseries of *T. cingulatiformis-cingulata* in northern Chile is likely to exhibit an accelerated evolution in shell form since 3 Ma (Fig. 8). No such accelerated evolution has been known from tropical seas in both the Pacific and Caribbean sides of the Isthmus of Panama since the time of its establishment at 3 Ma.

In the Neogene of New Zealand a similar accelerated evolution seems to occur in some turritellid gastropods during the interval of Pliocene and Pleistocene, considering their phyllogenic relationships and biochronologic positions (Marwick, 1957; Kotaka, 1978; Beu, 1983).

Conclusion

Two pronounced Pacific Neogene events notably evidenced in middle latitudes, the Mid-Neogene climatic optimum and the latest Neogene accelerated evolution, are discussed.

Based on evidence of benthic Mollusca, larger Foraminifera and associated planktonic Foraminifera in Japan, the Mid-Neogene climatic optimum culminated in the planktonic foraminiferal subzone N8b of the earliest Middle Miocene around 16 Ma, in southern Australia and New Zealand probably during the same interval, in Peru during the interval of late Subzone N8a and Subzone N8b, and in Chile during the interval of Subzones N8a, N8b, and the basal Zone N9, both around 16 Ma in South America. This marine climatic event is considered to be a prominent one occurring synchronously on a Pan-Pacific scale.

Based on marine molluscan biochronology, accelerated evolution in shell form is recognized in endemic forms of Japan through bioseries or successional species during the latest Neogene from 3 to 1 Ma. Considering changes in the composition of associated planktonic Foraminifera, the modification of shell form in these endemic Mollusca seems to occur in steps associated with episodes of cooling of seawater temperature. Similar accelerated evolution is likely to have happened in northern Chile and New Zealand, based on endemic bioseries of these areas during the latest Neogene, since 3 Ma. No such accelerated evolutionary change in shell form has been known in Mollusca of tropical seas on both sides of the Isthmus of Panama since 3 Ma. In middle latitudes, tropical and boreal water species appear and disappear in accordance with the rise and fall of seawater temperature, where endemic mollusca would survive by modifying their shell forms. An accelerated evolution may, therefore, be a characteristic biotic event in middle latitudes.

Marine faunal changes in middle latitudes are considered to be related causally to the rise and fall of seawater temperature. The above-mentioned two pronounced Neogene events may be notable examples of this in middle latitudes.

Acknowledgements

Dr. John A. Barron and Dr. Louie Marincovich, Jr., of the U.S. Geological Survey, Menlo Park, California, kindly read and reviewed the manuscript. The author is grateful to them for their critical suggestions.

References

Adams, C.G., Lee, D.E., and Rosen B.R., 1990. Conflicting isotopic and biotic evidence for tropical sea-surface temperatures during the Tertiary. *Palaeogeogr., Palaeoclimatol., Palaeoecol.*, 77: 289–313.

Benbow, M.C. and Lindsay, M.J., 1988. Discovery of *Lepidocyclina* (Foraminifera) in the Eucla Basin. *Quat. Geol. Notes, Geol. Surv. South Aust.*, 107: 1–8.

Berggren, W.A., Kent, D.V., Flynn, J.J., and Van Couvering, J. 1985. Cenozoic geochronology. *Geol. Soc. Amer. Bull.*, 96: 1407–1418.

Beu, A.G., 1983. Neogene geology – North Island east coast basin. *In:* N. de B. Hornibrook (Ed.), Guidebook for Tour B1, 15th Pacific Science Congress and 3rd International Meeting on Pacific Neogene Stratigraphy, Dunedin, 1983, pp. 1–28.

Chaproniere, G.C.H., 1980. Influence of plate tectonics on the distribution of late Paleogene to early Neogene larger foraminiferids in the Australasian region. *Palaeogeogr., Palaeoclimatol., Palaeoecol.*, 31: 299–317.

Chiji, M. and IGCP-246 National Working Group of Japan, 1990. Paleogeography and environments during the opening of the Sea of Japan. *In:* R. Tsuchi (Ed.), *Pacific Neogene Events*, University of Tokyo Press, pp. 161–169.

Herm, D., 1969. Marines Pliozan und Pleistozan in Nord- und Mittel-Chile unter besonderer Berucksichtigung der Entwicklung der Mollusken-Faunen. *Zitteliana*, 2: 1–159.

Hornibrook, N. de B., 1990. The Neogene of New Zealand — A basis for regional events. *In:* R. Tsuchi (Ed.), *Pacific Neogene Events*, Univ. of Tokyo Press, pp. 195–206.

Ibaraki, M., 1986. Neogene planktonic foraminiferal biostratigraphy of the Kakegawa area on the Pacific coast of central Japan. *Repts. Fac. Sci. Shizuoka Univ.*, 20: 39–173.

Ibaraki, M., 1990. Neogene planktonic foraminifera and events in Japan. *Palaeogeogr., Palaeoclimatol., Palaeoecol.*, 77: 335–343.

Ibaraki, M., 1992. Neogene planktonic foraminifera from the Camana Formation, Peru: their geologic ages and paleoceanographic implications. *In:* R. Tsuchi (Ed.), *Rep. Andean Studies, Shizuoka Univ., Spec. Vol. 4* (in press).

Kennett, J.P., Keller, G., and Srinivasan, M.S., 1985. Miocene planktonic foraminiferal biogeography and paleoceanographic development of the Indo-Pacific region. *In:* J.P. Kennett (Ed.), *The Miocene Ocean: Paleoceanography and Biogeography*, GSA Memoir 163, 197–236.

Kotaka, T., 1978. World-wide biostratigraphic correlation based on Turritellid phylogeny. *The Veliger*, 21: 189–196.

Marwick, J., 1957. New Zealand Genera of Turritellidae, and the species of *Stiracolpus*. New Zealand Geol. Surv., *Palaeont. Bull.*, 27: 1–55.

Suess, E., von Heune R., et al., 1988. *Proc. Ocean Dril. Prog. Init. Rep. Ocean Dril. Prog., College Station, TX* 112: 1015 p.

Tsuchi, R., 1986. Late Cenozoic molluscan faunas and their development in southwestern Japan. *In:* T. Kotaka (Ed.), Japanese Cenozoic Molluscs — Their Origin and Migration. *Paleontol. Soc. Jap. Spec. Pap.*, 29, pp. 33–46.

Tsuchi, R., 1987. Mid-Neogene migration of Tethyan tropical mollusca and larger foraminifera into northern Japan. *In:* K.G. McKenzie (Ed.), Shallow Tethys 2, *Proc. Int. Symp. Shallow Tetys 2*, Wagga Wagga, 15–17, Sept. 1986, pp. 455–459.

Tsuchi, R., 1990a. Neogene events in Japan and the Pacific. *Palaeogeogr., Palaeoclimatol., Palaeoecol.*, 77: 355–365.

Tsuchi R., 1990b. Accelerated Evolutionary Events in Japanese Endemic Mollusca during the latest Neogene. *In:* R. Tsuchi (Ed.), *Pacific Neogene Events*, Univ. of Tokyo Press, pp. 85–97.

Tsuchi, R. and IGCP-114 National Working Group of Japan, 1984. Neogene chronostratigraphy and bio-events in the Japanese Islands. *Palaeogeogr., Palaeoclimatol., Palaeoecol.*, 46: 37–51.

Tsuchi, R. and Ibaraki, M., 1988. Note on the Omma-Manganji molluscan fauna: Its geologic age and paleoceanographic implications *In:* Saito Ho-on Kai Spec. Publ. *Prof. T. Kotaka Commem. Vol.*, pp. 557–566.

Tsuchi, R., Shuto, T., Takayama, T., Fujiyoshi, A., Koizumi, I., Ibaraki, M., and Martinez-P.R., 1988. Fundamental data on Cenozoic biostratigraphy of Chile. *Rep. Andean Studies, Shizuoka Univ., Spec.,* 2: 71–95.

Tsuchi, R. and International Andean Studies Group, 1990. Trans-Pacific correlation of Neogene geologic events. *In:* R. Tsuchi (Ed.), *Rep. Andean Studies, Shizuoka Univ., Spec.,* 3: 1–7.

Yamanoi, T., 1984. Presence of sonneratiacous pollen in Middle Miocene sediments. *Rev. Paleobot. Palynol.,* 40: 347–357.

APPENDICES

1. Paleogeographic map of the Pacific region 16-17 m.y. ago

   Black arrow: Warm ocean current; White arrow: Cold ocean current.

The map was drafted in collaboration with L.T. Aguilar, J.A. Barron, G.C.H. Chaproniere, H. Duque-C., K. Hirooka, J.P. Kennett, L. Marincovich, Jr., R. Martinez-P., and S. Nishimura.

2. This time-space chart (T–S chart) is an attempt to make a view of various Neogene events in the Pacific region in time and space. The chart was drafted in collaboration with many of the IGCP-246 International Working Group members.

(The Editors)

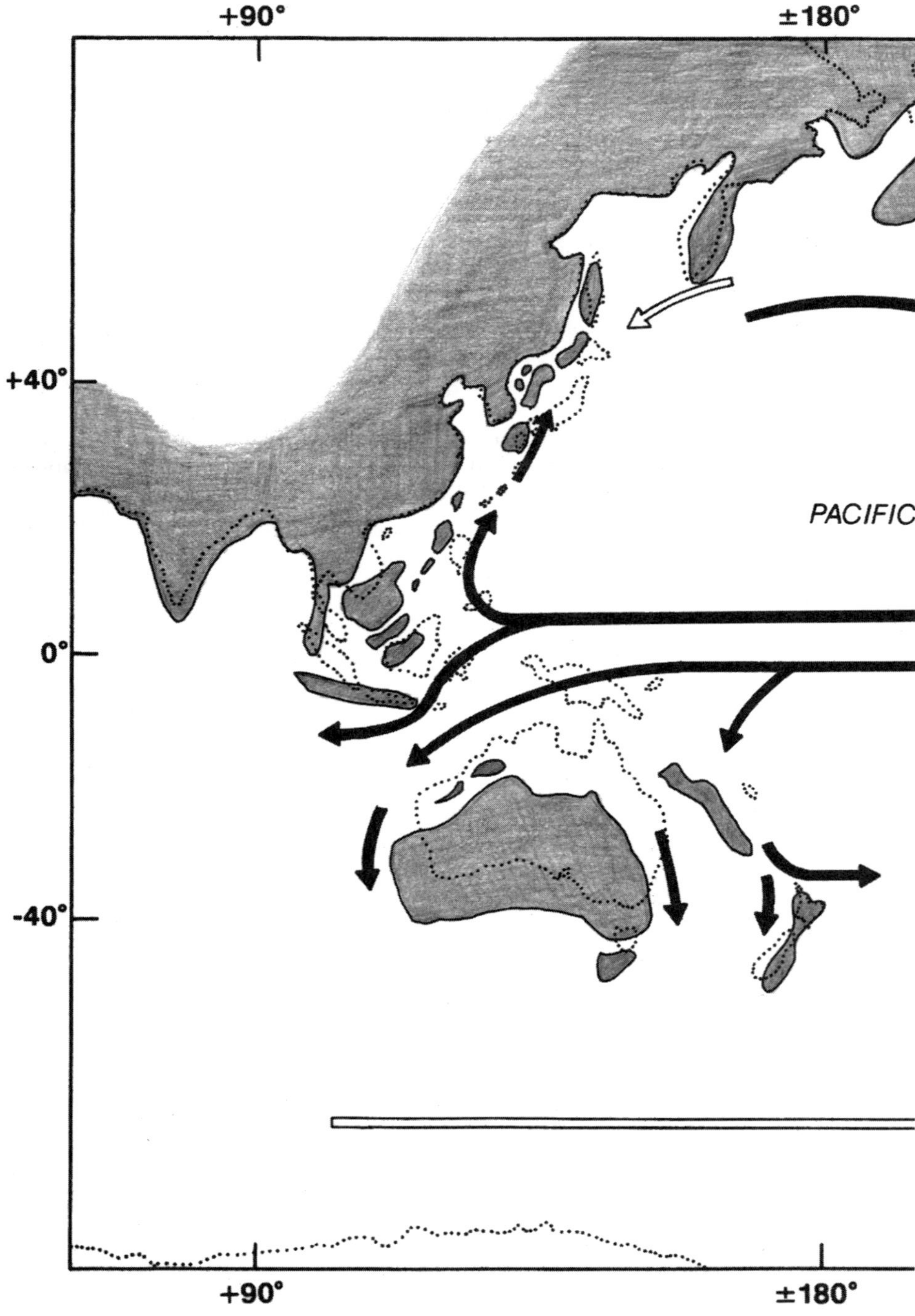

Neogene Paleogeographic Map of the Pacific Region around 16-17 Ma

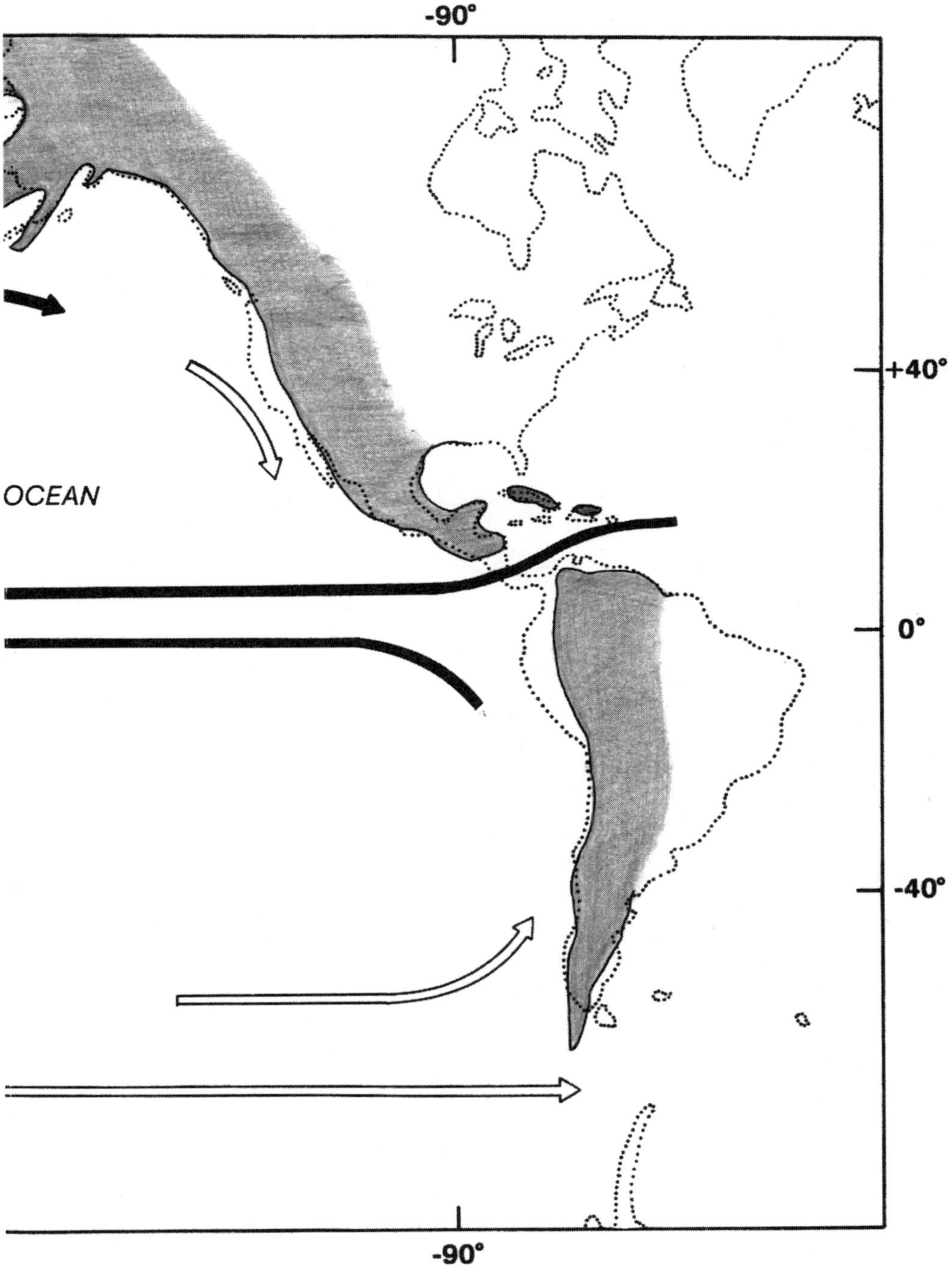
-90°
+40°
0°
-40°
-90°
OCEAN

Appendices

1. Time-Space Chart of Neogene Events in the NW Pacific.

	Ma	Climatic & Paleoceanographic	Biotic	Paleobiogeographic & Tectonic	Other events
PLEIS.	1				
PLIOCENE	2	W. palyn. flora ⟨TY⟩	Acc. evol. endemic moll. in Japan ⟨RT⟩	⊷ Collis. Izu Pen. ⟨RT⟩ −Sunda Str. & N. Makassar Sea open ⟨SN⟩ ⊷ Bering Strait open ⟨YG⟩ ⌐ Beringian transg. ⌐ _Astarte_ transg. ⟨KO⟩	
PLIOCENE	3				
PLIOCENE	4	W. clim. Kamchatka ⟨YG⟩			
PLIOCENE	5		Divers. dinofl. in N Japan ⟨KM⟩		
LATE MIOCENE	6				
LATE MIOCENE	7	W. clim. Kamchatka ⟨YG⟩ p.f. & dinofl. Japan ⟨SM⟩ ⟨YT⟩ ⟨KM⟩			
LATE MIOCENE	8				
LATE MIOCENE	9				
LATE MIOCENE	10				
MIDDLE MIOCENE	11	W. clim. Kamchatka ⟨YG⟩ Japan ⟨YT⟩			
MIDDLE MIOCENE	12	Init. equat. uhndercur. Intens. countercur. ⟨JK⟩			
MIDDLE MIOCENE	13				
MIDDLE MIOCENE	14		Neoteny of Pectinids in NE Japan ⟨KO⟩		
MIDDLE MIOCENE	15	Subtrop. clim. in Kamchatka ⟨YG⟩ Korea ⟨SY⟩	Subtrop.-subboreal moll in Kamchatka ⟨YG⟩ Trop. fauna in N. Japan moll., l.f., p.f. ⟨MI⟩⟨RT⟩	Japan Sea opening Rot. SW Japan ⟨KH⟩	
MIDDLE MIOCENE	16	⊷ Cimat. optimum culm. ⟨RT⟩			
EARLY MIOCENE	17				
EARLY MIOCENE	18				
EARLY MIOCENE	19	_Vicarya_ in centr. Japan ⟨KO⟩	Subboreal-boreal moll. in Kamchatka ⟨YG⟩		
EARLY MIOCENE	20				
EARLY MIOCENE	21				
EARLY MIOCENE	22				
EARLY MIOCENE	23				
OLIGO.	24				
OLIGO.	25				

Abbreviation:
w : Warm
c : Cold
moll: Mollusca
rad: Radiolaria
p.f.: Planktonic foraminifera
l.f.: Larger foraminifera

Author's abbreviation:
⟨JK⟩ J.P. Kennett
⟨KH⟩ K. Hirooka
⟨KM⟩ K. Matsuoka
⟨KO⟩ K. Ogasawara
⟨MI⟩ M. Ibaraki
⟨RT⟩ R. Tsuchi
⟨SM⟩ S. Maiya
⟨SN⟩ S. Nishimura
⟨TY⟩ T. Yamanoi
⟨YG⟩ Y.B. Gladenkov
⟨YT⟩ Y. Takayanagi

2. Time-Space Chart of Neogene Events in the NE Pacific.

Epoch	Ma	Climatic & Paleoceanographic	Biotic	Paleobiogeographic & Tectonic	Other events
PLEIS.	1			Uplift & folding of Neog. basins & margins ⟨JI⟩	
PLIOCENE	2				Cooling δ¹⁸O
	3			⟵ Bering Strait open ⟨LM⟩	
	4	Subtrop. p. f. Calif. ⟨JI⟩ ⟵ Poor diatom ⟨JB⟩			
	5	⟵ Cold p. f. Calif. ⟨JI⟩		Gulf Calif. opening ⟨JI⟩	
LATE MIOCENE	6	⟵ W. event diatom ⟨JB⟩ W. p. f. Calif. ⟨JI⟩			Negat. shift δ¹³C
	7	3rd clim. opt. diatom ⟨JB⟩ ⟵ W. dinofl. Aleutian ⟨KM⟩	Evol. cold diatom in Bering Sea ⟨JB⟩		
	8				Cooling δ¹⁸O
	9	⟵ Divers. c. dinofl. Bering Sea ⟨KM⟩	Diff. equat. & N. Pac. diatoms ⟨JB⟩		Cooling δ¹⁸O
	10				
MIDDLE MIOCENE	11	⟵ 2nd clim. opt. diatom ⟨JB⟩ Subtrop. p. f. Calif. ⟨JI⟩			
	12				No deposition in Bering Sea ⟨KM⟩
	13				
	14		Turnover of diatoms ⟨JB⟩		
	15				Cooling δ¹⁸O
	16	Brief appear. trop-w. temperate moll. & p. f. in Alaska ⟨LM⟩ Warm diatoms ⟨JB⟩		⟵ C. moll. in S. Alaska ⟨LM⟩	Glacial erratics in S. Alaska ⟨LM⟩ Low numbers diatom Crucidenticula ⟨JB⟩
	17	⟵ L. f. & subtrop. p. f. in Calif. ⟨JI⟩			
EARLY MIOCENE	18				⟵ Coast. mount. uplift in SE Alaska ⟨LM⟩
	19	W. dinofl. in Bering Sea ⟨KM⟩			
	20				
	21			⟵ Cool temperate moll. in Alaska ⟨LM⟩	
	22				
	23				
OLIGO.	24				(Oxygen and carbon isotope events in this column are worldwide)
	25				

Author's abbreviation:
⟨LM⟩ L. Marincovich Jr.
⟨JB⟩ J. A. Barron
⟨JI⟩ J. C. Ingle
⟨KM⟩ K. Matsuoka

3. Time-Space Chart of Neogene Events in the SE Pacific.

	Ma	Climatic & Paleoceanographic	Biotic evolutionary	Palaeobiogeographic & Tectonic	Otehr events
PLEIS.	1		Acc. evol. endemic moll. in N. Chile <RT>		
PLIOCENE	2–5			Onset of great Amer. intermingling. <HD> / Isthmus Panama closed <HD>	Diatomite deposition in N. Chile <MI>
LATE MIOCENE	6	Warm event diatom <JB> in Peru / Warm p.f. in Chile <MI>	Calif. & Carib. affinities in b.f. <HD>	Panama shallowing <HD>	
LATE MIOCENE	7–10				
MIDDLE MIOCENE	11–12		Calif. affinities in b.f. <HD>	Partial emergence of Panama <HD>	
MIDDLE MIOCENE	13–15				Significant Diatom. deposits in Peru <RM><MI>
MIDDLE MIOCENE	16	Clim. opt. in Peru–Chile <RM> <RT> <MI>	Rads disappear in the Caribbean <HD>	Shallow Panama sill <HD>	
EARLY MIOCENE	17				Diatom. deposits begin upwell. off Peru <MI>
EARLY MIOCENE	18–23				Drake Passage open
OLIGO.	24–25				

Abbreviation:
  b.f.: Benthic foraminifera
  p.f.: Planktonic foraminifera
  opt.: Optimum
  moll.: Mollusca

Author's abbreviation:
  <HD>   H. Duque
  <IK>   I. Koizumi
  <MI>   M. Ibaraki
  <RM>   R. Martinez
  <RT>   R. Tsuchi

4. Time-Space Chart of Neogene Events in the SW Pacific.

	Ma	Climatic & Paleoceanographic	Biotic	Paleobiogeographic & Tectonic	Other events
PLEIS.	1	Glacio-eustatic marine cyclothems in N.Z. ⟨NH⟩	Rapid moll. extinct. ⟨NH⟩		Volc. activities in central Vietnam & Thailand ⟨TD⟩ ⟨BR⟩
PLIOCENE	2–3	Rapid cooling in N.Z. ⟨NH⟩	Northward advance of c. moll in NZ ⟨NH⟩		$\delta^{18}O$ event ⟨AE⟩
PLIOCENE	5–6		Lowest moll. diversity & many extinct. in NZ ⟨NH⟩		Volc. activities in central Vietnam & Thai. ⟨TD·BR⟩; $\delta^{13}C$ shift ⟨AF⟩
LATE MIOCENE	7			Opening of N. Fiji Basin and tectonic pulse on Viti Levu platform ⟨PR⟩	Glacio-eustatic events ⟨AE⟩
LATE MIOCENE	10	Cooling in NZ ⟨NH⟩			
LATE MIOCENE	11		L. f. extinct. in NZ ⟨NH⟩		
MIDDLE MIOCENE	12–13			Clos. Indonesian Seaway ⟨JK⟩ ⟨SN⟩	Volc. activities in Thai ⟨BR⟩
MIDDLE MIOCENE	13.5	Ice vol. increase ⟨JK⟩			
MIDDLE MIOCENE	15	Bottom W. tem. drop ⟨JK⟩			Highest $\delta^{13}C$ $\delta^{18}O$ shift ⟨JK⟩
MIDDLE MIOCENE	16	Clim. opt. in Austr. N.Z. ⟨AC⟩ ⟨NH⟩	Peak of moll. divers. ⟨NH⟩	Rot. Timor, Seram, N. arm of Sulavesi ⟨SN⟩	
EARLY MIOCENE	17–18			Rot. Borneo, Celebes Sea Basin, E & W arm of Sulavesi ⟨SN⟩	Lowest $\delta^{18}O$ ⟨JK⟩
EARLY MIOCENE	19				Volc. activities in Thailand ⟨BR⟩
EARLY MIOCENE	20–21				Intrus. activities in Thailand ⟨BR⟩
EARLY MIOCENE	21			S. China Basin spreading occurred ⟨SN⟩	
OLIGO.	24				Highest transg. in Thailand ⟨BR⟩

Author's abbreviation:

⟨AC⟩	A. Carter	⟨NH⟩	N. de B. Hornibrook
⟨AE⟩	A. Edward	⟨PR⟩	P. Rodda
⟨BR⟩	B. Ratanasthien	⟨SN⟩	S. Nishimura
⟨JK⟩	J. Kennett	⟨TD⟩	T. Dzanh